TOPOLOGY

WILLIAM W. FAIRCHILD
Associate Professor of Mathematics
Union College

CASSIUS IONESCU TULCEA
Professor of Mathematics
Northwestern University

1971

W. B. SAUNDERS COMPANY PHILADELPHIA · LONDON · TORONTO

W. B. Saunders Company: West Washington Square
Philadelphia, Pa. 19105

12 Dyott Street
London, WC1A 1DB

1835 Yonge Street
Toronto 7, Ontario

Topology SBN 0-7216-3543-1

 Made in the United States of America. Press of W. B. Saunders Company. Library of Congress catalog card number 73-135323.

Print No.: 9 8 7 6 5 4 3 2 1

PREFACE

The purpose of this volume is to introduce the reader to general topology. Although no previous study of topology is assumed, familiarity with set theory (see [5]) and with the basic definitions of linear algebra and calculus is prerequisite.

Since non-metric spaces are often encountered in applications, we introduce general topological spaces from the beginning. However, metric spaces are treated in detail in Chapters 13 through 17. Chapters 24 through 29 contain a complete presentation of the main definitions and results concerning uniform spaces.

In Chapter 18 we discuss normed and Banach spaces and in Chapter 19 we prove the Stone-Weierstrass theorem. Partitions of unity are studied in Chapter 21. The presentation is made in such a way that the results may be applied directly to the case of continuous functions and to the case of differentiable functions on manifolds.

SOME REMARKS ABOUT NOTATION

1. Passages appearing between the triangles ▼ and ▲ may be omitted on a first reading.

2. The symbol Z is sometimes used to indicate a statement the understanding of which may cause difficulty.

3. The more difficult exercises are denoted by an asterisk.

CONTENTS

NOTATIONS and TERMINOLOGY xi

CHAPTER 1

TOPOLOGICAL SPACES 1

CHAPTER 2

NEIGHBORHOODS 8

CHAPTER 3

INTERIOR OF A SET and ADHERENCE OF A SET 15

CHAPTER 4

SUBSPACES 22

CHAPTER 5

CONTINUOUS FUNCTIONS 27

CHAPTER 6

INITIAL TOPOLOGIES 39

CHAPTER 7

PRODUCT SPACES 43

Appendix I—Projective Limits 56

CHAPTER 8

FURTHER RESULTS CONCERNING CONTINUOUS FUNCTIONS 60

CHAPTER 9

COMPACT SPACES.................................... 66

CHAPTER 10

LOCALLY COMPACT SPACES.................................... 80

CHAPTER 11

CONNECTED SPACES.................................... 89

Appendix I—Arcwise Connected Spaces.................................... 99

CHAPTER 12

FINAL TOPOLOGIES. QUOTIENT SPACES.................................... 103

CHAPTER 13

METRICS. METRIC SPACES.................................... 109

Appendix I—The Metric d_n on $\boldsymbol{R}^n$ ($n \in \mathbf{N}$).................................... 118

Appendix II—A Non-Metrizable Topological Space.................................... 122

CHAPTER 14

SEQUENCES.................................... 126

CHAPTER 15

COMPLETE METRIC SPACES.................................... 133

CHAPTER 16

COMPACT METRIZABLE SPACES.................................... 139

CHAPTER 17

UNIFORMLY CONTINUOUS FUNCTIONS.................................... 146

CHAPTER 18

NORMED SPACES.................................... 156

CHAPTER 19

THE STONE-WEIERSTRASS THEOREM.................................... 170

Appendix I—Complex Valued Functions.................................... 177

CHAPTER 20

NORMAL SPACES.................................... 180

Appendix I.................................... 190

CHAPTER 21

PARTITIONS OF UNITY 192

CHAPTER 22

BAIRE SPACES 198

CHAPTER 23

FILTERS. LIMITS. NETS 203

NOTATIONS AND TERMINOLOGY 216

CHAPTER 24

UNIFORM STRUCTURES. UNIFORM SPACES 218

CHAPTER 25

SEMI-METRICS AND UNIFORM SPACES 230

CHAPTER 26

COMPLETE UNIFORM SPACES 239

CHAPTER 27

UNIFORMLY CONTINUOUS FUNCTIONS 245

CHAPTER 28

ISOMORPHISMS. COMPLETION OF UNIFORM SPACES 255

CHAPTER 29

UNIFORMIZABLE SPACES. COMPLETELY REGULAR SPACES 259

BIBLIOGRAPHY 265

INDEX OF NOTATIONS 267

INDEX 269

Notations and Terminology

Although we assume that the reader is familiar with elementary set theory,* we shall introduce here certain notations and terminology that will be used throughout this volume.

If E is a set, then $\mathscr{P}(E)$ will denote the set of all subsets (or parts) of E including the void set $\varnothing$ and the set E itself. If E and F are two sets, then $E \subset F$ or $F \supset E$ means that E is a subset of F (E is contained in F); $E - F$ denotes the set $\{x \mid x \in E, x \notin F\}$. Whenever A is a subset of a set E, we may use $\complement A$ to denote the set $E - A$.

Let E and F be two sets. The notation $f: E \to F$ means that f is a function with domain E and range F. Whenever f is a function from E into F, we denote by $f(A)$ the set $\{f(x) \mid x \in A\}$, for each subset A of E, and by $f^{-1}(B)$ the set $\{x \mid f(x) \in B\}$, for each subset B of F. A mapping $f: E \to F$ is *onto* (or *surjective*) if $f(E) = F$, and *one to one* (or *injective*) if $x \neq y \Rightarrow f(x) \neq f(y)$ whenever x and y are in E.

Let I be a set and $(X_i)_{i \in I}$ a family of sets. We denote by $\prod_{i \in I} X_i$ the set $\{(t_i)_{i \in I} \mid t_i \in X_i, i \in I\}$, and call this the product of the family $(X_i)_{i \in I}$. If $I = \varnothing$, then $\prod_{i \in I} X_i$ contains exactly one element, whereas if $X_i \neq \varnothing$ for each i in I, then $\prod_{i \in I} X_i$ contains at least one element (Axiom of Choice). If $(X_i)_{i \in I}$ is considered to be a family of subsets of a set X, and $I = \varnothing$, then we may write $\bigcap_{i \in I} X_i = X$. For each subset J of I, pr_J denotes the mapping† $(t_i)_{i \in I} \mapsto (t_i)_{i \in J}$ of $\prod_{i \in I} X_i$ onto $\prod_{i \in J} X_i$, and is called the *projection* of $\prod_{i \in I} X_i$ onto $\prod_{i \in J} X_i$. If $J = \{\beta\}$, then we may write pr_β instead of pr_J.

A relation in a set E is an *order relation* if and only if it is reflexive, antisymmetric, and transitive. An order relation is frequently denoted by the symbol $\leq$. With this notation, the properties of an order relation are expressed as follows: (1) $x \leq x$ for all x in E (reflexivity); (2) $x \leq y$ and $y \leq x \Rightarrow x = y$ for all x, y, and z in E (antisymmetry); (3) $x \leq y$ and $y \leq z \Rightarrow x \leq z$ for all x, y, and z in E (transitivity). The notations $x \leq y$ and $y \geq x$ are equivalent. We write $x < y$ (or

* See, for instance, [5].

† A notation such as $x \mapsto \theta(x)$ or $f: x \mapsto \theta(x)$ describes a function that maps an element x in some domain set onto $\theta(x)$ in the range.

$y > x$) if $x \leq y$ and $x \neq y$. A set endowed with an order relation is called an *ordered set*.

Let E be an ordered set and $A \subset E$. An element y of E is an *upper bound* of A if $y \geq x$ for every x in A. If z is an upper bound of A such that $z \leq y$ for every upper bound y of A, then z is the *supremum* (or least upper bound) of A. In this case we write $z = \sup A$. An element m of E is called a *maximal element* if $x \in E$ and $x \geq m \Rightarrow x = m$ (that is, there is no x in E such that $m < x$). In a similar way, we define the *lower bounds*, the *infimum*, and the minimal elements. The subset A is said to be *totally ordered* if $x \in A$ and $y \in A$ imply $x \leq y$ or $y \leq x$. An ordered set is *inductive* if every totally ordered part has an upper bound. *Zorn's lemma states that every non-void inductive ordered set has a maximal element.*

Let E be an ordered set. A set $A \subset E$ is bounded if it has both a lower bound and an upper bound.

We denote by **N** the set of natural numbers 1, 2, 3, . . . , by **Z** we denote the set of all integers, by **Q**, the set of all rational numbers, and by **R**, the set of all real numbers. If $A \subset \mathbf{R}$, we write $A_+ = \{x \mid x \in A, x \geq 0\}$ and $A^* = \{x \mid x \in A, x \neq 0\}$. The complex numbers are denoted **C**.

Chapter I

Topological Spaces

In this chapter, we shall introduce the notion of topological space, and we shall give several elementary properties and several examples. We shall start with the:

1.1 Definition.—*Let X be a set and $\mathcal{T} \subset \mathcal{P}(X)$. We say that $\mathcal{T}$ is a topology on X if:*

(C_1) $\varnothing \in \mathcal{T}$ *and* $X \in \mathcal{T}$;

(C_2) *if A and B belong to $\mathcal{T}$, then $A \cap B \in \mathcal{T}$;*

(C_3) *if $(A_i)_{i \in I}$ is a family of elements of $\mathcal{T}$, then**

$$\bigcup_{i \in I} A_i \in \mathcal{T}.$$

From (C_2) we deduce, for instance, that if A, B, and C belong to $\mathcal{T}$, then $A \cap B \cap C = A \cap (B \cap C) \in \mathcal{T}$. More generally, we deduce that if $A_1, A_2, \ldots, A_n$ are sets belonging to $\mathcal{T}$, then

$$A_1 \cap \ldots \cap A_n \in \mathcal{T}.$$

Hence, if $(A_i)_{i \in I}$ is an *arbitrary finite*† family of elements of $\mathcal{T}$, then

$$\bigcap_{i \in I} A_i \in \mathcal{T}.$$

Notice that there is an essential difference between (C_2) and (C_3). The condition (C_3) asserts that the union of an *arbitrary family* of sets belonging to $\mathcal{T}$ belongs to $\mathcal{T}$. From (C_2), we deduce only that the intersection of an *arbitrary finite family* of sets belonging to $\mathcal{T}$ belongs to $\mathcal{T}$.

* Recall that the union of the void family is $\varnothing$.

† Recall that $\varnothing$ is a finite set and that $\bigcap_{i \in \varnothing} A_i = X$.

Example 1.—Let X be a set and let $\mathscr{D} = \mathscr{P}(X)$. Then $\mathscr{D}$ is a topology on X. It is called the *discrete topology* on X.

Example 2.—Let X be a set and let $\mathscr{W} = \{\varnothing, X\}$. Then $\mathscr{W}$ is a topology on X.

Let X be a set. Notice that if $\mathscr{D}$ and $\mathscr{W}$ are the topologies in Examples 1 and 2, and $\mathscr{T}$ is *any* topology on X, then we have $\mathscr{W} \subset \mathscr{T} \subset \mathscr{D}$.

Example 3.—Let a, b, and c be three distinct objects and let $X = \{a, b, c\}$ and

$$\mathscr{T} = \{\varnothing, \{a, b, c\}, \{b\}, \{a, b\}, \{c, b\}\}.$$

We leave it to the reader to verify that $\mathscr{T}$ is a topology on X.

Example 4.—Let $X \neq \varnothing$ be a set and let $a \in X$. Let $\mathscr{T}$ be the set of all $A \subset X$ such that:

$$\text{either} \quad A = \varnothing \quad \text{or} \quad A \ni a.$$

We leave it to the reader to verify that $\mathscr{T}$ is a topology on X.

Exercise 1.—Is the topology in Example 3 a particular case of the topology considered in Example 4?

Example 5.—Let X be a set. Let $\mathscr{T}$ be the set of all $A \subset X$ such that:

$$\text{either } A = \varnothing \quad \text{or} \quad \mathbf{C}A \text{ is finite.}$$

Then $\mathscr{T}$ is a topology on X.

In fact, (C_1) is obviously satisfied. Now let $A \in \mathscr{T}$ and $B \in \mathscr{T}$. If $A \cap B = \varnothing$, then $A \cap B \in \mathscr{T}$. If $A \cap B \neq \varnothing$, then $A \neq \varnothing$ and $B \neq \varnothing$. Since A and B belong to $\mathscr{T}$, we deduce that $\mathbf{C}A$ and $\mathbf{C}B$ are finite, whence

$$\mathbf{C}(A \cap B) = (\mathbf{C}A) \cup (\mathbf{C}B)$$

is finite; whence $A \cap B \in \mathscr{T}$. Hence, (C_2) is also satisfied. Now let $(A_i)_{i \in I}$ be a family of sets belonging to $\mathscr{T}$. If $\bigcup_{i \in I} A_i = \varnothing$, then $\bigcup_{i \in I} A_i \in \mathscr{T}$. If $\bigcup_{i \in I} A_i \neq \varnothing$, there is $i_0 \in I$ such that $A_{i_0} \neq \varnothing$. Then $\mathbf{C}A_{i_0}$ is finite. Since

$$\mathbf{C}(\textstyle\bigcup_{i \in I} A_i) \subset \mathbf{C}A_{i_0},$$

we deduce that $\mathbf{C}(\bigcup_{i \in I} A_i)$ is finite; whence $\bigcup_{i \in I} A_i \in \mathscr{T}$. Hence, (C_3) is also satisfied, and hence $\mathscr{T}$ is a topology on X.

Remark.—Let $X = \mathbf{R}$, and for each $n \in \mathbf{N}$, let

$$A_n = X - \{n\} = \mathbf{C}\{n\}.$$

If $\mathscr{T}$ is the topology introduced in Example 5, then $A_n \in \mathscr{T}$ for all $n \in \mathbf{N}$. However,

$$\bigcap_{n\in\mathbf{N}} A_n = \mathbf{CN}.$$

Since $\mathbf{N}$ is *not finite*, it follows that $\bigcap_{n\in\mathbf{N}} A_n \notin \mathscr{T}$.

1.2 Theorem.—*Let X be a set and let $\mathscr{E} \subset \mathscr{P}(X)$. Denote by $\mathscr{E}'$ the set of all parts of X that are intersections of finite* families of elements of $\mathscr{E}$. Denote by $\mathscr{E}''$ the set of all parts of X that are unions of families of elements of $\mathscr{E}'$. Then $\mathscr{E}''$ is a topology on X.*

A set A belongs to $\mathscr{E}'$ if and only if $A = \bigcap_{i\in I} A_i$, where $(A_i)_{i\in I}$ is a finite family of elements of $\mathscr{E}$. A set B belongs to $\mathscr{E}''$ if and only if $B = \bigcup_{j\in J} B_j$, where $(B_j)_{j\in J}$ is a family of elements of $\mathscr{E}'$.

Proof.—Since the union of the void family is $\varnothing$, $\mathscr{E}''$ satisfies (C_1).

Now suppose $U \in \mathscr{E}''$ and $V \in \mathscr{E}''$. Then $U = \bigcup_{h\in H} U_h$ and $V = \bigcup_{k\in K} V_k$ where $(U_h)_{h\in H}$ and $(V_k)_{k\in K}$ are families of elements of $\mathscr{E}'$. We deduce that

$$U \cap V = \left(\bigcup_{h\in H} U_h\right) \cap \left(\bigcup_{k\in K} V_k\right) = \bigcup_{(h,k)\in H\times K} (U_h \cap V_k).$$

Since $U_h \cap V_k \in \mathscr{E}'$ for every $(h, k) \in H \times K$, we conclude that $U \cap V \in \mathscr{E}''$. Thus, $\mathscr{E}''$ satisfies (C_2).

Now suppose $(U_i)_{i\in I}$ is a family of elements of $\mathscr{E}''$. Then for each $i \in I$, U_i is the union of a family of elements of $\mathscr{E}'$. Then, clearly, $\bigcup_{i\in I} U_i$ is the union of a family of elements of $\mathscr{E}'$, and hence it belongs to $\mathscr{E}''$. Thus $\mathscr{E}''$ satisfies (C_3) also. This completes the proof of Theorem 1.2.

Exercise 2.—Express $\bigcup_{i\in I} U_i$ explicitly as a union of a family of elements of $\mathscr{E}'$.

The topology $\mathscr{E}''$ is said to be the topology *spanned* (or *generated*) by $\mathscr{E}$. *In this book, it is often denoted* $\mathscr{T}_{\mathscr{E}}$. Notice that $\mathscr{E} \subset \mathscr{T}_{\mathscr{E}}$. If $\mathscr{T}$ is a topology on X containing $\mathscr{E}$, then $\mathscr{T}$ contains $\mathscr{E}'$ (use (C_1) and (C_2)); hence $\mathscr{T}$ contains $\mathscr{E}''$ also (use (C_3)). We conclude that $\mathscr{T}_{\mathscr{E}}$ is the smallest† topology on X containing $\mathscr{E}$.

---

* Recall that $\varnothing$ is a finite set, that the intersection of the void family of elements of $\mathscr{P}(X)$ is X, and that the union of the void family of elements of $\mathscr{P}(X)$ is $\varnothing$.

† If $\mathbf{T}$ is a set of topologies on X, we say that $\mathscr{T}_0$ is the smallest topology in $\mathbf{T}$ if $\mathscr{T}_0 \in \mathbf{T}$ and if $\mathscr{T}_0 \subset \mathscr{T}$ for every $\mathscr{T}$ in $\mathbf{T}$. Note that it is not necessarily true that a set of topologies *contains* a smallest element (this will be discussed later).

If $\mathscr{E} = \mathscr{E}'$, that is, $\bigcap_{i \in I} A_i \in \mathscr{E}$ whenever $(A_i)_{i \in I}$ is a finite family of elements of $\mathscr{E}$, then $\mathscr{T}_{\mathscr{E}}$ is the set of all parts of X that are unions of families of elements of $\mathscr{E}$.

If $\mathscr{T}$ is a topology on X, then we say that $\mathscr{B} \subset \mathscr{T}$ is a *basis* of $\mathscr{T}$ if every set in $\mathscr{T}$ is the union of a family of sets belonging to $\mathscr{B}$; we say that $\mathscr{C} \subset \mathscr{T}$ is a *subbasis* of $\mathscr{T}$ if every set in $\mathscr{T}$ is the union of a family of sets belonging to $\mathscr{C}'$. Theorem 1.2 shows that $\mathscr{E}$ is a subbasis of $\mathscr{E}''$ and that $\mathscr{E}'$ is a basis of $\mathscr{E}''$.

1.3 Theorem.—*Let X be a set and $\mathscr{E} \subset \mathscr{P}(X)$. For any set $A \subset X$, the following assertions are equivalent:*

1.4 $A \in \mathscr{T}_{\mathscr{E}}$;

1.5 *for every a in A there is B in $\mathscr{E}'$ such that $a \in B \subset A$;*

1.6 $A = \bigcup_{i \in I} A_i$, *where $(A_i)_{i \in I}$ is a family of elements of $\mathscr{E}'$.*

Proof of 1.4 ⇒ 1.5.—By the definition of $\mathscr{T}_{\mathscr{E}}$, $A = \bigcup_{i \in I} U_i$ where $U_i \in \mathscr{E}'$ for each i in I. Suppose that $a \in A$. Then there is $i_0 \in I$ such that $a \in U_{i_0}$. Since $U_{i_0} \in \mathscr{E}'$, A satisfies 1.5.

Proof of 1.5 ⇒ 1.6.—For each x in A, there exists a set B_x belonging to $\mathscr{E}$ such that $x \in B_x \subset A$. Since $A = \bigcup_{x \in A} B_x$, we deduce that A satisfies 1.6.

Proof of 1.6 ⇒ 1.4.—The fact that $A \in \mathscr{T}_{\mathscr{E}}$ follows from (C_3).

Let X be an arbitrary set and let $\mathscr{T}(X)$ be the set of all topologies on X. We may order $\mathscr{T}(X)$ by writing:

$$\mathscr{T}' \leq \mathscr{T}'' \quad \text{whenever} \quad \mathscr{T}' \subset \mathscr{T}''.$$

If $(\mathscr{T}_i)_{i \in I}$ is a family of elements of $\mathscr{T}(X)$, then $\bigcap_{i \in I} \mathscr{T}_i$ is again a topology on X, and it is easy to see that

1.7 $$\bigcap_{i \in I} \mathscr{T}_i = \inf_{i \in I} \mathscr{T}_i.$$

We leave it to the reader to prove that if $\mathscr{E} = \bigcup_{i \in I} \mathscr{T}_i$, then

1.8 $$\mathscr{T}_{\mathscr{E}} = \sup_{i \in I} \mathscr{T}_i$$

The set $\bigcup_{i \in I} \mathscr{T}_i$ is not necessarily a topology on X.

1.9 Definition.—*A couple $(X, \mathscr{T})$, where X is a set and $\mathscr{T}$ is a topology on X, is called a topological space.*

If $(X, \mathscr{T})$ is a topological space, we say that $\mathscr{T}$ is the topology of $(X, \mathscr{T})$. When there is no ambiguity as to what topology we consider on X, we shall often say the *topological space* X instead of the *topological space* $(X, \mathscr{T})$.

Let $(X, \mathscr{T})$ be a topological space. *A set* $U \subset X$ *is said to be open if and only if* $U \in \mathscr{T}$. From (C_1), it follows that $\varnothing$ and X are always open. From (C_2), it follows that the intersection of a *finite* family of open sets is open.

Example 6 (*The topological space* **R**).—For a and b in **R** satisfying $a < b$, we write*

1.10 $$(a, b) = \{x \mid x \in \mathbf{R}, a < x < b\}.$$

Let

$$\mathscr{E} = \{(a, b) \mid a \in \mathbf{R}, b \in \mathbf{R}, a < b\}$$

and let $\mathscr{U} = \mathscr{T}_{\mathscr{E}}$. Notice that $\mathscr{E}$ is a basis of $\mathscr{U}$; that is, every open set $U \subset \mathbf{R}$ is the union of a family of sets belonging to $\mathscr{E}$. Notice that $U \in \mathscr{U}$ if and only if for every $x \in U$, there are $a \in \mathbf{R}$, $b \in \mathbf{R}$, and $a < b$ such that

$$x \in (a, b) \subset U.$$

Unless we explicitly mention the contrary, whenever we consider **R** *as a topological space, we suppose that it is endowed with the topology* $\mathscr{U}$.

For $a \in \mathbf{R}$, let $U = \{x \mid x \in \mathbf{R}, x < a\}$; then $U \in \mathscr{U}$. In fact,

$$U = \bigcup_{n \in Z, n < a} (n, a).$$

For $b \in \mathbf{R}$, let $V = \{x \mid x \in \mathbf{R}, x > b\}$; then $V \in \mathscr{U}$. In fact,

$$V = \bigcup_{n \in Z, n > b} (b, n).$$

Let $(X, \mathscr{T})$ be a topological space. *A set* $F \subset X$ *is closed if*

$$X - F = \mathbf{C}F$$

is open.

Notice that $\varnothing$ and X are always closed. Hence, $\varnothing$ and X are both *open* and *closed.*

Example 7.—Consider the topological space **R**. For a and b in **R** satisfying $a < b$, we write

1.11 $$[a, b] = \{x \mid x \in \mathbf{R}, a \leq x \leq b\}.$$

Then $[a, b]$ is *closed*. In fact (with the notations of Example 6), we have

$$\mathbf{C}[a, b] = U \cup V,$$

* A set of the form 1.10 is called a bounded open interval. A set of the form 1.11 is called a bounded closed interval.

and as we have seen in Example 6, the sets U and V are open. Hence $U \cup V$ is open, and hence $[a, b]$ is closed.

1.12 Theorem.—*Let* $(X, \mathscr{T})$ *be a topological space. Then:*
(i) *if* $(F_i)_{i \in I}$ *is a finite family of closed sets,*

$$\bigcup_{i \in I} F_i \text{ is closed;}$$

(ii) *if* $(F_i)_{i \in I}$ *is an arbitrary family of closed sets,*

$$\bigcap_{i \in I} F_i \text{ is closed.}$$

Proof of (i).—Let $A = \bigcup_{i \in I} F_i$, with I finite. We have

$$\mathbf{C}A = \bigcap_{i \in I} \mathbf{C}F_i.$$

Since $\mathbf{C}F_i$ is open for all $i \in I$, it follows that $\bigcap_{i \in I} \mathbf{C}F_i$ is open, whence $\mathbf{C}A$ is open. Therefore, A is closed.

Proof of (ii).—Let $B = \bigcap_{i \in I} F_i$. We have

$$\mathbf{C}B = \bigcup_{i \in I} \mathbf{C}F_i.$$

Since $\mathbf{C}F_i$ is open for all $i \in I$, it follows that $\bigcup_{i \in I} \mathbf{C}F_i$ is open, whence $\mathbf{C}B$ is open. Therefore, B is closed.

Remark.—Consider the topological space $(\mathbf{R}, \mathscr{T})$, where $\mathscr{T}$ is the topology defined in Example 5. Then $B_n = \{n\}$ is closed for every $n \in \mathbf{N}$. However,

$$\bigcup_{n \in N} B_n = \mathbf{N}$$

is *not* closed.

Exercises for Chapter 1

1. Let a, b, *and* c be three distinct elements and let $X_1 = \{a\}$, $X_2 = \{a, b\}$, and $X_3 = \{a, b, c\}$. List all the topologies on X_1, X_2, and X_3.

2. Let X be a set and let $\mathscr{H}$ be the set of all subsets A of X such that $A = \varnothing$, or $\mathbf{C}A$ is finite or countable. Show that $\mathscr{H}$ is a topology on X.

3. In the topological space $(\mathbf{R}, \mathscr{U})$ (see Example 6), consider the family of open sets $(U_n)_{n \in \mathbf{N}}$ where $U_n = (-1, 1/n)$. Find

$$\bigcap_{n \in \mathbf{N}} U_n$$

and show that this set is not open.

4. Give an example of a family $(F_i)_{i \in I}$ of closed parts of $\mathbf{R}$ such that $\bigcup_{i \in I} F_i$ is not closed.

5. Let a, b, and c be three distinct objects and let $X = \{a, b, c\}$. Let

$$\mathscr{T}_1 = \{\varnothing, \{a, b, c\}, \{b\}, \{a, b\}, \{c, b\}\}$$

and

$$\mathscr{T}_2 = \{\varnothing, \{a, b, c\}, \{c\}, \{a, c\}, \{b, c\}\}.$$

Then $\mathscr{T}_1$ and $\mathscr{T}_2$ are topologies on X (see Example 3). The set $\mathscr{T}_1 \cup \mathscr{T}_2$ is not a topology on X.

6. Let X be a set and $\mathscr{B} = \{\{t\} \mid t \in X\}$. Then $\mathscr{T}_{\mathscr{B}} = \mathscr{P}(X)$.

7. Give an example of a topology $\mathscr{T}$ on $\mathbf{R}$ such that $\mathscr{T} \neq \mathscr{U}$, $\mathscr{T} \neq \{\varnothing, \mathbf{R}\}$, and $\mathscr{T} \neq \mathscr{P}(\mathbf{R})$.

8. Let $(\mathscr{T}_i)_{i \in I}$ be a family of topologies on a set X. Show that

$$\inf_{i \in I} \mathscr{T}_i = \bigcap_{i \in I} \mathscr{T}_i.$$

(See 1.7.)

9. Let $\mathscr{B}$ be a set of sets such that $A \in \mathscr{B}$, $B \in \mathscr{B}$, and $x \in A \cap B \Rightarrow$ there is $C \in \mathscr{B}$ such that $x \in C \subset A \cap B$. Then on $Y = \bigcup \mathscr{B}$, the topology $\mathscr{T}_{\mathscr{B}}$ is equal to $\{\bigcup \mathscr{G} \mid \mathscr{G} \subset \mathscr{B}\}$ (i.e., $\mathscr{B}$ is a basis of $\mathscr{T}_{\mathscr{B}}$).

10. Prove that if $(\mathscr{T}_i)_{i \in I}$ is a family of topologies on a set X, and $\mathscr{E} = \bigcup \{\mathscr{T}_i \mid i \in I\}$, then

$$\mathscr{T}_{\mathscr{E}} = \sup_{i \in I} \mathscr{T}_i.$$

(See 1.8.)

11. Let X be a set and $\leq$ an order relation on X. Show that $\mathscr{T} = \{U \mid U \in \mathscr{P}(X),\ x \in U \text{ and } y \leq x \Rightarrow y \in U\}$ is a topology on X.

Chapter 2

Neighborhoods

Let $(X, \mathscr{T})$ be a topological space. We shall now introduce the following:

2.1 Definition.—*A set* $V \subset X$ *is a neighborhood of* x *(in the space* $(X, \mathscr{T})$*) if there exists a set* $U \in \mathscr{T}$ *such that*

$$x \in U \subset V.$$

Hence, $V \subset X$ is a neighborhood of $x \in X$ if and only if V contains an open set containing x.

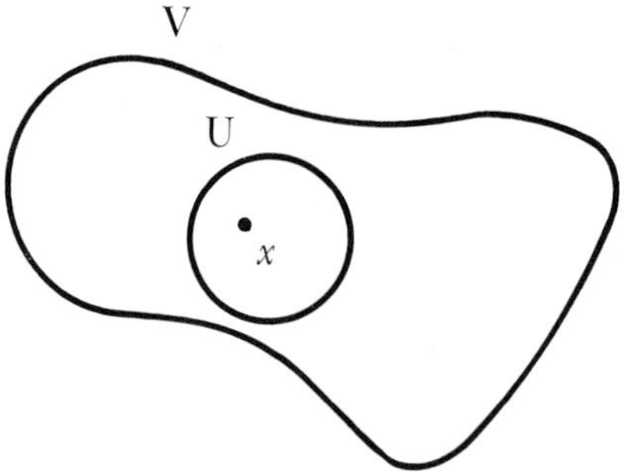

FIGURE 2.1

An open set containing x is clearly a neighborhood of x. Notice, however, that a neighborhood of x is not necessarily open. In any case, it contains an open set containing x.

For each $x \in X$, we denote by $\mathscr{N}(x)$ the set of all neighborhoods of x in the space $(X, \mathscr{T})$. Sometimes we shall write $\mathscr{N}_X(x)$, or $\mathscr{N}_{(X,\mathscr{T})}(x)$, or $\mathscr{N}_{\mathscr{T}}(x)$ instead of $\mathscr{N}(x)$.

2.2 Theorem.—*Let* $(X, \mathscr{T})$ *be a topological space and let* $x \in X$.

(a) *If* $V \in \mathscr{N}(x)$, *then* $V \ni x$.

(b) *If* $V_1 \in \mathscr{N}(x)$ *and* $V_1 \subset V_2$, *then* $V_2 \in \mathscr{N}(x)$.

(c) *If* V_1 *and* V_2 *belong to* $\mathcal{N}(x)$, *then* $V_1 \cap V_2 \in \mathcal{N}(x)$.

(d) *If* $V \in \mathcal{N}(x)$, *then there is* $W \in \mathcal{N}(x)$ *such that* $y \in W \Rightarrow V \in \mathcal{N}(y)$.

Proof.—The proofs of (a), (b), and (c) are left to the reader. To prove (d), we notice that since $V \in \mathcal{N}(x)$, there is an open set U such that $x \in U \subset V$. If we take $W = U$, then for $y \in W$ we have $y \in U \subset V$, whence $V \in \mathcal{N}(y)$.

We now introduce a very useful notion. Let $(X, \mathcal{T})$ be a topological space and let $t \in X$.

2.3 Definition.—*A set* $\mathcal{F} \subset \mathcal{N}(t)$ *is a fundamental system of* t *if for every* $V \in \mathcal{N}(t)$ *there exists* $W \in \mathcal{F}$ *such that* $W \subset V$.

Notice that if $\mathcal{F}$ is a fundamental system of $t \in X$, then every set $W \in \mathcal{F}$ is a neighborhood of t. However, a fundamental system of t does not necessarily contain all the neighborhoods of t.

Example 1.—Let $t \in X$ and let $\mathcal{F}$ be the set of all open parts of X containing t. It follows, from the definition of a neighborhood, that $\mathcal{F}$ is a fundamental system of t.

Remark.—Let $\mathcal{B}$ be a basis of the topology $\mathcal{T}$ and let $t \in X$. Since every set U in $\mathcal{T}$ is the union of sets belonging to $\mathcal{B}$, and since every neighborhood V of t contains an open set containing t, it follows that every neighborhood of t contains a set belonging to $\mathcal{B}$ and containing t. Thus, the set of all sets belonging to $\mathcal{B}$ and containing t is a fundamental system of t. In particular, if $\mathcal{E}$ generates $\mathcal{T}$, then the set of all sets belonging to $\mathcal{E}'$ and containing t is a fundamental system of t.

Let $(X, \mathcal{T})$ be a topological space, let $t \in X$, and let $\mathcal{F}$ be a fundamental system of t. We leave it to the reader to notice that:

(a′) if $W \in \mathcal{F}$, then $W \ni t$;
(b′) if $W_1 \in \mathcal{F}$ and $W_1 \subset W_2$, then $W_2 \in \mathcal{N}(t)$;
(c′) if W_1 and W_2 belong to $\mathcal{F}$, then $W_1 \cap W_2$ contains a set belonging to $\mathcal{F}$.

Example 2.—Consider the topological space **R** (introduced in Example 6 of Chapter 1). Then:

(i) the set $\mathcal{F}_1 = \{(a, b) \mid (a, b) \ni t\}$ is a fundamental system of t;

(ii) the set $\mathcal{F}_2 = \{[t - \varepsilon, t + \varepsilon] \mid \varepsilon > 0\}$ is a fundamental system of t;

(iii) the set $\mathcal{F}_3 = \{(t - \varepsilon, t + \varepsilon) \mid \varepsilon > 0\}$ is a fundamental system of t.

Proof of (i).—Since each set of the form (a, b) is open in $\mathbf{R}$, it follows that $\mathscr{F}_1 \subset \mathscr{N}(t)$. Let now $V \in \mathscr{N}(t)$. Then there exists U open (in $\mathbf{R}$) such that $t \in U \subset V$. By the definition of the topology of $\mathbf{R}$ (see Example 6 of Chapter 1), there exists (a, b) such that $t \in (a, b) \subset U$. Hence, there exists (a, b) such that $t \in (a, b) \subset V$. Since $V \in \mathscr{N}(t)$ was arbitrary, we deduce that $\mathscr{F}_1$ is a fundamental system of t.

Proof of (ii).—Since, for each $\varepsilon > 0$, $[t - \varepsilon, t + \varepsilon] \supset (t - \varepsilon, t + \varepsilon)$, it follows that $[t - \varepsilon, t + \varepsilon]$ is a neighborhood of t; whence $\mathscr{F}_2 \subset \mathscr{N}(t)$. Now let $V \in \mathscr{N}(t)$. By (i), there is a set of the form (a, b) such that $t \in (a, b) \subset V$. Then $a < t < b$. Let ε be the *smaller* of the two numbers $(t - a)/2$ and $(b - t)/2$; then

$$t + \varepsilon \leq t + (b - t)/2 < t + (b - t) = b$$

and

$$t - \varepsilon \geq t - (t - a)/2 > t - (t - a) = a.$$

If $x \in [t - \varepsilon, t + \varepsilon]$, then

$$a < t - \varepsilon \leq x \leq t + \varepsilon < b \Rightarrow a < x < b.$$

Hence, $[t - \varepsilon, t + \varepsilon] \subset (a, b) \subset V$. Since $V \in \mathscr{N}(t)$ was arbitrary, we deduce that $\mathscr{F}_2$ is a fundamental system of t.

Proof of (iii).—We leave this to the reader. (Hint: Use (ii).)

From (ii), we deduce that a set $V \in \mathscr{N}(t)$ if and only if there is $\varepsilon > 0$ such that $|x - t| \leq \varepsilon \Rightarrow x \in V$.

From (iii), we deduce that a set $V \in \mathscr{N}(t)$ if and only if there is $\varepsilon > 0$ such that $|x - t| < \varepsilon \Rightarrow x \in V$.

2.4 Theorem.—*Let $(X, \mathscr{T})$ be a topological space and let $A \subset X$. Then the following assertions are equivalent:*

(a) *A is open;*
(b) *$A \in \mathscr{N}(t)$ for every $t \in A$.*

Proof of (a) ⇒ (b).—The fact that (a) ⇒ (b) follows immediately from the definition of a neighborhood.

Proof of (b) ⇒ (a).—Let $t \in A$. Since $A \in \mathscr{N}(t)$, there is V_t open such that $t \in V_t \subset A$. Let

$$V = \bigcup_{t \in A} V_t.$$

By (C_3), V is *open*. Since $V_t \subset A$ for all $t \in A$, we have $V \subset A$. Since for every $t \in A$, $t \in V_t$, we have $A = V$. Hence A is open.

2.5 Theorem.—*Let X be a set and $\mathcal{T}_1$ and $\mathcal{T}_2$ be two topologies on X. Then the following assertions are equivalent:*

(a) $\mathcal{T}_1 = \mathcal{T}_2$;
(b) $\mathcal{N}_{\mathcal{T}_1}(x) = \mathcal{N}_{\mathcal{T}_2}(x)$ *for all* $x \in X$.

Proof.—Clearly, (a) $\Rightarrow$ (b). Conversely, suppose that (b) holds. Let $U \in \mathcal{T}_1$. By Theorem 2.4, $U \in \mathcal{N}_{\mathcal{T}_1}(x)$ for all $x \in U$. Since $\mathcal{N}_{\mathcal{T}_1}(x) = \mathcal{N}_{\mathcal{T}_2}(x)$ for all $x \in X$, we have $U \in \mathcal{N}_{\mathcal{T}_2}(x)$ for all $x \in U$. Hence, $U \in \mathcal{T}_2$, and hence $\mathcal{T}_1 \subset \mathcal{T}_2$. In the same way, we show that $\mathcal{T}_2 \subset \mathcal{T}_1$. Hence (b) $\Rightarrow$ (a).

The following result is very useful in applications (particularly when defining topologies for topological vector spaces or topological groups).

2.6 Theorem.—*Let X be a set, and for each x in X let $\mathcal{B}(x)$ be a set of parts of X. Suppose that:*

(a) *if* $V \in \mathcal{B}(x)$, *then* $V \ni x$;
(b) *if* $V_1 \in \mathcal{B}(x)$ *and* $V_1 \subset V_2 \subset X$, *then* $V_2 \in \mathcal{B}(x)$;
(c) *if* V_1 *and* V_2 *belong to* $\mathcal{B}(x)$, *then* $V_1 \cap V_2 \in \mathcal{B}(x)$;
(d) *if* $V \in \mathcal{B}(x)$, *then there is* $W \in \mathcal{B}(x)$ *such that* $y \in W \Rightarrow V \in \mathcal{B}(y)$.

Then there is one, and only one, topology $\mathcal{T}$ on X such that $\mathcal{N}_{(X,\mathcal{T})}(x) = \mathcal{B}(x)$ for every x in X.

We have shown in Theorem 2.2 that if $(X, \mathcal{T})$ is a topological space, then the sets $\mathcal{N}(x)$, for $x \in X$, satisfy the conditions (a), (b), (c), and (d) of Theorem 2.6. Theorem 2.6 shows, *conversely*, that if, for each x in X, a set $\mathcal{B}(x)$ is given such that (a), (b), (c), and (d) are satisfied, then there is a unique topology $\mathcal{T}$ such that, for each x in X, the set of all neighborhoods of x in the space $(X, \mathcal{T})$ coincides with $\mathcal{B}(x)$.

Proof.—By Theorem 2.5, there can be at most one topology $\mathcal{T}$ on X such that $\mathcal{N}_{(X,\mathcal{T})}(x) = \mathcal{B}(x)$ for all x in X. Thus, $\mathcal{T}$ is unique (if it exists). The existence of $\mathcal{T}$ still must be proved.

Let $\mathcal{T}$ be the set of all subsets U of X such that $y \in U \Rightarrow U \in \mathcal{B}(y)$. It is easy to see that $\mathcal{T}$ satisfies the conditions (C_1), (C_2), and (C_3); hence $\mathcal{T}$ is a topology on X.

We still need to show that $\mathcal{N}_{(X,\mathcal{T})}(x) = \mathcal{B}(x)$ for all x in X. Suppose that $x \in X$ and $V \in \mathcal{N}_{(X,\mathcal{T})}(x)$. Then there is U in $\mathcal{T}$ such that $x \in U \subset V$. By the definition of $\mathcal{T}$, $U \in \mathcal{B}(x)$. We deduce (using (b)) that $V \in \mathcal{B}(x)$, and hence that $\mathcal{N}_{(X,\mathcal{T})}(x) \subset \mathcal{B}(x)$.

Now, suppose that $V \in \mathcal{B}(x)$ and let

$$U = \{y \mid y \in V, V \in \mathcal{B}(y)\}.$$

Notice that $x \in U$, since by our hypothesis $V \in \mathcal{B}(x)$. We will now show that $U \in \mathscr{T}$: Let z be an arbitrary element of U. By the definition of U, $V \in \mathscr{B}(z)$. By (d), there is a set W in $\mathscr{B}(z)$ such that $s \in W \Rightarrow V \in \mathscr{B}(s)$. We deduce from (a) and from the definition of U again that $z \in W \subset U$. Using (b), we obtain $U \in \mathscr{B}(z)$. Thus, $z \in U \Rightarrow U \in \mathscr{B}(z)$, that is, $U \in \mathscr{T}$. Since $x \in U$ and $U \subset V$, we conclude that $V \in \mathscr{N}_{(X,\mathscr{T})}(x)$. Therefore, $\mathscr{B}(x) \subset \mathscr{N}_{(X,\mathscr{T})}(x)$. Since x was arbitrary, we deduce that $\mathscr{B}(x) = \mathscr{N}_{(X,\mathscr{T})}(x)$ for all x in X.

2.7 Definition.—*A topological space $(X, \mathscr{T})$ is said to be separated (or Hausdorff) if for every x and y in X, $x \neq y$, there exist $V_x \in \mathscr{N}(x)$ and $V_y \in \mathscr{N}(y)$ satisfying*

$$V_x \cap V_y = \varnothing.$$

Notice that in Definition 2.7 we may suppose V_x and V_y to be *open*.

If $(X, \mathscr{T})$ is separated, $x \in X$, and $\mathscr{F}$ is a fundamental system of x, then

$$\bigcap_{V \in \mathscr{F}} V = \{x\}.$$

Example 3.—The topological space **R** is separated.

In fact, take x and y in **R**, $x \neq y$. Suppose, for instance, that $x < y$ and let $\varepsilon = (y - x)/2$. Then

$$(x - \varepsilon, x + \varepsilon) \cap (y - \varepsilon, y + \varepsilon) = \varnothing.$$

Hence (see Example 2), **R** is separated.

Example 4.—(a) The topological space $(X, \mathscr{P}(X))$ is separated. (b) The topological space considered in Example 5, Chapter 1, is *not* separated if X is infinite.

2.8 Definition.—*A topological space $(X, \mathscr{T})$ is said to be regular if for every t in X there exists a fundamental system of t consisting of closed sets.*

By (ii) of Example 2, $\mathscr{F}_2 = \{[t - \varepsilon, t + \varepsilon] \mid \varepsilon > 0\}$ is a fundamental system of t in **R**. Since for each $\varepsilon > 0$ the set $[t - \varepsilon, t + \varepsilon]$ is closed, it follows that $(\mathbf{R}, \mathscr{U})$ is regular.

Let $(X, \mathscr{T})$ be a topological space. A family $(A_i)_{i \in I}$ of parts of X is said to be *locally finite* if for every $x \in X$ there is $V \in \mathscr{N}(x)$ such that

$$\{i \mid V \cap A_i \neq \varnothing\}$$

is finite.

2.9 Theorem.—*Let* $(X, \mathscr{T})$ *be a topological space and* $(A_i)_{i \in I}$ *a locally finite family of closed parts of* X. *Then* $A = \bigcup_{i \in I} A_i$ *is closed.*

Proof.—Let $x \in \mathbf{C}A$. Let $V \in \mathscr{N}(x)$ open, such that

$$J = \{i \mid V \cap A_i \neq \varnothing\}$$

is finite and let $A_J = \bigcup_{i \in J} A_i$. Then $U = V \cap \mathbf{C}\, A_J \in \mathscr{N}(x)$ and

$$U \cap A = A \cap (V \cap \mathbf{C}A_J) = A_J \cap (V \cap \mathbf{C}A_J) = \varnothing .$$

Hence $U \subset \mathbf{C}A$, so that $\mathbf{C}A \in \mathscr{N}(x)$. Since $x \in \mathbf{C}A$ was arbitrary, $\mathbf{C}A$ is open (see 2.4). Hence, A is closed.

Remark.—Notice that if $(A_i)_{i \in I}$ is an *arbitrary* family of closed parts of a topological space, then $\bigcup_{i \in I} A_i$ is not necessarily closed (see the remark following Theorem 1.12).

Exercises for Chapter 2

1. Consider the topological space $\mathbf{R}$ and let $t \in \mathbf{R}$. Then the sets $\mathscr{H}_1$, $\mathscr{H}_2$, $\mathscr{H}_3$, and $\mathscr{H}_4$ defined below are fundamental systems of t:

$$\mathscr{H}_1 = \{[t - \varepsilon, t + 2\varepsilon] \mid \varepsilon > 0\}; \qquad \mathscr{H}_2 = \{[t - 2\varepsilon, t + \varepsilon] \mid \varepsilon > 0\};$$

$$\mathscr{H}_3 = \{(t - \varepsilon, t + 2\varepsilon) \mid \varepsilon > 0\}; \qquad \mathscr{H}_4 = \{(t - 2\varepsilon, t + \varepsilon) \mid \varepsilon > 0\}.$$

2. In Example 1 of Chapter 1, for each $t \in X$, $\{\{t\}\}$ is a fundamental system of t.

3. Consider Examples 2, 3, and 5 of Chapter 1. In each case, write a simple fundamental system of $t \in X$.

4. The topological space $(X, \mathscr{T})$ in Example 2, Chapter 1 is separated if and only if either $X = \varnothing$ or X contains only one element.

5. The topological space $(X, \mathscr{T})$ in Example 3, Chapter 1 is not separated. The topological space $(X, \mathscr{T})$ in Example 4, Chapter 1 is not separated if X contains at least two elements.

6. Let $(X, \mathscr{T})$ be a topological space, let $t \in X$, and let $\mathscr{F}_1$, $\mathscr{F}_2$ be two fundamental systems of t. Show that

$$\bigcap_{V \in \mathscr{F}_1} V = \bigcap_{U \in \mathscr{F}_2} U.$$

7. A topological space $(X, \mathscr{T})$ is a T_1 space if, given any distinct elements s and t of X, there is $U \in \mathscr{N}(s)$ such that $t \notin U$. Show that $(X, \mathscr{T})$ is a T_1 space if and only if $\{t\}$ is closed for every t in X.

8. If $(X, \mathscr{T})$ is a topological space and $S \subset X$, we denote by $\mathscr{N}(S)$ the set of all neighborhoods of S; i.e.,

$$\mathscr{N}(S) = \{V \mid \text{there is } U \text{ in } \mathscr{T} \text{ such that } S \subset U \subset V\}.$$

Show that if $(X, \mathscr{T})$ is a T_1 space, and $S \subset X$, then $\bigcap \mathscr{N}(S) = S$. (The reader may also define a fundamental system of S ($S \subset X$) in an obvious way and show that if $(X, \mathscr{T})$ is a T_1 space, $S \subset X$, and $\mathscr{F}$ a fundamental system of S, then $\bigcap \mathscr{F} = S$.)

9. Show that the space in Example 3, Chapter 1, is not a T_1 space. Show that the space in Example 5, Chapter 1, is a T_1 space.

10. Let $(X, \mathscr{T})$ be a regular topological space. Suppose that $\mathscr{G} \subset \mathscr{T} \Rightarrow \bigcap \mathscr{G} \in \mathscr{T}$; that is, an arbitrary *intersection* of open sets is open. Then every open subset of X is closed.

11. Let $(X, \mathscr{T})$ be a topological space. A subset $\mathscr{B}$ of $\mathscr{T}$ is a basis of $\mathscr{T}$ if and only if $\mathscr{B}_x = \{B \mid B \in \mathscr{B} \text{ and } x \in B\}$ is a fundamental system at x for every x in X.

12. Let $(X, \mathscr{T})$ be a topological space. Show that $(X, \mathscr{T})$ is regular if and only if, given any closed subset F of X and $t \in X - F$, there are open sets U and V such that $t \in U$, $F \subset V$, and $U \cap V = \varnothing$.

13. Give an example of a topological space that is not regular.

14. On $\mathbf{R}$, consider the topology generated by

$$\mathscr{E} = \{(a, b) \mid a \in \mathbf{R}, b \in \mathbf{R}, a < b\} \cup \{\mathbf{Q}\}.$$

Show that $(\mathbf{R}, \mathscr{T}_{\mathscr{E}})$ is separated but not regular.

Chapter 3

Interior of a Set
Adherence of a Set

We denote by $(X, \mathscr{T})$ a topological space.

3.1 Definition.—*Let $A \subset X$. We say that $x \in X$ is an interior point of A if A is a neighborhood of x.*

Clearly, $x \in X$ is an interior point of A if and only if there is a neighborhood V_x of x such that $V_x \subset A$ (use 2.2 (b)).

The set of interior points of A is called the *interior* of A and is denoted $\mathring{A}$. Notice that $\mathring{X} = X$ and $\mathring{\varnothing} = \varnothing$.

3.2 Theorem.—*For each $A \subset X$, the set $\mathring{A}$ is open.*

Proof.—Let $x \in \mathring{A}$. Then A is a neighborhood of x. Hence, there exists $V_x \in \mathscr{T}$ such that

$$x \in V_x \subset A.$$

Since $x \in \mathring{A}$ was arbitrary, if we denote $W = \bigcup_{x \in \mathring{A}} V_x$, we obtain

$$\mathring{A} \subset W \subset A.$$

By (C_3), W is open; hence, if $x \in W$, then A is a neighborhood of x. We deduce that $W \subset \mathring{A}$. We conclude that $\mathring{A} = W$; thus $\mathring{A}$ is open.

3.3 Corollary.—*A set $A \subset X$ is open if and only if $A = \mathring{A}$.*

Proof.—If $A = \mathring{A}$, then by 3.2, A is open. If A is open, then A is a neighborhood of x for every $x \in A$. By Definition 3.1, $A \subset \mathring{A}$. Since $\mathring{A} \subset A$, we conclude that $A = \mathring{A}$.

We shall now list several further properties of the interior of a set. Their proofs are relatively easy and will be left as exercises for the reader.

We denote by $(X, \mathscr{T})$ a topological space and by A and B two parts of X; then:

3.4 $A \subset B \Rightarrow \mathring{A} \subset \mathring{B}$;

3.5 A open and $A \subset B \Rightarrow A \subset \mathring{B}$;

3.6 $\overset{\circ}{\widehat{A \cap B}} = \mathring{A} \cap \mathring{B}$;

3.7 $\overset{\circ}{\widehat{A \cup B}} \supset \mathring{A} \cup \mathring{B}$.

3.8 Show that it is not necessarily true that $\overset{\circ}{\widehat{A \cup B}} = \mathring{A} \cup \mathring{B}$. (Hint: In Example 3, Chapter 1, take $A = \{a\}$ and $B = \{b\}$.)

Exercise 1.—Let $A \subset X$ and let $\mathscr{A}$ be the set of all open parts $B \subset A$. Then

$$\mathring{A} = \bigcup_{B \in \mathscr{A}} B.$$

Hence, $\mathring{A}$ is the "largest" open set contained in A.

Denote by $(X, \mathscr{T})$ a topological space.

3.9 Definition.—*Let $A \subset X$. We say that $x \in X$ is adherent to A if every neighborhood of x has a non-void intersection with A.*

The set of all points adherent to A is called the adherence of A (or the *closure* of A); it is denoted by $\bar{A}$. Notice that $\bar{X} = X$ and $\bar{\varnothing} = \varnothing$. It is also obvious that

3.10 $A \subset \bar{A}$

for every set $A \subset X$.

With the notations we have just introduced, we may say that $x \in \bar{A}$ *if and only if $V \cap A \neq \varnothing$ for every $V \in \mathscr{N}(x)$.*

For applications, it is important to notice that *if $\mathscr{F}$ is a fundamental system of x, then $x \in \bar{A}$ if and only if $W \cap A \neq \varnothing$ for every $W \in \mathscr{F}$.*

Let $x \in \bar{A}$. If $W \in \mathscr{F}$, then $W \in \mathscr{N}(x)$, whence $W \cap A \neq \varnothing$. Conversely, let $V \in \mathscr{N}(x)$. Then there is $W \in \mathscr{F}$ such that $W \subset V$. Since $W \cap A \neq \varnothing$, we obtain $V \cap A \neq \varnothing$. Since $V \in \mathscr{N}(x)$ was arbitrary, we conclude that $x \in \bar{A}$.

Example 1.—Consider the topological space $\mathbf{R}$. Let $A \subset \mathbf{R}$ and $t \in \mathbf{R}$. Then the following assertions are equivalent:

(i) $t \in \bar{A}$;
(ii) for every $\varepsilon > 0$, $[t - \varepsilon, t + \varepsilon] \cap A \neq \varnothing$;
(iii) for every $\varepsilon > 0$, $(t - \varepsilon, t + \varepsilon) \cap A \neq \varnothing$.

The equivalence of these three assertions follows from the fact that $\{[t - \varepsilon, t + \varepsilon] \mid \varepsilon > 0\}$ and $\{(t - \varepsilon, t + \varepsilon) \mid \varepsilon > 0\}$ are fundamental systems of t (see Example 2, Chapter 2).

Example 2.—If $A \subset \mathbf{R}$ is a bounded set, then

$$\inf A \in \bar{A} \quad \text{and} \quad \sup A \in \bar{A}.$$

Let $\alpha = \inf A$. Then $\alpha \leq x$ for all $x \in A$. Now let $\varepsilon > 0$. Since $\alpha + \varepsilon > \alpha$, it follows that $\alpha + \varepsilon$ is *not* a minorant of A. Hence, there is $x_\varepsilon \in A$ such that $x_\varepsilon < \alpha + \varepsilon$. Hence,

$$(\alpha - \varepsilon, \alpha + \varepsilon) \cap A \ni x_\varepsilon,$$

and hence

$$(\alpha - \varepsilon, \alpha + \varepsilon) \cap A \neq \varnothing.$$

Since $\varepsilon > 0$ was arbitrary, we deduce that $\alpha \in \bar{A}$ (see (iii) of Example 1). In a similar way, we prove that $\sup A \in \bar{A}$.

Example 3.—Consider the topological space $(X, \mathscr{T})$ where $X = \{a, b, c\}$ and where $\mathscr{T}$ is the topology introduced in Example 3 of Chapter 1. If $A = \{a, b\}$, then $c \in \bar{A}$. In fact, if $V \in \mathscr{N}(c)$, then $V = \{b, c\}$ or $V = \{a, b, c\}$ (these are the only neighborhoods of c in $(X, \mathscr{T})$) and we have

$$\{b, c\} \cap A = \{b\} \neq \varnothing \quad \text{and} \quad \{a, b, c\} \cap A = \{a, b\} \neq \varnothing.$$

Hence, $c \in \bar{A}$. Notice that in this example, $\bar{A} = X$.

Example 4.—Let $(X, \mathscr{T})$ be the topological space considered in Example 3 above. If $B = \{a, c\}$, then $b \notin \bar{B}$. In fact, $\{b\}$ is a neighborhood of b (why?) and $\{b\} \cap B = \varnothing$. Notice that in this example, $\bar{B} = B$.

Example 5.—Consider the topological space $(X, \mathscr{P}(X))$ (see Example 1, Chapter 1). Then $A = \bar{A}$ for every $A \subset X$ (we leave the proof to the reader).

Example 6.—Consider the topological space $(X, \mathscr{T})$ where X is an *infinite* set and $\mathscr{T}$ is the topology introduced in Example 5 of Chapter 1. Then for every infinite part $A \subset X$, we have $\bar{A} = X$. In fact, let $x \in X$ and let $V \in \mathscr{N}(x)$. Then there is an open set W such that $V \supset W \ni x$. Since W is open, $\mathbf{C}W$ is finite; hence $\mathbf{C}V$ is finite. Since A is infinite, $A \not\subset \mathbf{C}V$; therefore $A \cap V \neq \varnothing$. Since $V \in \mathscr{N}(x)$ was arbitrary, we conclude that $x \in \bar{A}$. Since x was arbitrary, $X = \bar{A}$.

Let $(X, \mathscr{T})$ be a topological space.

3.11 Theorem.—*If* $A \subset X$, *then* $\mathbf{C}\bar{A} = \overset{\circ}{\widehat{\mathbf{C}A}}$.

Proof.—Let $x \in \mathbf{C}\bar{A}$. Then there is $V_x \in \mathscr{N}(x)$ such that $V_x \cap A = \varnothing$; hence, $V_x \subset \mathbf{C}A$, and hence $x \in \overset{\circ}{\widehat{\mathbf{C}A}}$. Since $x \in \mathbf{C}\bar{A}$ was arbitrary,

(i) $\mathbf{C}\bar{A} \subset \overset{\circ}{\widehat{\mathbf{C}A}}$.

Conversely, let $x \in \overset{\circ}{\widehat{\mathbf{C}A}}$. Then there is $V_x \in \mathscr{N}(x)$ such that $V_x \subset \mathbf{C}A$. Then $V_x \cap A = \varnothing$, and thus $x \notin \bar{A}$; that is, $x \in \mathbf{C}\bar{A}$. Since $x \in \overset{\circ}{\widehat{\mathbf{C}A}}$ was arbitrary, we have

(ii) $\overset{\circ}{\widehat{\mathbf{C}A}} \subset \mathbf{C}\bar{A}$.

Comparing (i) and (ii), we obtain 3.11.

3.12 Corollary.—*If* $A \subset X$, *then* $\bar{A}$ *is closed.*

Proof.—By 3.2, the set $\overset{\circ}{\widehat{\mathbf{C}A}}$ is open. Hence, $\mathbf{C}\bar{A}$ is open, whence $\bar{A}$ is closed.

3.13 Theorem.—*A set* $A \subset X$ *is closed if and only if* $A = \bar{A}$.

Proof.—If $A = \bar{A}$, then by 3.12, A is closed. Conversely, suppose that A is closed. Then $\mathbf{C}A$ is open, whence by 3.3, $\mathbf{C}A = \overset{\circ}{\widehat{\mathbf{C}A}}$. By 3.11, $\mathbf{C}\bar{A} = \mathbf{C}A$; that is,

$$\bar{A} = \mathbf{CC}\bar{A} = \mathbf{CC}A = A.$$

3.14 Corollary.—*If* $A \subset X$, *then* $\bar{\bar{A}} = \bar{A}$.

Remark.—If $(X, \mathscr{T})$ is separated, then $\overline{\{x\}} = \{x\}$ for every $x \in X$. In fact, if $y \neq x$, then there exists $V \in \mathscr{N}(y)$ such that $x \notin V$. Hence, $\{x\}$ is closed for every $x \in X$ if $(X, \mathscr{T})$ is separated.

Example 7.—If $A \subset \mathbf{R}$ is a bounded closed set, then

$$\inf A \in A \quad \text{and} \quad \sup A \in A.$$

This follows immediately from Example 2 and from the fact that A is closed if and only if $A = \bar{A}$.

More generally, if $B \subset A$, then

$$\inf B \in A \quad \textit{and} \quad \sup B \in A.$$

In fact, by Example 2, $\inf B \in \bar{B} \subset A$ and $\sup B \in \bar{B} \subset A$.

We shall now list several further properties of the adherence of a set. Their proofs are relatively easy and will be left as exercises for the reader.

We denote by $(X, \mathscr{T})$ a topological space and by A and B two parts of X; then:

3.15 $A \subset B \Rightarrow \bar{A} \subset \bar{B}$;

3.16 B closed and $A \subset B \Rightarrow \bar{A} \subset B$;

3.17 $\overline{A \cup B} = \bar{A} \cup \bar{B}$;

3.18 $\bar{A} \cap \bar{B} \supset \overline{A \cap B}$.

3.19 Show that it is not necessarily true that

$$\bar{A} \cap \bar{B} = \overline{A \cap B}.$$

Exercise 2.—Let $B \subset X$ and let $\mathscr{F}$ be the set of all closed sets containing B. Then

$$\bar{B} = \bigcap_{A \in \mathscr{F}} A.$$

Hence, $\bar{B}$ is the "smallest" closed set containing B.

Example 8.—Consider the topological space **R**. Then

3.20 $\overline{(a, b)} = [a, b]$.

By Example 7, Chapter 1 $[a, b]$ is closed. By 3.16, $\overline{(a, b)} \subset [a, b]$. Now let $\varepsilon > 0$. Then $(a - \varepsilon, a + \varepsilon) \cap (a, b) \neq \varnothing$. (Why?) Since $\{(a - \varepsilon, a + \varepsilon) \mid \varepsilon > 0\}$ is a fundamental system of a, we conclude that $a \in \overline{(a, b)}$. In the same way, we show that $b \in \overline{(a, b)}$. Since $(a, b) \subset \overline{(a, b)}$, we conclude that 3.20 is true.

3.21 Definition.—*Let $A \subset X$ and $B \subset X$. We say that A is dense in B if $\bar{A} \supset B$.*

In particular, A is *dense* in X if $\bar{A} = X$. Notice that if $A \subset X$, then A is dense in $\bar{A}$.

Example 9.—(a) In the topological space **R**, (a, b) is dense in $[a, b]$ (see 3.20). (b) In the topological space **R**, **Q** is dense in **R**.

(iii) In the topological space $(X, \mathscr{T})$ considered in Example 6, every *infinite* part of X is dense in X.

3.22 Definition.—*Let $A \subset X$. We say that $x \in X$ is a frontier point* of A if every neighborhood of x has a non-void intersection with both A and $\mathbf{C}A$.*

Hence, x is a frontier point of A if and only if

$$x \in \bar{A} \cap \overline{\mathbf{C}A}.$$

The set of all frontier points of A is denoted $Fr(A)$ and called the *frontier of A*.

We shall now list several properties of the frontier of a set. Their relatively easy proofs will be left as exercises for the reader.

We denote by $(X, \mathscr{T})$ a topological space and by A, B, and C three parts of X; then:

3.23 $Fr(A) = \bar{A} - \mathring{A}$;

3.24 a set $B \subset X$ is closed $\Leftrightarrow B \supset Fr(B)$;

3.25 a set $C \subset X$ is both open and closed $\Leftrightarrow Fr(C) = \varnothing$;

3.26 if $A \subset X$, then $Fr(A)$ is a closed set.

Exercises for Chapter 3

1. In the topological space $\mathbf{R}$, show that if $a \in \mathbf{R}$, $b \in \mathbf{R}$, and $a < b$, then

$$\overset{\circ}{\widehat{[a, b]}} = (a, b).$$

(See 1.10 and 1.11 for the notations.) Show that if $t \in \mathbf{R}$, then $\{t\} = \overline{\{t\}}$ and $\overset{\circ}{\widehat{\{t\}}} = \varnothing$. Show that if $F \subset \mathbf{R}$ is finite, then $F = \bar{F}$ and $\mathring{F} = \varnothing$.

2. Let $(X, \mathscr{T})$ be the topological space in Example 4, Chapter 1. Show that if $B \ni a$, then $\bar{B} = X$. Show that if $D \not\ni a$, then $\mathring{D} = \varnothing$.

3. Prove assertions 3.4 through 3.8.

4. Prove assertions 3.15 through 3.19.

5. Let $(X, \mathscr{T})$ be a topological space and suppose that $U \in \mathscr{T}$, $A \in \mathscr{P}(X)$. Then $U \cap A = \varnothing \Leftrightarrow U \cap \bar{A} = \varnothing$.

6. Let $(X, \mathscr{T})$ be a topological space and suppose that $A \subset X$. Then A is dense in X if and only if $A \cap S \neq \varnothing$ for every $S \in \mathscr{T} - \{\varnothing\}$.

---

* We say *frontier* instead of *boundary* since the term boundary is used with a different meaning in the theory of manifolds.

7. Let $(X, \mathscr{T})$ be a topological space. For each $A \in \mathscr{P}(X)$,

$$\bar{A} = A \cup Fr(A).$$

8. Let $(X, \mathscr{T})$ be a topological space. A subset S of X is closed if and only if $\bar{A} \cup S = \overline{A \cup S}$ for every A in $\mathscr{P}(X)$.

9. Let $(X, \mathscr{T})$ be a topological space. A subset A of X is said to be nowhere dense if and only if $\mathring{\bar{A}} = \varnothing$. Let F be a closed subset of X. Then:

(i) F is nowhere dense $\Leftrightarrow$ $\mathbf{C}F$ is dense in X;
(ii) $Fr(F)$ is nowhere dense.

10. Let $(X, \mathscr{T})$ be a topological space. For each t in X, define $\mathscr{A}(t) = \{A \mid A \subset X, t \in \bar{A}\}$; that is, $A \in \mathscr{A}(t) \Leftrightarrow$ for every V in $\mathscr{N}(t)$, $V \cap A \neq \varnothing$. Show that $U \in \mathscr{N}(t) \Leftrightarrow$ for every A in $\mathscr{A}(t)$, $U \cap A \neq \varnothing$.

11. Prove the assertions 3.23–3.26.

12. Let $(X, \mathscr{T})$ be a topological space, $A \subset X$ open and $Z \subset X$. Then

$$A \cap \bar{Z} \subset \overline{A \cap Z}.$$

Chapter 4

Subspaces

In this section we shall define and discuss the notion of a *subspace* of a topological space.

Let $(X, \mathscr{T})$ be a topological space and let $S \subset X$. Let

$$\mathscr{T}_S = \{U \cap S \mid U \in \mathscr{T}\}.$$

Then $\mathscr{T}_S$ *is a topology on* S. In fact, we have

$$\varnothing = \varnothing \cap S \quad \text{and} \quad S = X \cap S.$$

Since $\varnothing \in \mathscr{T}$ and $X \in \mathscr{T}$, we deduce that:

(C_1) $\varnothing \in \mathscr{T}_S$ *and* $S \in \mathscr{T}_S$.

Now let A and B be sets belonging to $\mathscr{T}_S$. Then there are U and V in $\mathscr{T}$ such that

$$A = U \cap S \quad \text{and} \quad B = V \cap S.$$

Hence

$$A \cap B = (U \cap S) \cap (V \cap S) = (U \cap V) \cap S.$$

Since $U \cap V \in \mathscr{T}$, we deduce that $A \cap B \in \mathscr{T}_S$. Hence:

(C_2) *if* A *and* B *belong to* $\mathscr{T}_S$, *then* $A \cap B \in \mathscr{T}_S$.

Finally, let $(A_i)_{i \in I}$ be a family of elements of $\mathscr{T}_S$. For each $i \in I$, there is $U_i \in \mathscr{T}$ such that $A_i = U_i \cap S$. Hence

$$\bigcup_{i \in I} A_i = \bigcup_{i \in I} (U_i \cap S) = \left(\bigcup_{i \in I} U_i\right) \cap S.$$

Since $\bigcup_{i \in I} U_i \in \mathscr{T}$, we deduce that $\bigcup_{i \in I} A_i \in \mathscr{T}_S$. Hence:

(C_3) *if* $(A_i)_{i \in I}$ *is a family of elements of* $\mathscr{T}_S$, *then*

$$\bigcup_{i \in I} A_i \in \mathscr{T}_S.$$

We conclude that $\mathscr{T}_S$ satisfies the conditions (C_1), (C_2), and (C_3), and hence that *$\mathscr{T}_S$ is a topology on S*. We usually say that $\mathscr{T}_S$ is the topology induced on S by $\mathscr{T}$.

This way we define the topological space $(S, \mathscr{T}_S)$. Any topological space of this form will be called a *subspace* of $(X, \mathscr{T})$. We shall often say the *subspace S of X*, instead of the *subspace $(S, \mathscr{T}_S)$ of $(X, \mathscr{T})$*.

For further reference, let us state here the following result, which follows immediately from the definition of $\mathscr{T}_S$.

4.1 Theorem. —*A set $V \subset S$ is open in the subspace S (of $(X, \mathscr{T})$) if and only if there is a set $U \subset X$, open (in the space $(X, \mathscr{T})$) such that $V = S \cap U$.*

Remarks.—Let $\mathscr{E} \subset \mathscr{T}$ be such that* $\mathscr{T}_{\mathscr{E}} = \mathscr{T}$. It easily follows, then, that if $\mathscr{E}_S = \{U \cap S \mid U \in \mathscr{E}\}$, then $\mathscr{T}_{\mathscr{E}_S} = \mathscr{T}_S$. If $\mathscr{B} \subset \mathscr{T}$ is a basis of $\mathscr{T}$, and if $\mathscr{B}_S = \{U \cap S \mid U \in \mathscr{B}\}$, then $\mathscr{B}_S$ is a basis of $\mathscr{T}_S$.

We shall now give several elementary results concerning subspaces. We denote by $(X, \mathscr{T})$ a topological space, by S a subset of X, and by $\mathscr{T}_S$ the topology induced on S by $\mathscr{T}$.

4.2 *A set $L \subset S$ is closed in the subspace S (of $(X, \mathscr{T})$) if and only if there is a set $F \subset X$, closed (in the space $(X, \mathscr{T})$) such that $L = S \cap F$.*

Proof.—Suppose that F is closed in $(X, \mathscr{T})$. Then $\mathbf{C}F$ is open in $(X, \mathscr{T})$ and, if $L = F \cap S$, we have

$$S - L = S \cap (X - F).$$

Since $X - F$ is open in $(X, \mathscr{T})$, it follows that $S - L$ is open in the subspace S, whence L is closed in the subspace S.

Conversely, suppose that $L \subset S$ is closed in the subspace S. Then $S - L$ is open in the subspace S. Hence, there is a set $U \subset X$, open in $(X, \mathscr{T})$, such that

$$S - L = S \cap U.$$

Then

$$L = S - (S \cap U) = S \cap (X - U).$$

Since $X - U$ is closed in $(X, \mathscr{T})$, Assertion 4.2 is proved.

4.3 *For each $t \in S$, we have*

$$\mathscr{N}_S(t) = \{V \cap S \mid V \in \mathscr{N}_X(t)\}.$$

Proof.—Let $V \in \mathscr{N}_X(t)$. Then there is U open in $(X, \mathscr{T})$ such that $t \in U \subset V$. Since $t \in U \cap S \subset V \cap S$ and $U \cap S$ is open in S, we

* The notation $\mathscr{T}_{\mathscr{E}}$, for $\mathscr{E}$ a set of parts, and the notation $\mathscr{T}_S$, for S a set have different meanings.

deduce that $V \cap S \in \mathcal{N}_S(t)$. Hence

$$\{V \cap S \mid V \in \mathcal{N}_X(t)\} \subset \mathcal{N}_S(t).$$

Conversely, let $W \in \mathcal{N}_S(t)$. Then there is A open in the subspace S such that $t \in A \subset W$. Since A is open in S, there exists B open in $(X, \mathcal{T})$ such that $A = B \cap S$. Now, $V = W \cup B \supset B \ni t$, and hence $V \in \mathcal{N}_X(t)$. Moreover,

$$V \cap S = (W \cap S) \cup (B \cap S) = W \cup A = W.$$

Hence

$$\mathcal{N}_S(t) \subset \{V \cap S \mid V \in \mathcal{N}_X(t)\}.$$

We conclude that 4.3 is proved.

4.4 *Let $t \in S$ and let $\mathcal{F}$ be a fundamental system of t in the space $(X, \mathcal{T})$. Then*

$$\mathcal{F}_S = \{V \cap S \mid V \in \mathcal{F}\}$$

is a fundamental system of t in the subspace S.

Proof.—We leave the proof to the reader.

4.5 *If $A \subset S$, then the adherence of A in the subspace S is the intersection of S with the adherence of A in $(X, \mathcal{T})$.*

Proof.—Denote by $\hat{A}$ the adherence of A in S by $\bar{A}$ the adherence of A in X. Since $\bar{A}$ is closed in $(X, \mathcal{T})$, $\bar{A} \cap S$ is closed in S. Since $A \subset \bar{A} \cap S$, we deduce that

$$\hat{A} \subset \bar{A} \cap S.$$

Conversely, let $x \in \bar{A} \cap S$. Let $W \in \mathcal{N}_S(x)$ and let $V \in \mathcal{N}_X(x)$ such that $V \cap S = W$. Then

$$W \cap A = V \cap S \cap A = (V \cap A) \cap S.$$

Since $x \in \bar{A}$, we have $V \cap A \neq \varnothing$. Since $A \subset S$, we deduce that $(V \cap A) \cap S \neq \varnothing$, and thus $W \cap A \neq \varnothing$. Since $W \in \mathcal{N}_S(x)$ was arbitrary, we conclude that $x \in \hat{A}$. Hence

$$\bar{A} \cap S \subset \hat{A},$$

and hence 4.5 is proved.

4.6 *The subspace $(S, \mathcal{T}_S)$ is separated if $(X, \mathcal{T})$ is.*

Proof.—Let x and y be in S, $x \neq y$. Since $(X, \mathscr{T})$ is separated, there are $V_x \in \mathscr{N}_X(x)$ and $V_y \in \mathscr{N}_X(y)$ such that $V_x \cap V_y = \varnothing$. Let $U_x = V_x \cap S$ and $U_y = V_y \cap S$. Then $U_x \in \mathscr{N}_S(x)$, $V_y \in \mathscr{N}_S(y)$, and $U_x \cap U_y = \varnothing$. Since x and y were arbitrary in S such that $x \neq y$, we conclude that $(S, \mathscr{T}_S)$ is separated.

4.7 *The subspace $(S, \mathscr{T}_S)$ is regular if $(X, \mathscr{T})$ is.*

Proof.—Let $x \in S$ and let $U \in \mathscr{N}_S(x)$. Then there is $V \in \mathscr{N}_X(x)$ such that $U = V \cap S$. Since $(X, \mathscr{T})$ is regular, there is a closed $W \in \mathscr{N}_X(x)$ contained in V. Then $W \cap S$ is closed and belongs to $\mathscr{N}_S(x)$, and $W \cap S \subset U$. Since x and U were arbitrary, we deduce that $(S, \mathscr{T}_S)$ is regular.

4.8 Definition.—*Let $(X, \mathscr{T})$ be a topological space. A set $Z \subset X$ is locally closed if for every $z \in Z$ there exists $V \in \mathscr{N}_X(x)$ such that* $Z \cap V$ is closed in the subspace $(V, \mathscr{T}_V)$.*

4.9 Theorem.—*Let $(X, \mathscr{T})$ be a topological space and $Z \subset X$. Then the following assertions are equivalent:*

(a) *the set Z is locally closed;*
(b) *there is an open set $U \subset X$ and a closed set $F \subset X$ such that $Z = U \cap F$.*

Proof of (a) $\Rightarrow$ (b).—Let $z \in Z$ and $V(z) \in \mathscr{N}_X(z)$, open, such that $Z \cap V(z)$ is closed in $V(z)$. Then $Z \cap V(z) = \bar{Z} \cap V(z)$ for all $z \in Z$ (see 4.5 and Exercise 12, Chapter 3). Let

$$U = \bigcup_{z \in Z} V(z) \quad \text{and} \quad F = \bar{Z}.$$

Then

$$Z \subset U \cap F = U \cap \bar{Z} \subset \left(\bigcup_{z \in Z} V(z)\right) \cap \bar{Z}$$
$$= \bigcup_{z \in Z}(V(z) \cap \bar{Z}) = \bigcup_{z \in Z}(V(z) \cap Z) \subset Z;$$

hence $Z = U \cap F$.

Proof of (b) $\Rightarrow$ (a).—For each $z \in Z$, let $V(z) = U$. Then

$$Z \cap V(z) = F \cap U \cap V(z) = F \cap V(z)$$

is closed in $V(z)$. Hence, Z is locally closed.

* Notice that we may assume V to be open.

Exercises for Chapter 4

1. Let $(X, \mathscr{T})$ be a topological space and suppose $Y \subset X$ and $Z \subset Y$. Show that $\mathscr{T}_Z = (\mathscr{T}_Y)_Z$. (Hence, the subspace $(Z, \mathscr{T}_Z)$ of $(X, \mathscr{T})$ is the same as the subspace $(Z, (\mathscr{T}_Y)_Z)$ of the subspace $(Y, \mathscr{T}_Y)$ of $(X, \mathscr{T})$.)

2. Let $(X, \mathscr{T})$ be a topological space and suppose that S is an open subset of X and T a closed subset of X. Then:

(i) a subset A of S is open in the space $(S, \mathscr{T}_S)$ if and only if A is open in $(X, \mathscr{T})$;

(ii) a subset B of T is closed in the space $(T, \mathscr{T}_T)$ if and only if B is closed in $(X, \mathscr{T})$.

4. Consider the subspace $([0, 1], \mathscr{U}_{[0,1]})$ of $\mathbf{R}$. Is

$$\{(1 - \varepsilon, 1] \mid 0 \leq \varepsilon \leq 1\}$$

a fundamental system of 1? Is

$$\{[0, \varepsilon) \mid 0 < \varepsilon \leq 1\}$$

a fundamental system of 0?

5. Let X and Y be sets such that $X \subset Y$. Suppose $\mathscr{T}$ is a topology on X. Then $\mathscr{W} = \mathscr{T} \cup \{Y\}$ is a topology on Y and $\mathscr{W}_X = \mathscr{T}$.

Chapter 5

Continuous Functions

In this chapter, we shall introduce the notion of *continuous function* and we shall give several properties.

Below we denote by $(X, \mathscr{T})$ and $(Y, \mathscr{T})$ two topological spaces.

5.1. Definition.—*Let $f: X \to Y$. We say that f is continuous at $c \in X$ if for every neighborhood V of $f(c)$ there exists a neighborhood U of c such that $f(x) \in V$ whenever $x \in U$.*

Notice that U *depends* on V. Some notation that indicates this is sometimes used; for instance, we may write U_V instead of U. Actually, U also depends upon f and c; however, we usually do not use additional notation to indicate this.

If $f: X \to Y$ is continuous at c, and if $V \in \mathscr{N}_Y(f(c))$, then by definition there exists $U \in \mathscr{N}_X(c)$ such that $f(x) \in V$ if $x \in U$. Hence, $f^{-1}(V) \supset U$, and hence $f^{-1}(V) \in \mathscr{N}_X(c)$. Conversely, if for every $V \in \mathscr{N}_Y(f(c))$ the set $f^{-1}(V) \in \mathscr{N}_X(c)$, we have $f(x) \in V$ if $x \in U = f^{-1}(V)$; hence f is continuous at c. We conclude therefore, that *$f: X \to Y$ is continuous at $c \in X$ if and only if*

$$V \in \mathscr{N}_Y(f(c)) \Rightarrow f^{-1}(V) \in \mathscr{N}_X(c).$$

Now let $S \subset X$, and consider the topological space $(S, \mathscr{T}_S)$ (see Chapter 4). A function $f: S \to Y$ is *continuous at* $c \in S$ if, considered as a mapping "of the *subspace* $(S, \mathscr{T}_S)$ into $(Y, \mathscr{T})$," it is continuous in the sense of Definition 5.1. It follows that *$f: S \to Y$ is continuous at $c \in S$ if and only if for every $V \in \mathscr{N}_Y(f(c))$ there exists $U \in \mathscr{N}_X(c)$ such that*

$$x \in U \cap S \Rightarrow f(x) \in V.$$

Example 1.—Let $S \subset X$ and let $j_{S,X}: S \to X$ be defined by $j_{S,X}(x) = x$ for $x \in S$. If $c \in S$, then $j_{S,X}$ *is continuous at* c. In fact, let

$V \in \mathscr{N}(j_{S,X}(c))$ and let $U = V$. Then

$$x \in U \cap S \Rightarrow x \in U = V;$$

whence

$$x \in U \cap S \Rightarrow j_{S,X}(x) \in V.$$

Since $V \in \mathscr{N}(j_{S,X}(c))$ was arbitrary, we deduce that $j_{S,X}$ is continuous at c.

Example 2.—Let $S \subset X$, $\alpha \in Y$, and let $f: S \to Y$ be defined by $f(x) = \alpha$ for $x \in S$. If $c \in S$, then f *is continuous at* c. In fact, let $V \in \mathscr{N}(f(c))$ (notice that $f(c) = \alpha$). Then

$$x \in S \Rightarrow f(x) = \alpha \in V;$$

whence

$$x \in X \cap S \Rightarrow f(x) \in V.$$

Since $V \in \mathscr{N}(f(c))$ was arbitrary, and since $X \in \mathscr{N}(c)$, we deduce that f is continuous at c.

In this example, for every $V \in \mathscr{N}(f(c))$ we may take* $U = X$; this is due to the special form of the function f.

Example 3.—Let $h: \mathbf{R} \to \mathbf{R}$ be defined by

$$h(x) = \begin{cases} 1 & \text{if} \quad x \geq 0 \\ 0 & \text{if} \quad x < 0. \end{cases}$$

Then h is *not* continuous at 0 (we leave the proof of this assertion to the reader).

5.2 Theorem.—*Let $T \subset S \subset X$ and let $c \in T$. Let $f: S \to Y$ be a function continuous at c. Then $f \mid T$ is continuous at c.*

Proof.—In fact, let $V \in \mathscr{N}((f \mid T)(c))$. Since $(f \mid T)(c) = f(c)$, and since f is continuous at c, there is $U \in \mathscr{N}(c)$ such that $x \in U \cap S$ implies $f(x) \in V$. Since $U \cap T \subset U \cap S$, we deduce that, for $x \in U \cap T$,

$$(f \mid T)(x) = f(x) \in V.$$

Since $V \in \mathscr{N}(f(c))$ was arbitrary, we deduce that $f \mid T$ is continuous at c.

Theorem 5.2 asserts that the continuity of f at $c \in T \subset S$ implies the continuity of $f \mid T$ at c. The converse is not necessarily true, as

---

* In fact, we could have taken for U *any* neighborhood of c.

the following example shows: Let $X = \mathbf{R}$, $S = X$, $T = \{x \mid x \geq 0\}$, and let h be the function defined in Example 3. Then $h \mid T$ is continuous at 0 (see Example 2), whereas h is not continuous at 0. We have, however, the following theorem.

5.3 Theorem.—*Let $S \subset X$, $c \in S$, and $f: S \to Y$ a mapping. Let $W \in \mathcal{N}(c)$, let $T = S \cap W$, and suppose that $f \mid T$ is continuous at c. Then f is continuous at c.*

Proof.—Notice that $(f \mid T)(c) = f(c)$ and let $V \in \mathcal{N}(f(c))$. Since $f \mid T$ is continuous at c, there is $U \in \mathcal{N}(c)$ such that

$$(f \mid T)(x) \in V$$

if $x \in U \cap T$. But $U \cap T = (U \cap W) \cap S$ and $U \cap W \in \mathcal{N}(c)$. Hence

$$f(x) = (f \mid T)(x) \in V$$

if $x \in (U \cap W) \cap S$. Since $V \in \mathcal{N}(f(c))$ was arbitrary, we deduce that f is continuous at c.

The continuity of $f \mid T$ at $c \in T \subset S$ implies that f is continuous at c because of the fact that T is of the form $S \cap W$, where W *is a neighborhood of c.*

Let $S \subset X$ and $c \in S$. Let $f: S \to Y$ be a mapping. Then

5.4 Theorem.—*Let $\mathscr{C}$ be a fundamental system of c and $\mathscr{F}$ a fundamental system of $f(c)$. Then the following assertions are equivalent:*

(a) *the function f is continuous at c;*

(b) *for every $V \in \mathscr{F}$, there exists $U \in \mathscr{C}$ such that $f(x) \in V$ whenever $x \in U \cap S$.*

Proof of (a) $\Rightarrow$ (b).—Let $V \in \mathscr{F}$; then $V \in \mathcal{N}(f(c))$. Since f is continuous at c, there exists $W \in \mathcal{N}(c)$ such that $f(x) \in V$ whenever $x \in W$. Since $\mathscr{C}$ is a fundamental system of c, there exists $U \in \mathscr{C}$, $U \subset W$. Clearly, then, $f(x) \in V$ whenever $x \in U \cap S$ (since $U \cap S \subset W \cap S$).

Proof of (b) $\Rightarrow$ (a).—Let $V \in \mathcal{N}(f(c))$. Since $\mathscr{F}$ is a fundamental system of $f(c)$, there exists $V' \in \mathscr{F}$, $V' \subset V$. By (b), there is $U \in \mathscr{C}$ such that $f(x) \in V'$ whenever $x \in U \cap S$. Hence, $f(x) \in V$ whenever $x \in U \cap S$. Since $V \in \mathcal{N}(f(c))$ was arbitrary, we deduce (a).

Theorem 5.4 shows that, to establish the continuity of f at c, it is not necessary to consider *all* the neighborhoods V of $f(c)$. It is enough to consider only the neighborhoods in a *given* fundamental system of $f(c)$.

**Exercise* 1.—For any topological space $(T, \mathscr{T})$, define

$$\mathscr{A}(t) = \{A \mid A \subset T, t \in \bar{A}\}$$

for every $t \in T$. Let X and Y be topological spaces, $c \in X$, and $f: X \to Y$. Show that f is continuous at c if and only if

$$A \in \mathscr{A}(c) \Rightarrow f(A) \in \mathscr{A}(f(c)).$$

We have shown in Example 2, Chapter 2, that if $c \in \mathbf{R}$, then

$$\{[c - \varepsilon, c + \varepsilon] \mid \varepsilon > 0\} \quad \text{and} \quad \{(c - \varepsilon, c + \varepsilon) \mid \varepsilon > 0\}$$

are fundamental systems of c. We also know that if $\varepsilon > 0$, then

$$x \in [c - \varepsilon, c + \varepsilon] \Leftrightarrow |x - c| \leq \varepsilon$$

and

$$x \in (c - \varepsilon, c + \varepsilon) \Leftrightarrow |x - c| < \varepsilon.$$

From these remarks, and from Theorem 5.4, we deduce immediately the following results.

5.5 *Let $(X, \mathscr{T})$ be a topological space; let $S \subset X$ and $c \in S$. Let $f: S \to \mathbf{R}$ be a mapping. Then the following assertions are equivalent:*

(i) *f is continuous at c;*
(ii) *for every $\varepsilon > 0$, there is $V_\varepsilon \in \mathscr{N}(c)$ such that*

$$|f(x) - f(c)| \leq \varepsilon$$

if $x \in V_\varepsilon \cap S$;
(iii) *for every $\varepsilon > 0$, there is $U_\varepsilon \in \mathscr{N}(c)$ such that*

$$|f(x) - f(c)| < \varepsilon$$

if $x \in U_\varepsilon \cap S$.

5.6 *Let $S \subset \mathbf{R}$ and $c \in S$. Let $f: S \to \mathbf{R}$ be a mapping. Then the following assertions are equivalent:*

(i) *f is continuous at c;*
(ii) *for every $\varepsilon > 0$, there is $\delta_\varepsilon > 0$ such that*

$$|f(x) - f(c)| \leq \varepsilon$$

if $x \in S$ and $|x - c| \leq \delta_\varepsilon$;
(iii) *for every $\varepsilon > 0$, there is $\gamma_\varepsilon > 0$ such that*

$$|f(x) - f(c)| < \varepsilon$$

if $x \in S$ and $|x - c| < \gamma_\varepsilon$.

Let $(X, \mathscr{T})$, $(Y, \mathscr{I})$, and $(Z, \mathscr{W})$ be three topological spaces. The following is an important result with many applications.

5.7 Theorem.—*Let* $S \subset X$ *and* $f: S \to Y$ *and let* $T \subset Y$ *and* $g: T \to Z$. *Let* $c \in S$. *Suppose that:*

(a) $x \in S \Rightarrow f(x) \in T$;

(b) f *is continuous at* c *and* g *is continuous at* $f(c)$. *Then* $g \circ f$ *is continuous at* c.

The hypothesis (a) allows us to define $g \circ f$.

Proof.—We must show that for *every* $V \in \mathscr{N}(g \circ f(c))$ there exists $U \in \mathscr{N}(c)$ such that $x \in U \cap S \Rightarrow g \circ f(x) \in V$.

Hence, let $V \in \mathscr{N}(g \circ f(c)) = \mathscr{N}(g(f(c)))$. Since g is continuous at $f(c)$, there exists $W \in \mathscr{N}(f(c))$ such that

(i) $x \in W \cap T \Rightarrow g(x) \in V$.

Since $W \in \mathscr{N}(f(c))$ and since f is continuous at c, there exists $U \in \mathscr{N}(c)$ such that

(ii) $x \in U \cap S \Rightarrow f(x) \in W$.

Hence (use (i) and (ii))

$$x \in U \cap S \Rightarrow f(x) \in W \cap T \Rightarrow g(f(x)) \in V.$$

Hence

$$x \in U \cap S \Rightarrow g \circ f(x) \in V.$$

Since $V \in \mathscr{N}(g \circ f(c))$ was arbitrary, we conclude that $g \circ f$ is continuous at c.

Again, denote by $(X, \mathscr{T})$ and $(Y, \mathscr{I})$ two topological spaces.

5.8 Definition.—*Let* $f: X \to Y$. *We say that* f *is continuous on* X *if* f *is continuous at every point* $c \in X$.

Whenever we use the expression, "the mapping $f: X \to Y$ is continuous," we mean that f is continuous on X in the sense of Definition 5.8.

Now let $S \subset X$ and consider the topological space $(S, \mathscr{T}_S)$. A function $f: S \to X$ is *continuous on* S if, considered as a mapping "of the subspace $(S, \mathscr{T}_S)$ into $(Y, \mathscr{I})$," it is continuous on S in the sense of Definition 5.8. This means that $f: S \to Y$ is continuous on S if it is continuous at every point $c \in S$.

Whenever we use the expression, "the mapping $f: S \to Y$ is continuous," we mean that f is continuous on S in the sense defined above.

Notice that if $f: S \to Y$ is continuous on S, and $T \subset S$, then $f \mid T$ is continuous on T (see Theorem 5.2).

Example 4.—The function $j_{S,X}$ considered in Example 1 is continuous on S. The function f considered in Example 2 is continuous on S. In fact, it is enough to note that we proved in Examples 1 and 2 that $j_{S,X}$ and f were continuous at c for *arbitary* $c \in S$.

Consider a topological space $(X, \mathscr{T})$ and the topological space $\mathbf{R}$ (see Example 6, Chapter 1).

5.9 Definition.—*For each* $S \subset X$, *we denote by* $C_{\mathbf{R}}(S)$, *the set of all* $f: S \to \mathbf{R}$ *continuous on* S.

We shall now discuss several examples of continuous functions.

Example 5.—Let $S \subset \mathbf{R}$ and $f: S \to \mathbf{R}$. We say that f is a *Lipschitz function* if there is $L \geq 0$ such that

$$|f(x) - f(y)| \leq L\,|x - y|$$

for all x and y in S.

We shall show that if f is a Lipschitz function, then $f \in C_{\mathbf{R}}(S)$.

In fact, let $c \in S$ and let $V \in \mathscr{N}(f(c))$. Then there is $\varepsilon > 0$ such that $(f(c) - \varepsilon, f(c) + \varepsilon) \subset V$ (see Example 2, Chapter 2). Now let $\delta = \varepsilon/(1 + L)$ and let $U = (c - \delta, c + \delta)$. Then $U \in \mathscr{N}(c)$ and

$$x \in U \Rightarrow c - \delta < x < c + \delta \Rightarrow |x - c| < \delta \leq \varepsilon/(1 + L).$$

Hence

$$x \in U \cap S \Rightarrow |f(x) - f(c)| \leq L\,|x - c| \leq L\frac{\varepsilon}{1 + L} < \varepsilon.$$

Thus

$$x \in U \cap S \Rightarrow f(x) \in (f(c) - \varepsilon, f(c) + \varepsilon) \subset V.$$

Since $V \in \mathscr{N}(f(c))$ was arbitrary, we deduce* that f is continuous at c. Since $c \in S$ was arbitrary, we deduce that $f \in C_{\mathbf{R}}(S)$.

Example 6.—Let $S \subset \mathbf{R}$. Then the function $\beta: x \mapsto |x|$ on S to $\mathbf{R}$ belongs to $C_{\mathbf{R}}(S)$.

It is enough to remark that

$$|\beta(x) - \beta(y)| = ||x| - |y|| \leq |x - y|$$

for all $x \in S, y \in S$ and to use the result in Example 5.

---

* We may use 5.6.

Example 7.—Let $a \in \mathbf{R}$, $b \in \mathbf{R}$. Let $S \subset \mathbf{R}$ and let $h: S \to \mathbf{R}$ be defined by $h(x) = ax + b$ for $x \in S$. Then $h \in C_{\mathbf{R}}(S)$.

It is enough to remark that

$$|h(x) - h(y)| = |(ax + b) - (ay + b)| = |ax - ay| = |a|\,|x - y|$$

for all $x \in S, y \in S$ and to use the result in Example 5.

Example 8.—Let $S \subset \mathbf{R}^* = \mathbf{R} - \{0\}$, and let $k: S \to \mathbf{R}$ be defined by $k(x) = 1/x$ for $x \in S$. Then $k \in C_{\mathbf{R}}(S)$.

In fact, let $c \in S$ and $V \in \mathscr{N}(k(c))$. Then there is $\varepsilon > 0$ such that $(k(c) - \varepsilon, k(c) + \varepsilon) \subset V$. Let

$$\delta = \inf\{|c|/2, |c|^2 \varepsilon/2\}.$$

Recall that $x \in (c - \delta, c + \delta) \Leftrightarrow |x - c| < \delta$. Now

$$|x - c| < \delta \Rightarrow ||x| - |c|| < \delta \Rightarrow |x| > |c| - \delta \geq |c| - |c|/2 = |c|/2.$$

Hence, $x \in (c - \delta, c + \delta) \cap S$ implies that

$$|k(x) - k(x)| = \left|\frac{1}{x} - \frac{1}{c}\right| = \frac{|x - c|}{|x|\,|c|} \leq \frac{2}{|c|^2}|x - c| < \frac{2}{|c|^2}|c|^2 \varepsilon/2 = \varepsilon.$$

Therefore,

$$x \in (c - \delta, c + \delta) \cap S \Rightarrow |k(x) - k(c)| < \varepsilon$$

$$\Rightarrow k(x) \in (k(c) - \varepsilon, k(c) + \varepsilon) \subset V.$$

Since $V \in \mathscr{N}(k(c))$ was arbitrary, we deduce* that k is continuous at c, and since $c \in S$ was arbitrary, we deduce that k is continuous on S.

Let $(X, \mathscr{T})$, $(Y, \mathscr{S})$, and $(Z, \mathscr{W})$ be three topological spaces. From Theorem 5.7, we deduce immediately the following theorem.

5.10 Theorem.—*Let $S \subset X$ and $f: S \to Y$ and let $T \subset Y$ and $g: T \to Z$. Suppose that:*

(a) $x \in S \Rightarrow f(x) \in T$;

(b) *f is continuous on S and g is continuous on T. Then $g \circ f$ is continuous on S.*

In the following examples, we denote by $(X, \mathscr{T})$ a topological space, and by S, a part of X.

---

* We may use 5.6.

Example 9.—Let $f \in C_{\mathbf{R}}(S)$ and let $|f|$ be the mapping $x \mapsto |f(x)|$ of S into $\mathbf{R}$. Then $|f| \in C_{\mathbf{R}}(S)$.

In fact, if $\beta(x) = |x|$ for $x \in \mathbf{R}$, then β is continuous on $\mathbf{R}$ (see Example 6) and $|f| = \beta \circ f$. By Theorem 5.10, $|f| \in C_{\mathbf{R}}(S)$.

Example 10.—Let $f \in C_{\mathbf{R}}(S)$; let $a \in \mathbf{R}$, $b \in \mathbf{R}$, and let $af + b$ be the mapping $x \mapsto af(x) + b$ of S into $\mathbf{R}$. Then $af + b \in C_{\mathbf{R}}(S)$.

In fact, if $h(x) = ax + b$ for $x \in \mathbf{R}$, then h is continuous on $\mathbf{R}$ (see Example 7) and $af + b = h \circ f$. By Theorem 5.10, $af + b \in C_{\mathbf{R}}(S)$.

Note that af is the mapping $x \mapsto af(x)$. We also write $f = 1 \cdot f$ and $-f = (-1)f$.

Example 11.—Let $f \in C_{\mathbf{R}}(S)$. Suppose that $f(x) \neq 0$ for $x \in S$ and let $1/f$ be the mapping $x \mapsto 1/f(x)$ of S into $\mathbf{R}$. Then $1/f \in C_{\mathbf{R}}(S)$.

In fact, if $k(x) = 1/x$ for $x \in \mathbf{R}^*$, then k is continuous on $\mathbf{R}^*$ (see Example 7). Since $f(x) \neq 0 \Leftrightarrow f(x) \in \mathbf{R}^*$, we deduce that $k \circ f$ is defined. Now, clearly, $k \circ f = 1/f$. By Theorem 5.10, the function $1/f \in C_{\mathbf{R}}(S)$.

5.11 Theorem.—*Let $(X, \mathscr{T})$ and $(Y, \mathscr{I})$ be two topological spaces and let $f: X \to Y$. Then the following assertions are equivalent:*

(a) *f is continuous on X;*
(b) *$E \subset Y$ open $\Rightarrow f^{-1}(E)$ open;*
(c) *$F \subset Y$ closed $\Rightarrow f^{-1}(F)$ closed;*
(d) *$f(\bar{A}) \subset \overline{f(A)}$ for every $A \subset X$.*

Proof of (a) $\Rightarrow$ (b).—Let $E \subset Y$ be open and let $c \in f^{-1}(E)$. Since $c \in f^{-1}(E)$, it follows that $f(c) \in E$. Since E is open, $E \in \mathscr{N}(f(c))$. Since f is continuous, there is $U \in \mathscr{N}(c)$ such that $f(x) \in E$ for $x \in U$. We deduce that $c \in U \subset f^{-1}(E)$. Hence, $f^{-1}(E)$ is a neighborhood of c. Since $c \in f^{-1}(E)$ was arbitrary, we deduce that $f^{-1}(E)$ is open.

Proof of (b) $\Rightarrow$ (a).—Let $c \in X$ and let $V \in \mathscr{N}(f(c))$. Let $U = f^{-1}(\mathring{V})$. By (b), the set U is open, and clearly $U \ni c$. Since $x \in U \Rightarrow f(x) \in \mathring{V} \subset V$, it follows that f is continuous at c. Since $c \in X$ was arbitrary, (b) $\Rightarrow$ (a).

Proof of (b) $\Rightarrow$ (c).—Let $F \subset Y$ be closed. Then $\mathbf{C}F$ is open and, by (b), $\mathbf{C}f^{-1}(F) = f^{-1}(\mathbf{C}F)$ is open. Hence, $f^{-1}(F)$ is closed.

Proof of (c) $\Rightarrow$ (b).—Let $U \subset Y$ be open. Then $\mathbf{C}U$ is closed and, by (c), $\mathbf{C}f^{-1}(U) = f^{-1}(\mathbf{C}U)$ is closed. Hence, $f^{-1}(U)$ is open.

Proof of (a) $\Rightarrow$ (d).—Let $y \in f(\bar{A})$. Then there exists $x \in \bar{A}$ such that $f(x) = y$. Let $V \in \mathscr{N}_Y(y)$. Since f is continuous on X, there exists $U \in \mathcal{N}_X(x)$ such that $f(U) \subset V$. Since $x \in \bar{A}$, there is $x' \in U \cap A$. Then $f(x') \in V$, and thus $V \cap f(A) \neq \varnothing$. Since $V \in \mathcal{N}_Y(y)$ was arbitrary, we deduce that $y \in \overline{f(A)}$. Since $y \in f(\bar{A})$ was arbitrary, we obtain $f(\bar{A}) \subset \overline{f(A)}$.

Proof of (d) $\Rightarrow$ (a).—Let $F \subset Y$ be closed and let $E = f^{-1}(F)$. Then

$$f(\bar{E}) \subset \overline{f(E)} = \overline{f(f^{-1}(F))} \subset \bar{F} = F;$$

that is,

$$\bar{E} \subset f^{-1}(F) = E.$$

Since $E \subset \bar{E}$, we deduce that $E = \bar{E}$. Hence, $f^{-1}(F)$ is closed if $F \subset X$ is closed, and hence (use the fact that (c) $\Rightarrow$ (a)) f is continuous on X.

Exercise 2.—Use Exercise 1 to show that (a) $\Leftrightarrow$ (d).

For applications, it is useful to have the following theorem.

5.12 Theorem.—*Let $\mathscr{E} \subset \mathscr{I}$ be such that $\mathscr{T}_{\mathscr{E}} = \mathscr{I}$ (see Chapter* 1) *and let $f: X \to Y$. Then the following assertions are equivalent:*

(a) *f is continuous on X;*
(b) *$E \in \mathscr{E} \Rightarrow f^{-1}(E)$ open.*

Proof of (a) $\Rightarrow$ (b).—Since $\mathscr{E} \subset \mathscr{I}$, this implication follows from Theorem 5.10 (implication (a) $\Rightarrow$ (b)).

Proof of (b) $\Rightarrow$ (a).—Let $\mathscr{W}$ be the set of all $A \in \mathscr{P}(Y)$ such that $f^{-1}(A) \in \mathscr{T}$. It is immediate that $\mathscr{W}$ is a topology on Y. Since $\mathscr{E} \subset \mathscr{W}$, we deduce that $\mathscr{T}_{\mathscr{E}} \subset \mathscr{W}$; that is, $\mathscr{I} \subset \mathscr{W}$. Hence, $E \in \mathscr{I} \Rightarrow f^{-1}(E) \in \mathscr{T}$. By Theorem 5.10 (implication (b) $\Rightarrow$ (a)), f is continuous.

We denote again by $(X, \mathscr{T})$ and $(Y, \mathscr{I})$ two topological spaces.

5.13 Definition.—*We call a homeomorphism of $(X, \mathscr{T})$ onto $(Y, \mathscr{I})$ any bijection $f: X \to Y$ such that both f and f^{-1} are continuous.*

When there is no ambiguity, we shall simply say *homeomorphism of X onto Y*, instead of *homeomorphism of $(X, \mathscr{T})$ onto $(Y, \mathscr{I})$*.

Notice that a homeomorphism is necessarily a bijection. Notice also that f is a homeomorphism of X onto Y if and only if f^{-1} is a homeomorphism of Y onto X (observe that $(f^{-1})^{-1} = f$).

Two topological spaces are said to be *homeomorphic* if there exists a homeomorphism of one onto the other.

Example 13.—Let $a \in \mathbf{R}^*$ and $b \in \mathbf{R}$ and let $h(x) = ax + b$ for $x \in \mathbf{R}$. Then f is a homeomorphism of $\mathbf{R}$ onto $\mathbf{R}$.

Let $x \in \mathbf{R}, y \in \mathbf{R}$, and $x \neq y$. Then $ax \neq ay$ and $ax + b \neq ay + b$; hence $h(x) \neq h(y)$, and hence f is an injection. If $z \in \mathbf{R}$, then $h((1/a)z - b/a) = z$. Hence, f is a bijection of $\mathbf{R}$ onto $\mathbf{R}$. Clearly, h^{-1} is the mapping $x \mapsto (1/a)x - b/a$ of $\mathbf{R}$ onto $\mathbf{R}$. By the result in Example 7, the mappings h and h^{-1} are continuous. Hence h is a homeomorphism.

Example 14.—The mapping $x \mapsto x/(1 + |x|)$ of $\mathbf{R}$ onto $(-1, +1)$ is a homeomorphism of $\mathbf{R}$ onto $(-1, +1)$, considered as a subspace of $\mathbf{R}$. We leave the proof of this assertion to the reader.

Other examples of homeomorphisms will be given later in this book.

A *continuous bijection* of X onto Y is not necessarily a homeomorphism. For instance, the mapping* $j_{\mathbf{R}}$ of $(\mathbf{R}, \mathscr{D})$ onto $(\mathbf{R}, \mathscr{U})$ (see Example 1, Chapter 1) is a continuous bijection, but $j_{\mathbf{R}}^{-1}$ is *not* continuous. Hence, $j_{\mathbf{R}}$ is *not* a homeomorphism of $(\mathbf{R}, \mathscr{D})$ onto $(\mathbf{R}, \mathscr{U})$.

Let $(X, \mathscr{T})$, $(Y, \mathscr{I})$, and $(Z, \mathscr{W})$ be three topological spaces. Then we get the following theorem.

5.14 Theorem.—*If f is a homeomorphism of X onto Y, and g is a homeomorphism of Y onto Z, then $g \circ f$ is a homeomorphism of X onto Z.*

Proof.—Since f is a bijection of X onto Y and g is a bijection of Y onto Z, we deduce that $g \circ f$ is a bijection of X onto Z. By Theorem 5.10, $g \circ f$ is continuous. Since $(g \circ f)^{-1} = f^{-1} \circ g^{-1}$, we deduce, using Theorem 5.10 again, that $(g \circ f)^{-1}: Z \to X$ is continuous. Hence, $g \circ f$ is a homeomorphism.

Let $(X, \mathscr{T})$ and $(Y, \mathscr{I})$ be two topological spaces. A mapping $f: X \to Y$ is *open* if $f(A)$ is open for every open set $A \subset X$. A mapping $f: X \to Y$ is *closed* if $f(A)$ is closed for every closed set $A \subset X$.

Let $f: X \to Y$ be a bijection. We leave it to the reader to prove that *the following assertions are equivalent:*

5.15 *f is a homeomorphism of X onto Y;*

5.16 *f is continuous and open;*

5.17 *f is continuous and closed;*

5.18 $\mathscr{N}_Y(f(x)) = \{f(V) \mid V \in \mathscr{N}_X(x)\}$ *for all x in X;*

5.19 *f is continuous and there exists $\mathscr{E} \subset \mathscr{T}$ such that $\mathscr{T}_{\mathscr{E}} = \mathscr{T}$ and $f(A)$ is open for each A in $\mathscr{E}$.*

Exercises for Chapter 5

1. Consider the function h in Example 3. Let $V = (\frac{1}{2}, \frac{3}{2}) \in \mathscr{N}(h(0))$ and show that there is no $U \in \mathscr{N}(0)$ such that $f(U) \subset V$.

---

* If X is a set, j_X is the bijection $x \mapsto x$ of X onto X.

2. Let $(X, \mathscr{T})$ and $(Y, \mathscr{I})$ be topological spaces. Let $S \subset Y$ and suppose $f: X \to Y$ is such that $f(X) \subset S$. Then f is continuous if and only if $f: X \to S$ is continuous (here S is endowed with the topology $\mathscr{I}_S$).

3. Let X and Y be sets and $f: X \to Y$. Show that if X is endowed with the topology $\mathscr{P}(X)$ and Y is endowed with any topology, then f is continuous on X. Show that if Y is endowed with the topology $\{Y, \varnothing\}$ and X is endowed with any topology, then f is continuous on X.

4. Let X be a set and let $j(x) = x$ for $x \in X$. Let $\mathscr{T}_1$ and $\mathscr{T}_2$ be two topologies on X. Show that j, as a mapping of X endowed with $\mathscr{T}_1$ into X endowed with $\mathscr{T}_2$, is continuous if and only if $\mathscr{T}_2 \subset \mathscr{T}_1$.

5. Let $S \subset \mathbf{R}$ be a set and let $f: S \to \mathbf{R}$ and $g: S \to \mathbf{R}$ be two Lipschitz functions. Show that $\lambda f + \mu g$ is a Lipschitz function for every $\lambda \in \mathbf{R}$, $\mu \in \mathbf{R}$.

6. Show that if $S \subset \mathbf{R}$ and if $f: S \to \mathbf{R}$ and $g: S \to \mathbf{R}$ are two bounded† Lipschitz functions, then fg is a bounded Lipschitz function.

7. Let $S \subset \mathbf{R}$ be a bounded set and let $u_n (n \in \mathbf{N})$ be the mapping $x \mapsto x^n$ of S into S. Show that u_n is a Lipschitz function.

8. Show that the mapping u_n on $\mathbf{R}$ to $\mathbf{R}$ (defined as in Exercise 7) is a Lipschitz function if and only if $n = 1$.

9. Show that if $f: S \to \mathbf{R}$ is a Lipschitz function, then $|f|: S \to \mathbf{R}$ is also a Lipschitz function.

10. Let $(X, \mathscr{T})$ be a topological space and let $(Y, \mathscr{I})$ be a separated topological space. Let $A \subset X$ and suppose that $f: A \to Y$ is continuous on A. Show that there is at most one function $g: \bar{A} \to Y$ such that g is continuous on $\bar{A}$, and $g \mid A = f$.

11. Let $k(x) = 1/x$ for $x \in \mathbf{R}^*$ (see Example 8). Show that there is no continuous function $g: \mathbf{R} \to \mathbf{R}$ such that $g \mid \mathbf{R}^* = k$.

12. Let $a \in \mathbf{R}$, $a \neq 0$, and $b \in \mathbf{R}$, and let $S \subset \mathbf{R} - \{-b/a\}$. Show that the mapping $x \mapsto 1/(ax + b)$ of S into $\mathbf{R}$ is continuous.

* 13. Let $(X, \mathscr{T})$ be a topological space. Let $C_X(X)$ be the set of all continuous mappings of X into X. Then $C_X(X)$ is a stable‡ part

---

† A function $h: S \to \mathbf{R}$ is bounded if there is $M > 0$ such that $|h(x)| \leq M$ for all x in S.

‡ A set $S \subset \mathscr{F}(X, X)$ is stable if $f \in S$ and $g \in S$ imply $g \circ f \in S$.

of $\mathscr{F}(X, X)$ for the law of composition $(f, g) \mapsto f \circ g$. Let $\mathscr{H}(X)$ be the set of all homeomorphisms of X. Show that $\mathscr{H}(X)$ is a group when it is endowed with the law of composition $(f, g) \mapsto f \circ g$.

14. Let $(X, \mathscr{T})$ and $(Y, \mathscr{I})$ be two topological spaces and suppose that $h: X \to Y$ is a homeomorphism. If $\mathscr{F}$ is a fundamental system of $t \in X$, then $\{h(A) \mid A \in \mathscr{F}\}$ is a fundamental system of $h(t)$.

Chapter 6

Initial Topologies

Let X be a set, $(Y, \mathscr{S})$ a topological space, and f a mapping of X into Y. Let

$$\mathscr{T} = \{f^{-1}(U) \mid U \in \mathscr{S}\}.$$

Then *$\mathscr{T}$ is a topology on X.*

In fact,

$$\varnothing = f^{-1}(\varnothing) \quad \text{and} \quad X = f^{-1}(Y);$$

since $\varnothing \in \mathscr{T}$ and $Y \in \mathscr{T}$, we deduce that $\mathscr{T}$ satisfies (C_1) of Chapter 1. If A and B belong to $\mathscr{T}$, there exist U and V in $\mathscr{S}$ such that

$$A = f^{-1}(U) \quad \text{and} \quad B = f^{-1}(V).$$

We deduce that

$$A \cap B = f^{-1}(U) \cap f^{-1}(V) = f^{-1}(U \cap V) \in \mathscr{T},$$

since $U \cap V \in \mathscr{S}$. Hence, $\mathscr{T}$ satisfies (C_2) of Chapter 1. If $(A_i)_{i \in I}$ is a family of elements of $\mathscr{T}$, there exists a family $(U_i)_{i \in I}$ of elements of $\mathscr{S}$ such that

$$A_i = f^{-1}(U_i)$$

for all $i \in I$. We deduce that

$$\bigcup_{i \in I} A_i = \bigcup_{i \in I} f^{-1}(U_i) = f^{-1}(\bigcup_{i \in I} U_i) \in \mathscr{T},$$

since $\bigcup_{i \in I} U_i \in \mathscr{S}$. Hence, $\mathscr{T}$ satisfies (C_3) of Chapter 1 also.

The topology $\mathscr{T}$ is called *the inverse image of $\mathscr{S}$* by f (and may be denoted $f^{-1}(\mathscr{S})$). Notice that *$\mathscr{T}$ is the smallest topology on X such that $f: X \to Y$ is continuous.* In fact, first it is clear that f is continuous when X is endowed with $\mathscr{T}$. If $\mathscr{T}'$ is a topology on X such that $f: X \to Y$

is continuous when X is endowed with $\mathscr{T}'$, then $f^{-1}(U) \in \mathscr{T}'$ for all $U \in \mathscr{I}$ (see 5.11). Hence, $\mathscr{T} \subset \mathscr{T}'$, and hence our assertion is proved.

We shall now generalize the above construction.

Let X be a set and $((Y_i, \mathscr{I}_i))_{i \in I}$ a family of topological spaces, and for each $i \in I$ let f_i be a mapping of X into Y_i.

Denote by $\mathscr{E}$ the set of all parts of X that are of the form $f_i^{-1}(U)$, for some $i \in I$ and $U \in \mathscr{I}_i$.

Hence, $A \in \mathscr{E}$ if and only if there exists $i \in I$ and $U \in \mathscr{I}_i$ such that $A = f_i^{-1}(U)$. Notice that if $I = \{i_0\}$, then $\mathscr{E}$ is the inverse image of $\mathscr{I}_{i_0}$ by f_{i_0}.

6.1 Definition.—*The topology $\mathscr{T}_{\mathscr{E}}$ generated by $\mathscr{E}$ is called the initial topology on X associated with the families $((Y_i, \mathscr{I}_i))_{i \in I}$ and $(f_i)_{i \in I}$.*

If $I = \{i_0\}$, then we shall usually say *the initial topology associated with $(Y_{i_0}, \mathscr{I}_{i_0})$ and f_{i_0}* instead of *the initial topology associated with the families $((Y_i, \mathscr{I}_i))_{i \in I}$ and $(f_i)_{i \in I}$*.

If $Y_i = Y$ for all i in I, we shall often say that $\mathscr{T}_{\mathscr{E}}$ is the initial topology associated with Y and $(f_i)_{i \in I}$.

When there is no ambiguity, we shall say simply that $\mathscr{T}_{\mathscr{E}}$ is the initial topology on X instead of saying that $\mathscr{T}_{\mathscr{E}}$ is the initial topology on X associated with the families $((X_i, \mathscr{I}_i))_{i \in I}$ and $(f_i)_{i \in I}$.

We shall now indicate several properties of *initial topologies*. We shall use the notations introduced in Definition 6.1.

6.2 Theorem.—*The topology $\mathscr{T}_{\mathscr{E}}$ is the smallest topology on X such that all the functions f_i, $i \in I$, are continuous.*

Proof.—Let $i \in I$ and $U \in \mathscr{I}_i$. Then $f_i^{-1}(U) \in \mathscr{E}$, and hence $f_i^{-1}(U) \in \mathscr{T}_{\mathscr{E}}$. Hence, $f_i : X \to Y$ is continuous if X is endowed with the topology $\mathscr{T}_{\mathscr{E}}$.

Suppose now that $\mathscr{T}'$ is a topology on X such that all the functions f_i, $i \in I$, are continuous when X is endowed with $\mathscr{T}'$. Let $A \in \mathscr{E}$. Then $A = f_i^{-1}(U)$ for some $i \in I$ and $U \in \mathscr{I}_i$. Since f_i is continuous when X is endowed with $\mathscr{T}'$, we deduce that $A = f_i^{-1}(U) \in \mathscr{T}'$. Since A was an arbitrary element of $\mathscr{E}$, $\mathscr{E} \subset \mathscr{T}'$, whence $\mathscr{T}_{\mathscr{E}} \subset \mathscr{T}'$.

6.3 Theorem.—*Let $(Z, \mathscr{W})$ be a topological space, $a \in Z$, and $f : Z \to X$. Then f is continuous at a (when X is endowed with $\mathscr{T}_{\mathscr{E}}$) if and only if $f_i \circ f : Z \to X$ is continuous at a for each i in I.*

Proof.—If f is continuous at a, then $f_i \circ f$ is continuous at a for each i in I (see 5.7). Conversely, suppose that $f_i \circ f$ is continuous at a for every i in I. Let $V \in \mathscr{N}_X(f(a))$ and pick U in $\mathscr{E}'$ such that $f(a) \in U \subset V$. Since $U \in \mathscr{E}'$, we have $U = \bigcap_{j \in J} f_j^{-1}(W_j)$ where J is a finite subset of I and $W_j \in \mathscr{I}_j$ for each j in J. Since $f(a) \in f_j^{-1}(W_j)$

for each j in J, it follows that $f_j \circ f(a) \in W_j$ for each j in J, and hence $W_j \in \mathcal{N}_{Y_j}(f_j \circ f(a))$ for each j in J. Since $f_j \circ f$ is continuous at a for each j in J, we deduce that $(f_j \circ f)^{-1}(W_j) \in \mathcal{N}_Z(a)$ for each j in J. We obtain

$$\begin{aligned} f^{-1}(U) &= f^{-1}(\textstyle\bigcap_{j\in J} f_j^{-1}(W_j)) = \bigcap_{j\in J} f^{-1}(f_j^{-1}(W_j)) \\ &= \textstyle\bigcap_{j\in J} (f_j \circ f)^{-1}(W_j) \in \mathcal{N}_Z(a); \end{aligned}$$

since $V \supset U$ it follows that $f^{-1}(V) \supset f^{-1}(U)$, whence $f^{-1}(V) \in \mathcal{N}_Z(a)$. Since V in $\mathcal{N}_X(f(a))$ was arbitrary, we conclude that f is continuous at a.

6.4 Corollary.—*Let $(Z, \mathscr{W})$ be a topological space and $f: Z \to X$. Then f is continuous on Z if and only if $f_i \circ f: Z \to Y_i$ is continuous on Z for each i in I.*

An important particular case of initial topologies (the case of product spaces) will be discussed in the next chapter. We shall now give two simple examples.

Example 1.—Let X be a set, $(Y, \mathscr{I})$ a topological space and $f: X \to Y$. The inverse image of $\mathscr{I}$ by f is the initial topology associated with $(Y, \mathscr{I})$ and f. In fact, with the notations of Definition 6.1, we have $\mathscr{E} = \{f^{-1}(U) \mid U \in \mathscr{I}\}$, and hence $\mathscr{E} = \mathscr{T}_{\mathscr{E}}$.

Example 2.—Let $(Y, \mathscr{I})$ be a topological space and let $T \subset Y$. Then $\mathscr{I}_T = \{U \cap T \mid U \in \mathscr{I}\}$. Since for each A in $\mathscr{P}(Y)$ we have* $A \cap T = j_{T,Y}^{-1}(A)$, we deduce that $\mathscr{I}_T = \{j_{T,Y}^{-1}(U) \mid U \in \mathscr{I}\}$. Hence, $\mathscr{I}_T$ is the initial topology on T associated with $(Y, \mathscr{I})$ and $j_{T,Y}$.

Example 3.—Let X be a set and let $(\mathscr{T}_i)_{i\in I}$ be a family of topologies on X. Consider the family of topological spaces $((Y_i, \mathscr{T}_i))_{i\in I}$ where $Y_i = X$ for each i in I. For each i in I, let f_i be the mapping $j_X: x \mapsto x$ of X onto Y_i. Then, if $\mathscr{E} = \bigcup_{i\in I} \mathscr{T}_i$, we have† $\mathscr{T}_{\mathscr{E}} = \sup_{i\in I} \mathscr{T}_i$. Hence, $\sup_{i\in I} \mathscr{T}_i$ is the initial topology associated with the families $((Y_i, \mathscr{T}_i))_{i\in I}$ and $(f_i)_{i\in I}$.

6.5 Theorem.—*Suppose that:* (i) *for each $i \in I$ the space $(Y_i, \mathscr{I}_i)$ is separated, and* (ii) *for every x and y in X, $x \neq y$, there is i in I such that $f_i(x) \neq f_i(y)$. Then $(X, \mathscr{T}_{\mathscr{E}})$ is separated.*

---

* Recall that $j_{T,Y}$ is the mapping $x \mapsto x$ of T into Y.

† The set of all topologies on X is *ordered* by inclusion (hence $\mathscr{T}' \leq \mathscr{T}''$ if and only if $\mathscr{T}' \subset \mathscr{T}''$).

Proof.—Let x and y be in X, $x \neq y$, and let i in I be such that $f_i(x) \neq f_i(y)$. Since $(Y_i, \mathscr{I}_i)$ is separated, there are V_x in $\mathscr{N}_{Y_i}(f_i(x))$ and V_y in $\mathscr{N}_{Y_i}(f_i(y))$ such that $V_x \cap V_y = \varnothing$. Then $f_i^{-1}(V_x) \cap f_i^{-1}(V_y) = f_i^{-1}(V_x \cap V_y) = \varnothing$. Since $f_i: X \to Y_i$ is continuous, $f_i^{-1}(V_x) \in \mathscr{N}_X(x)$ and $f_i^{-1}(V_y) \in \mathscr{N}_X(y)$. Since x and y were arbitrary, except for being distinct, we conclude that $(X, \mathscr{T}_{\mathscr{E}})$ is separated.

Exercises for Chapter 6

1. Let X be a set and $((Y_i, \mathscr{I}_i))_{i \in I}$ a family of topological spaces, and for each i in I let f_i be a mapping of X into Y_i. For each i in I, denote by $\mathscr{T}_i$ the topology $f_i^{-1}(\mathscr{I}_i)$. Then the initial topology on X associated with the families $((Y_i, \mathscr{I}_i))_{i \in I}$ and $(f_i)_{i \in I}$ is $\sup_{i \in I} \mathscr{T}_i$ (see Example 3).

2. Let $X = \{a, b, c\} = Y_1 = Y_2$. Let $\mathscr{T}_1$ and $\mathscr{T}_2$ be the topologies on X given in Exercise 5 at the end of Chapter 1. Let $f_1 = f_2 = j_X$. Construct the initial topology on X associated with the families $((Y_i, \mathscr{I}_i))_{i \in \{1,2\}}$ and $(f_i)_{i \in \{1,2\}}$.

3. Let X be a set and $((Y_i, \mathscr{I}_i))_{i \in I}$ a family of topological spaces, and for each i in I let f_i be a mapping of X into Y_i. Let $\mathscr{T}$ denote the initial topology on X associated with the families $((Y_i, \mathscr{I}_i))_{i \in I}$ and $(f_i)_{i \in I}$. If $S \subset X$, then $\mathscr{T}_S$ (the induced topology on S) is the initial topology associated with the families $((Y_i, \mathscr{I}_i))_{i \in I}$ and $(f_i \mid S)_{i \in I}$.

4. Let $(X, \mathscr{T})$ be a topological space and $((Y_i, \mathscr{I}_i))_{i \in I}$ a family of topological spaces, and for each i in I let f_i be a mapping of X into Y_i. Suppose that for each topological space $(Z, \mathscr{W})$ and mapping f of Z into X we have f continuous if and only if $f_i \circ f$ is continuous for every i in I. Then $\mathscr{T}$ is the initial topology on X associated with the families $((Y_i, \mathscr{I}_i))_{i \in I}$ and $(f_i)_{i \in I}$.

5. Consider the mapping $x \mapsto x^3$ of $\mathbf{R}$ into $\mathbf{R}$. Describe the weakest topology on $\mathbf{R}$ that makes this mapping continuous (when the range space is endowed with the usual topology).

Chapter 7

Product Spaces

Let $((X_i, \mathscr{T}_i))_{i \in I}$ be a family of topological spaces and let

$$X = \prod_{i \in I} X_i.$$

For each k in I, let pr_k be the *projection* $(x_i)_{i \in I} \mapsto x_k$ of X onto X_k.

7.1 Definition.—*The initial topology $\mathscr{T}$ on X associated with the families $((X_i, \mathscr{T}_i))_{i \in I}$ and $(pr_i)_{i \in I}$ is called the product topology (on X) of the family of topologies $(\mathscr{T}_i)_{i \in I}$. The topological space $(X, \mathscr{T})$ is called the product of the family $((X_i, \mathscr{T}_i))_{i \in I}$.*

When there is no ambiguity, we shall often say that X is the product space $\prod_{i \in I} X_i$ instead of saying that $(X, \mathscr{T})$ is the product of the family $((X_i, \mathscr{T}_i))_{i \in I}$.

Let us recall here that the initial topology $\mathscr{T}$ on X associated with the families $((X_i, \mathscr{T}_i))_{i \in I}$ and $(pr_i)_{i \in I}$ is constructed as follows:

We denote by $\mathscr{E}$ the set of all parts of X that are of the form $pr_k^{-1}(U)$, for some k in I and $U \subset X_k$ *open*. Notice that

$$pr_k^{-1}(U) = \prod_{i \in I} A_i$$

where $A_i = U$ if $i = k$ and $A_i = X_i$ if $i \neq k$. We consider then the set $\mathscr{E}'$ of all parts of X that are intersections of finite families of sets belonging to $\mathscr{E}$. Hence, the sets belonging to $\mathscr{E}'$ are of the form

$$\bigcap_{k \in J} pr_k^{-1}(U_k)$$

where $J \subset I$ is finite and where $U_k \subset X_k$ is *open* for each $k \in J$. Notice that

$$\bigcap_{k \in J} pr_k^{-1}(U_k) = \prod_{i \in I} A_i$$

where $A_i = U_i$ if $i \in J$ and $A_i = X_i$ if $i \notin J$. Finally, $\mathscr{E}''$ is the set of all parts of X that are unions of families of sets belonging to $\mathscr{E}'$.

The elements of $\mathscr{E}'$, that is, the parts of X that can be written in the form

7.2 $$\prod_{i \in I} A_i$$

with $A_i \subset X_i$ *open*, for each $i \in I$, and $\{i \mid A_i \neq X_i\}$ *finite*, are called *elementary sets.*

Hence, *a part of X is open if and only if it is the union of a family of elementary sets.*

Hence, the elementary parts of X form a *basis* of the product topology.

If I is finite, then a part of X is elementary if and only if it is of the form

7.3 $$\prod_{i \in I} V_i$$

with $V_i \in \mathscr{T}_i$ for all $i \in I$.

If $I = \{1, \ldots, n\}$, we *identify* a product of the form $\prod_{i \in I} A_i$ with $A_1 \times \ldots \times A_n$. In this case, the elementary parts of X are of the form

$$U_1 \times \ldots \times U_n$$

where $U_1 \in \mathscr{T}_1, \ldots, U_n \in \mathscr{T}_n$.

If $I = \{1, \ldots, n\}$ we shall often say that $\mathscr{T}$ is the product of the topologies $\mathscr{T}_1, \ldots, \mathscr{T}_n$ and that $(X, \mathscr{T})$ is the product of the topological spaces $(X_1, \mathscr{T}_1), \ldots, (X_n, \mathscr{T}_n)$.

By Theorem 6.2, the mapping $pr_k : X \to X_k$ is *continuous* for each $k \in I$. If $U = \prod_{i \in I} A_i$ is a non-void elementary set, then

$$pr_k(U) = A_k \in \mathscr{T}_k.$$

Hence, $pr_k(U) \subset X_k$ is open if U is an elementary set. If $U \subset X$ is an arbitrary open set, then $U = \bigcup_{i \in J} U_j$ for some family $(U_j)_{j \in J}$ of elementary parts of X. We deduce that

$$pr_k(U) = pr_k(\bigcup_{j \in J} U_j) = \bigcup_{j \in J} pr_k(U_j) \in \mathscr{T}_k.$$

We conclude that pr_k is *continuous and open.*

The mappings pr_k are not necessarily closed (see Exercise 2, Chapter 8).

7.4 Theorem.—*For each $a \in X$, the set of all elementary parts of X containing a is a fundamental system of a.*

Proof.—We have noticed that the elementary parts of X form a basis of the product topology on X. The theorem then follows from the remark following Definition 2.3.

7.5 Theorem.—*Let* $a = (a_i)_{i\in I} \in X$ *and let* $V \subset X$. *Then the following assertions are equivalent:*

(a) $V \in \mathcal{N}_X(a)$;

(b) *there is a family* $(V_i)_{i\in I}$ *such that* $V_i \in \mathcal{N}_{X_i}(a_i)$ *for all* $i \in I$, $\{i \mid V_i \neq X_i\}$ *is finite, and* $\prod_{i\in I} V_i \subset V$.

Proof of (a) $\Rightarrow$ (b).—Since $V \in \mathcal{N}_X(a)$, there exists an elementary set $\prod_{i\in I} V_i$ such that

$$a \in \prod_{i\in I} V_i \subset V.$$

Then V_i is open and $a_i \in V_i$ for all $i \in I$; whence $V_i \in \mathcal{N}_X(a_i)$. Moreover, since $\prod_{i\in I} V_i$ is elementary, we deduce that the set $\{i \mid V_i \neq X_i\}$ is finite.

Proof of (b) $\Rightarrow$ (a).—Let $J = \{i \mid V_i \neq X_i\}$; by hypothesis J is finite. For each $i \in J$, let D_i be an open set such that $a_i \in D_i \subset V_i$. Let $D_i = X_i$ for $i \in I - J$. Then $\prod_{i\in I} D_i$ is an elementary set and

$$\prod_{i\in I} D_i \subset \prod_{i\in I} V_i \subset V.$$

Hence, $V \in \mathcal{N}_X(a)$.

Remark.—If $I = \{1, \ldots, n\}$, we deduce that $V \in \mathcal{N}_X(a)$ if and only if there are $V_1 \in \mathcal{N}_{X_1}(a_1), \ldots, V_n \in \mathcal{N}_{X_n}(a_n)$ such that

$$V_1 \times \ldots \times V_n \subset V.$$

We deduce the following from 7.5:

7.6 *If* $V_i \in \mathcal{N}_{X_i}(a)$ *for all* $i \in I$, *and if* $\{i \mid V_i \neq X_i\}$ *is finite, then*

$$\prod_{i\in I} V_i \in \mathcal{N}_X(a).$$

7.7 *If, for each* i *in* I, $\mathcal{F}_i$ *is a fundamental system of* a_i, *then the set of all parts of* X *of the form*

$$\prod_{i\in I} A_i,$$

where either $A_i \in \mathcal{F}_i$ *or* $A_i = X_i$ *and where* $\{i \mid A_i \neq X_i\}$ *is finite, is a fundamental system of* a.

Remarks.—Suppose $I = \{1, \ldots, n\}$. It follows from 7.6 and 7.7 that:

7.6′ *If* $V_1 \in \mathcal{N}_{X_1}(a_1), \ldots, V_n \in \mathcal{N}_{X_n}(a_n)$, *then*

$$V_1 \times \ldots \times V_n \in \mathcal{N}_X(a).$$

7.7′ *If* $\mathcal{F}_i$ *is, for each* $i = 1, \ldots, n$, *a fundamental system of* a_i, *then*

$$\mathcal{F} = \{V_1 \times \ldots \times V_n \mid V_1 \in \mathcal{F}_1, \ldots, V_n \in \mathcal{F}_n\}$$

is a fundamental system of a.

Example 1.—Let $X_1, \ldots, X_n$ be $n(>1)$ sets and consider the topological spaces $(X_1, \mathcal{P}(X_1)), \ldots, (X_n, \mathcal{P}(X_n))$. Let $\mathcal{T}$ be the product topology of the topologies $\mathcal{P}(X_1), \ldots, \mathcal{P}(X_n)$. Then $\mathcal{T} = \mathcal{P}(X_1 \times \ldots \times X_n)$.

To simplify the writing, we shall prove the assertion only for $n = 2$.

Clearly, $\mathcal{T} \subset \mathcal{P}(X_1 \times X_2)$. Conversely, let $(a_1, a_2) \in X_1 \times X_2$. Then $\{a_1\}$ is open in X_1 and $\{a_2\}$ is open in X_2, whence

$$\{(a_1, a_2)\} = \{a_1\} \times \{a_2\} \in \mathcal{T}.$$

For $A \subset X_1 \times X_2$ arbitrary, we have

$$A = \bigcup_{(x_1, x_2) \in A} \{(x_1, x_2)\}$$

Since $\{(x_1, x_2)\} \in \mathcal{T}$, and since $\mathcal{T}$ (being a topology) satisfies condition (C_3), we deduce that $A \in \mathcal{T}$. Since $A \subset X_1 \times X_2$ was arbitrary, we deduce that $\mathcal{P}(X_1 \times X_2) \subset \mathcal{T}$. We conclude that $\mathcal{T} = \mathcal{P}(X_1 \times X_2)$. Hence, the product of the topological spaces $(X_1, \mathcal{P}(X_1))$ and $(X_2, \mathcal{P}(X_2))$ is $(X_1 \times X_2, \mathcal{P}(X_1 \times X_2))$.

This result does not extend to the case of an arbitrary family $((X_i, \mathcal{P}(X_i))_{i \in I}$. We leave it to the reader to justify the assertion.

Example 2.—Unless we explicitly mention the contrary, whenever we consider* $\mathbf{R}^n (n > 1)$ as a topological space, we assume that $\mathbf{R}^n$ is endowed with the product topology of the topologies $\mathcal{U}_1, \ldots, \mathcal{U}_n$ where $\mathcal{U}_1 = \ldots = \mathcal{U}_n = \mathcal{U}$ (see Example 6, Chapter 1 for the definition of $\mathcal{U}$). This product topology will be denoted $\mathcal{U}^n$. To unify the writing, we shall often write $\mathbf{R}^1 = \mathbf{R}$ and $\mathcal{U}^1 = \mathcal{U}$.

7.8 Theorem.—*If* $((X_i, \mathcal{T}_j))_{i \in I}$ *is a family of separated topological spaces, then the product* $(X, \mathcal{T})$ *of the family* $((X_i, \mathcal{T}_i))_{i \in I}$ *is separated.*

---

* $\mathbf{R}^n = \mathbf{R} \times \ldots \times \mathbf{R}$ (n-times).

Proof.—Let $x = (x_i)_{i \in I}$ and $y = (y_i)_{i \in I}$ in X, $x \neq y$. Then there is $k \in I$ such that $x_k \neq y_k$. Hence,

$$pr_k(x) = x_k \neq y_k = pr_k(y).$$

Since x and y in X, $x \neq y$, were arbitrary, we deduce from Theorem 6.5 that $(X, \mathscr{T})$ is separated.

Example 3.—The topological space $(\mathbf{R}^n, \mathscr{U}^n)$ $(n \in \mathbf{N})$ is separated (see Example 3, Chapter 2).

Exercise 1.—If $X \neq \varnothing$ and $(X, \mathscr{T})$ is separated, then every $(X_i, \mathscr{T}_i)$ $(i \in I)$ is separated.

We shall now state and prove the following theorem, which gives a useful characterization of separated spaces.

7.9 Theorem.—*Let* $(X, \mathscr{T})$ *be a topological space and let** $\Delta = \{(x, x) \mid x \in X\}$. *Then the following assertions are equivalent:*

(a) *the space* $(X, \mathscr{T})$ *is separated;*
(b) *the set* Δ *is closed in* $X \times X$.

Proof of (a) $\Rightarrow$ (b).—Let $(s, t) \notin \Delta$. Then $s \neq t$; hence there are $U \in \mathscr{N}_X(s)$ and $V \in \mathscr{N}_X(t)$ such that $U \cap V = \varnothing$. Then $(U \times V) \cap \Delta = \varnothing$. Since $U \times V \in \mathscr{N}_{X \times X}((s, t))$, we conclude that $(s, t) \notin \bar{\Delta}$. Hence, $\Delta = \bar{\Delta}$, and thus Δ is closed in $X \times X$.

Proof of (b)$\Rightarrow$(a).—Let s and t be in X, $s \neq t$. Then $(s, t) \notin \Delta$. Since $\mathbf{C}\Delta$ is open, there are $U \in \mathscr{N}_X(s)$ and $V \in \mathscr{N}_X(t)$ such that $U \times V \subset \mathbf{C}\Delta$. If $U \cap V \neq \varnothing$, and if $z \in U \cap V$, then $(z, z) \in U \times V$; that is, $(U \times V) \cap \Delta \neq \varnothing$. Since this leads to a contradiction, it follows that $U \cap V = \varnothing$. Since s and t in X, $s \neq t$, were arbitrary, we deduce that $(X, \mathscr{T})$ is separated.

7.10 Theorem.—*Let* $A_i \subset X_i$ *for each* $i \in I$. *Then*

$$\overline{\prod_{i \in I} A_i} = \prod_{i \in I} \bar{A}_i$$

Proof.—Let $x = (x_i)_{i \in I} \in \overline{\prod_{i \in I} A_i}$. Let $k \in I$ and let $V_k \in \mathscr{N}_{X_k}(x_k)$. Define

$$V = \prod_{i \in I} D_i$$

where $D_i = X_i$ if $i \neq k$ and $D_k = V_k$. By 7.6, $V \in \mathscr{N}_X(x)$ and hence

* We often write $\Delta_X = \{(x, x) \mid x \in X\}$.

$V \cap (\prod_{i \in I} A_i) \neq \varnothing$. But

$$V \cap (\textstyle\prod_{i \in I} A_i) = \prod_{i \in I} (D_i \cap A_i).$$

Since $\prod_{i \in I} (D_i \cap A_i) \neq \varnothing$, we have in particular $D_k \cap A_k \neq \varnothing$; that is, $V_k \cap A_k \neq \varnothing$. Since $V_k \in \mathscr{N}_{X_k}(x_k)$ was arbitrary, we deduce that $x_k \in \bar{A}_k$. Since $k \in I$ was arbitrary, we obtain $x = (x_i)_{i \in I} \in \prod_{i \in I} \bar{A}_i$. Since $x \in \overline{\prod_{i \in I} A_i}$ was arbitrary, we conclude that

$$(1) \qquad \overline{\textstyle\prod_{i \in I} A_i} \subset \prod_{i \in I} \bar{A}_i.$$

Conversely, let $x = (x_i)_{i \in I} \in \prod_{i \in I} \bar{A}_i$. Let $V = \prod_{i \in I} D_i$ be an elementary set containing x. Then, for each $i \in I$, D_i is a neighborhood of x_i, whence $D_i \cap A_i \neq \varnothing$ for each $i \in I$. We deduce that

$$(2) \qquad V \cap (\textstyle\prod_{i \in I} A_i) = \prod_{i \in I} (D_i \cap A_i) \neq \varnothing.$$

Since the elementary sets containing x form a fundamental system of x, and since V was arbitrary, we deduce that $x \in \overline{\prod_{i \in I} A_i}$. Since $x \in \prod_{i \in I} \bar{A}_i$ was arbitrary, we conclude that

$$\textstyle\prod_{i \in I} \bar{A}_i \subset \overline{\prod_{i \in I} A_i}.$$

The conclusion of the theorem follows from (1) and (2).

7.11 Corollary.—*Let $A_i \subset X_i$ for each $i \in I$. Then:*

(a) *if A_i is closed for each $i \in I$, then $\prod_{i \in I} A_i$ is closed;*
(b) *if $\prod_{i \in I} A_i$ is non-void and closed, the set A_i is closed for each $i \in I$.*

Proof of (a).—We have $A_i = \bar{A}_i$ for each $i \in I$. By Theorem 7.10,

$$\overline{\textstyle\prod_{i \in I} A_i} = \prod_{i \in I} \bar{A}_i = \prod_{i \in I} A_i,$$

whence $\prod_{i \in I} A_i$ is closed.

Proof of (b).—We have

$$\textstyle\prod_{i \in I} \bar{A}_i = \overline{\prod_{i \in I} A_i} = \prod_{i \in I} A_i \neq \varnothing,$$

whence* $\bar{A}_i = A_i$ for each $i \in I$. Therefore, A_i is closed for each $i \in I$.

---

* Notice that $\prod_{i \in I} D_i = \prod_{i \in I} C_i \neq \varnothing$ implies that $D_i = C_i$ for all $i \in I$.

Remarks.—Suppose $I = \{1, \ldots, n\}$. It follows from Theorem 7.10 and from Corollary 7.11 (a) that:

7.10′ *if* $A_1 \subset X_1, \ldots, A_n \subset X_n$, *then*

$$\overline{A_1 \times \ldots \times A_n} = \overline{A}_1 \times \ldots \times \overline{A}_n;$$

7.11′(a) *if* $A_1 \subset X_1, \ldots, A_n \subset X_n$ *are closed, then* $A_1 \times \ldots \times A_n$ *is closed;*

7.11′(b) *if* $A_1 \subset X_1, \ldots, A_n \subset X_n$, *and if* $A_1 \times \ldots \times A_n$ *is non-void and closed, then* $A_1, \ldots, A_n$ *are closed.*

7.12 Corollary.—*If* $((X_i, \mathscr{T}_i))_{i\in I}$ *is a family of regular topological spaces, then its product,* $(X, \mathscr{T})$, *is regular.*

Proof.—Let $a = (a_i)_{i\in I} \in X$ and let $V \in \mathscr{N}_X(a)$. By Theorem 7.5, there is a family $(V_i)_{i\in I}$ such that $V_i \in \mathscr{N}_{X_i}(a_i)$ for all $i \in I$, $J = \{i \mid V_i \neq X_i\}$ is finite and $\prod_{i\in I} V_i \subset V$. For every $i \in J$, let $D_i \in \mathscr{N}_{X_i}(a_i)$, *closed* and such that $a_i \in D_i \subset V_i$. Let $D_i = X_i$ if $i \notin J$. Then $\prod_{i\in I} D_i$ is a closed neighborhood of a contained in $\prod_{i\in I} V_i$ and hence in V. Since $a \in X$ and $V \in \mathscr{N}_X(a)$ were arbitrary, $(X, \mathscr{T})$ is regular.

Example 4.—Let $n > 1$ and let $(a_1, \ldots, a_n)$ and $(b_1, \ldots, b_n)$ in $\mathbf{R}^n$. Suppose that $a_1 < b_1, \ldots, a_n < b_n$. Let

7.13 $$((a_1, \ldots, a_n), (b_1, \ldots, b_n)) = \{(x_1, \ldots, x_n) \in \mathbf{R}^n \mid a_1 < x_1 < b_1, \ldots, a_n < x_n < b_n\}$$

and

7.14 $$[(a_1, \ldots, a_n), (b_1, \ldots, b_n)] = \{(x_1, \ldots, x_n) \in R^n \mid a_1 \leq x_1 \leq b_1, \ldots, a_n \leq x_n \leq b_n\}.$$

A set of the form 7.13 is called (bounded n-dimensional) *open interval;* a set of the form 7.14 is called (bounded n-dimensional) *closed interval.*

Notice that

$$((a_1, \ldots, a_n), (b_1, \ldots, b_n)) = \textstyle\prod_{i=1}^n (a_i, b_i)$$

and

$$[(a_1, \ldots, a_n), (b_1, \ldots, b_n)] = \textstyle\prod_{i=1}^n [a_i, b_i].$$

From the definition of the topology of $\mathbf{R}^n$, it follows that an open interval is an elementary set (and hence an open set). A closed interval is a closed set.

Let $a = (a_1, \ldots, a_n) \in \mathbf{R}^n$. We deduce from Example 2, Chapter 2, and from 7.7′ that:

7.15 *the set* $\{((a_1 - \varepsilon, \ldots, a_n - \varepsilon)(a_1 + \varepsilon, \ldots, a_n + \varepsilon)) \mid \varepsilon > 0\}$ *is a fundamental system of* a;

7.16 *the set* $\{[(a_1 - \varepsilon, \ldots, a_n - \varepsilon), (a_1 + \varepsilon, \ldots, a_n + \varepsilon)] \mid \varepsilon > 0\}$ *is a fundamental system of* a.

7.17 Theorem.—*Let* $(Z, \mathscr{W})$ *be a topological space,* $S \subset Z$, $a \in S$, *and* $f: S \to \prod_{i \in I} X_i$. *Then the following assertions are equivalent:*

(a) *the mapping* f *is continuous at* a;
(b) *the mapping* $pr_i \circ f$ *is continuous at* a *for each* $i \in I$.

Proof.—Since the topology of $\prod_{i \in I} X_i$ is the initial topology associated with the families $((X_i, \mathscr{T}_i))_{i \in I}$ and $(pr_i)_{i \in I}$, the theorem is a particular case of Theorem 6.3.

7.18 Corollary.—*Let* $(Z, \mathscr{W})$ *be a topological space,* $S \subset Z$, *and* $f: S \to \prod_{i \in I} X_i$. *Then the following assertions are equivalent:*

(a) *the mapping* f *is continuous on* S;
(b) *the mapping* $pr_i \circ f$ *is continuous on* S *for each* $i \in I$.

Proof.—The corollary follows from Theorem 7.17 or from Corollary 6.4.

7.19 Theorem.—*Let* $(Z, \mathscr{W})$ *be a topological space,* $S \subset Z$, $a \in S$, *and* $f_i: S \to X_i$ *for every* $i \in I$. *Then the following assertions are equivalent:*

(a) *the mapping* $\varphi: x \mapsto (f_i(x))_{i \in I}$ *of* S *into* $\prod_{i \in I} X_i$ *is continuous at* a;
(b) *The mapping* $f_i: S \to X_i$ *is continuous at* a *for every* $i \in I$.

Proof.—We have

$$f_i = pr_i \circ \varphi$$

for every $i \in I$. By Theorem 7.17, the mapping f_i is continuous at a, for every $i \in I$, if and only if φ is continuous at a.

7.20 Corollary.—*Let* $(Z, \mathscr{W})$ *be a topological space,* $S \subset Z$, *and* $f_i: S \to X_i$ *for every* $i \in I$. *Then the following assertions are equivalent:*

(a) *the mapping* $\varphi: x \mapsto (f_i(x))_{i \in I}$ *of* S *into* $\prod_{i \in I} X_i$ *is continuous on* S;
(b) *the mapping* $f_i: S \to X_i$ *is continuous on* S for each $i \in I$.

Example 5.—Let X be a topological space and $S \subset X$, and let k be the mapping $x \mapsto (x, x)$ of S into $X \times X$. Then k is continuous on S.

In connection with various applications, we shall also state and prove the following.

7.21 Corollary.—*Let* $(E, \mathscr{T})$ *and* $(F, \mathscr{O})$ *be two topological spaces and let* $f: E \to F$ *and* $g: E \to F$ *be two continuous mappings. Suppose* $(F, \mathscr{O})$ *separated. Then the set* $C = \{x \mid f(x) = g(x)\}$ *is closed.*

Proof.—By 7.20, the mapping $(f, g): x \mapsto (f(x), g(x))$ is continuous on X, and by 7.9 and our hypothesis $\triangle = \{(x, x) \mid x \in F\}$ is closed. Since

$$C = \{x \mid (f, g)(x) \in \triangle\},$$

we deduce from Theorem 5.11(c) that C is closed.

Let $(Z, \mathscr{W})$ and $(Y, \mathscr{I})$ be two topological spaces. Recall that we denoted by $(X, \mathscr{T})$ the product of the family $((X_i, \mathscr{T}_i))_{i \in I}$.

7.22 Theorem.—*Let* $S \subset Z$ *and* $a \in S$ *and let* $T \subset X$. *Let* $f_i: S \to X_i$, *for every* $i \in I$, *and let* $g: T \to Y$. *Suppose that:*

(a) $x \in S \Rightarrow (f_i(x))_{i \in I} \in T$;
(b) $f_i: S \to X_i$ *is continuous at* a *for every* $i \in I$;
(c) $g: T \to Y$ *is continuous at* $(f_i(a))_{i \in I}$.

Then the mapping $h: x \mapsto g((f_i(x))_{i \in I})$ *is continuous at* a.

By (a), the mapping $h: x \mapsto g((f_i(x))_{i \in I})$ can be defined.

Proof.—By Theorem 7.19, the mapping $\varphi: x \mapsto (f_i(x))_{i \in I}$, of S into $\prod_{i \in I} X_i$, is continuous at a. Since by (c), g is continuous at $\varphi(a) = (f_i(a))_{i \in I}$, we deduce (see Theorem 5.7) that $k = g \circ \sigma$ is continuous at a.

7.23 Corollary.—*Let* $S \subset Z$ *and let* $T \subset X$. *Let* $f_i: S \to X_i$ *for every* $i \in I$, *and let* $g: T \to Y$. *Suppose that:*

(a) $x \in S \Rightarrow (f_i(x))_{i \in I} \in T$;
(b) $f_i: S \to X_i$ *is continuous on* S *for every* $i \in I$;
(c) $g: T \to Y$ *is continuous on* T.

Then the mapping $h: x \mapsto g((f_i(x))_{i \in I})$ *is continuous on* S.

Proof.—By Theorem 7.22, h is continuous at every $c \in S$, whence h is continuous on S.

Suppose $I = \{1, \ldots, n\}$. Let $S \subset Z$, $a \in S$, and $T \subset X_1 \times \ldots \times X_n$. Let $f_1: S \to X_1, \ldots, f_n: S \to X_n$, and let $g: T \to Y$. Assume that

$$y \in S \Rightarrow (f_1(y), \ldots, f_n(y)) \in T.$$

Let $h = g(f_1, \ldots, f_n)$ be the mapping

$$y \mapsto g(f_1(y), \ldots, f_n(y))$$

of S into Y. We deduce from 7.22 and 7.23 that:

7.22′ *if $f_1, \ldots, f_n$ are continuous at a and g is continuous at $(f_1(a), \ldots, f_n(a))$, then $h = g(f_1, \ldots, f_n)$ is continuous at a;*

7.23′ *if $f_1, \ldots, f_n$ are continuous on S and g is continuous on T, then $h = g(f_1, \ldots, f_n)$ is continuous on S.*

Let $X = \prod_{i \in I} X_i$ and let $a = (a_i)_{i \in I}$. Let $k \in I$. For each $i \in I$, $i \neq k$, let u_i be the (constant) mapping $x \mapsto a_i$ of X_k into X_i; let $u_k = j_{X_k}$ (recall that j_{X_k} is the mapping $x \mapsto x$ of X_k onto X_k). Then u_i is continuous on X_k for all $i \in I$, and hence the mapping $u^{(k)}: x \mapsto (u_i(x))_{i \in I}$ of X_k into X is continuous on X_k. Notice that $u^{(k)}(a_k) = a$. If $f: X \to Y$ is continuous at a, we deduce that $f \circ u^{(k)}$ is continuous at a_k. If $f: X \to Y$ is continuous on X, we deduce that $f \circ u^{(k)}$ is continuous on X_k.

It follows from the above that if $I = \{1, \ldots, n\}$, and if $f: X \to Y$ is continuous at $(a_1, \ldots, a_n) \in X_1 \times \ldots \times X_n$, then the n mappings

7.24
$$\begin{cases} x_1 \mapsto f(x_1, a_2, a_3, \ldots, a_n) \\ x_2 \mapsto f(a_1, x_2, a_3, \ldots, a_n) \\ \cdots\cdots\cdots\cdots\cdots \\ x_n \mapsto f(a_1, a_2, a_3, \ldots, x_n) \end{cases}$$

are continuous* at $a_1, a_2, \ldots, a_n$ respectively.

It is not true that the continuity of the functions in 7.24 implies that of f at $(a_1, \ldots, a_n)$. This is shown in Example 6, below.

Example 6.—Let $f: \mathbf{R}^2 \to \mathbf{R}$ be defined by

$$f((x, y)) = \begin{cases} 0 & \text{if } (x, y) = (0, 0) \\ \dfrac{2xy}{(x^2 + y^2)} & \text{if } (x, y) \neq (0, 0). \end{cases}$$

Then f is *not* continuous at (0, 0).

In fact, let h be the mapping $x \mapsto (x, x)$ of $\mathbf{R}$ into $\mathbf{R}^2$. By Example 5, h is continuous on $\mathbf{R}$ and $h(0) = (0, 0)$. If f were continuous at

* If f is continuous on X, then the n functions given in 7.24 are continuous on $X_1, X_2, \ldots, X_n$ respectively.

(0, 0), then $f \circ h$ would be continuous at (0, 0). But

$$f \circ h(x) = \begin{cases} 0 & \text{if} \quad x = 0 \\ 1 & \text{if} \quad x \neq 0, \end{cases}$$

and it is easy to see that $f \circ h$ is *not* continuous at 0. However, $f(x, 0) = f(0, x) = 0$ for all $x \in \mathbf{R}$, and hence

$$x \mapsto f(x, 0) \quad \text{and} \quad x \mapsto f(0, x)$$

are continuous on **R**.

▼ We shall close this section with four more results that are sometimes useful. Their proofs are only sketched; the details are left to the reader.

We denote by $((X_i, \mathscr{T}_i))_{i \in I}$ a family of topological spaces and by $(X, \mathscr{T})$ their product.

7.25 Theorem.—*Let φ be a permutation of I. Then the mapping*

$$u: (x_i)_{i \in I} \mapsto (x_{\varphi(i)})_{i \in I}$$

is a homeomorphism of $\prod_{i \in I} X_i$ onto $\prod_{i \in I} X_{\varphi(i)}$.

7.26 Theorem.—*Let $(I(j))_{j \in J}$ be a partition of I. Then the mapping*

$$v: (x_i)_{i \in I} \mapsto ((x_i)_{i \in I(j)})_{j \in J}$$

is a homeomorphism of $\prod_{i \in I} X_i$ onto $\prod_{j \in J} (\prod_{i \in I(j)} X_i)$.

Theorems 7.25 and 7.26 can be proved as follows: It is clear first that u and v are bijections. Using Corollary 7.20, we deduce that they are continuous. Finally, we notice that if A is an elementary part of $\prod_{i \in I} X_i$, then $u(A)$ and $v(A)$ are elementary parts of $\prod_{i \in I} X_{\varphi(i)}$ and $\prod_{j \in J} (\prod_{i \in I(j)} X_i)$ respectively. The equivalence 5.15 ⇔ 5.19 (see Chapter 5) shows that u and v are homeomorphisms.

It follows from 7.26 that, for instance,

$$\mathbf{R}^3, \qquad \mathbf{R} \times \mathbf{R}^2, \quad \text{and} \quad \mathbf{R}^2 \times \mathbf{R}$$

are homeomorphic. ▲

Let $((Y_i, \mathscr{S}_i))_{i \in I}$ be a second family of topological spaces. For each $i \in I$, let $f_i: X_i \to Y_i$ be continuous at $a_i \in X_i$.

7.27 Theorem.—*The mapping*

$$\prod_{i \in I} f_i : (x_i)_{i \in I} \mapsto (f_i(x_i))_{i \in I}$$

is continuous at $(a_i)_{i \in I}$.

To prove the theorem, we notice that, for each $k \in I$,

$$pr_k \circ \left(\prod_{i \in I} f_i\right)$$

is the mapping $(x_i)_{i \in I} \mapsto f_k(x_k)$ of $\prod_{i \in I} X_i$ onto Y_k. This mapping is continuous at a_k, since it is the composition $f_k \circ pr_k$ (we denote in the same way the projection mappings corresponding to $\prod_{i \in I} X_i$ and $\prod_{i \in I} Y_i$). The continuity of $\prod_{i \in I} f_i$ at $(a_i)_{i \in I}$ is then a consequence of 7.17.

It follows from 7.27 that if, for each $i \in I, f_i : X_i \to Y_i$ is continuous on X_i, then $\prod_{i \in I} f_i$ is continuous on $\prod_{i \in I} X_i$. It also follows that if, for each $i \in I, f_i$ is a homeomorphism of X_i onto Y_i, then $\prod_{i \in I} f_i$ is a homeomorphism of $\prod_{i \in I} X_i$ onto $\prod_{i \in I} Y_i$.

7.28 Theorem.—*For each* $i \in I$, *let* $A_i \subset X_i$ *and let* $A = \prod_{i \in I} A_i$. *Then the topology* $\mathscr{T}_A$ *and the topology* $\mathscr{A}$, *product of the family* $((\mathscr{T}_i)_{A_i})_{i \in I}$, *are identical.*

To prove the theorem, we notice that $\mathscr{T}_A$ is generated by

$$\{U \cap A \mid U \text{ elementary in } \prod_{i \in I} X_i\}$$

and that every elementary set B in the product $\prod_{i \in I} A_i$ can be written $U \cap A$, with U elementary in $\prod_{i \in I} X_i$.

Exercises for Chapter 7

1. Consider the topological space $\mathbf{R}^2$. Let a, b, c, d be real numbers such that $a < c$ and $b < d$. Show that

$$\overline{((a, b), (c, d))} = [(a, b), (c, d)]$$

and

$$\overset{\circ}{\overbrace{[(a, b), (c, d)]}} = ((a, b), (c, d)).$$

Determine $Fr\,((a, b), (c, d))$.

2. In $\mathbf{R}^2$, define

$$S(a) = \{(x, y) \mid y \in \mathbf{R}, x < a\}$$

for each $a \in \mathbf{R}$. Then

$$\overline{S(a)} = \{(x, y) \mid y \in \mathbf{R}, x \leq a\}$$

and $\overset{\circ}{\overline{S(a)}} = \widehat{\overset{\circ}{S(a)}} = S(a)$. Determine $Fr\,(S(a))$.

3. Let $(X, \mathscr{T})$ be a topological space and $(Y, \mathscr{I})$ a separated topological space. Suppose that $f: X \to Y$ is continuous. Then $\{(x, f(x)) \mid x \in X\}$ is a closed subset of the product space $X \times Y$.

4. Show that the mapping $u^{(k)}$ (see the paragraph following 7.23′) is a homeomorphism of X_k onto the subspace $u^{(k)}(X_k)$ of X.

5. Let $(X_1, \mathscr{T}_1)$ and $(X_2, \mathscr{T}_2)$ be topological spaces and $(X, \mathscr{T})$ their product. For $R \subset X_1 \times X_2$ and $A \subset X_1$, define

$$R(A) = \{y \mid \text{there is } x \text{ in } A \text{ such that } (x, y) \in R\}.$$

Show that if $U \in \mathscr{T}$ and $A \subset X_1$ then $U(A)$ is open in X_2.

6. Let $\mathscr{T}_i$ be a topology on a set X for each i in an index set I. Define $X_i = X$ for each i in I. Then $f: X \to \prod_{i \in I} X_i$, given by $(f(x))_i = x$ for each i in I, is a homeomorphism of $(X, \sup_{i \in I} \mathscr{T}_i)$ with $f(X)$, considered as a subspace of the topological space product of the family $((X_i, \mathscr{T}_i))_{i \in I}$.

▼ Projective Limits*

A relation in a set E is a *preorder relation* if it is reflexive and transitive. If the preorder relation we consider is denoted by the symbol $\prec$, then its properties are expressed as follows: (1) $x \prec x$ for all $x \in E$, and (2) $x \prec y$ and $y \prec z \Rightarrow x \prec z$ for all x, y, and z in E. A set endowed with a preorder relation is called a preordered set.

Now let I be a preordered set, and let

$$G = \{(i, j) \mid (i, j) \in I \times I, i \prec j\}$$

be its graph. Let $(X_i)_{i \in I}$ be a family of sets, and for every $(i, j) \in G$ let $f_{ij} : X_j \to X_i$. Assume that:

(PL_1) $i \prec j \prec k \Rightarrow f_{ij} \circ f_{jk} = f_{ik}$;

(PL_2) for every $i \in I, f_{ii} = j_{X_i}$. †

Let $X = \Pi_{i \in I} X_i$ and let E be the set of all elements $x = (x_i)_{i \in I}$ of X such that

$$x_i = f_{ij}(x_j)$$

whenever $(i, j) \in G$. The set E is said to be *the projective limit of* $(X_i)_{i \in I}$ *with respect to* $(f_{ij})_{(i,j) \in G}$. We write

$$E = \lim_{\longleftarrow} ((X_i)_{i \in I}, (f_{ij})_{(i,j) \in G}),$$

or

$$E = \lim_{\longleftarrow} (X_i)_{i \in I}$$

if there is no ambiguity. For each $i \in I$, we write $f_i = pr_i \mid E$ and call f_i the canonical mapping of E *into* X_i. Notice that

$$f_i = f_{ij} \circ f_j$$

for all $(i, j) \in G$.

---

* The definitions and results of this appendix are not used in the rest of the volume. Hence they may be omitted in a first reading.

† If Y is a set, j_Y is the mapping $x \mapsto x$ of Y onto Y.

Sometimes we say that $((X_i)_{i \in I}, (f_{ij})_{(i,j) \in G})$ is a *projective system* (if $(f_{ij})_{(i,j) \in G}$ satisfies Conditions (PL_1) and (PL_2)) and that E is the projective limit of this system.

Example 7.—Let I be an arbitrary set and let $G = \{(x, x) \mid x \in I\}$. Then G is an order relation on I. Let $(X_i)_{i \in I}$ be a family of sets, and for every $(i, i) \in G$ let $f_{ii} = j_{X_i}$. Then

$$\lim_{\leftarrow} ((X_i)_{i \in I}, (f_{ij})_{(i,j) \in G}) = \prod_{i \in I} X_i.$$

Example 8.—Let T be a separated topological space and let $\mathscr{K}$ be the set of all compact* parts of T endowed with the order relation

$$G = \{(K, L) \mid K \subset L\}.$$

If $(K, L) \in G$ and $f \in C_R(K)$, we denote by $f': L \to R$ the mapping defined by $f'(x) = f(x)$ if $x \in K$ and $f'(x) = 0$ if $x \in L - K$.

For each $K \in \mathscr{K}$, let $M(K)$ be the vector space of all (Radon) measures on K. If $(K, L) \in G$ and if $\mu \in M(L)$, we denote by μ_K the measure on K defined by

$$\mu_K(f) = \mu(f')$$

for $f \in C_R(K)$. For $(K, L) \in G$, let $f_{KL}: M(L) \to M(K)$ be defined by

$$f_{KL}(\mu) = \mu_K$$

for $\mu \in M(L)$. It is easy to see that $(f_{KL})_{(K,L) \in G}$ satisfies Conditions (PL_1) and (PL_2). Hence

$$((M(K))_{K \in \mathscr{K}}, (f_{KL})_{(K,L) \in G})$$

is a projective system. The elements of

$$\lim_{\leftarrow} (M(K))_{K \in \mathscr{K}}$$

are called *premeasures* on T. Notice that μ is a premeasure on T if and only if $\mu = (\mu_K)_{K \in \mathscr{K}}$ where $\mu_K \in M(K)$ for every $K \in \mathscr{K}$ and $(\mu_L)_K = \mu_K$ for every $(K, L) \in G$.

Let I be a preordered set, G its graph, and let

$$\mathscr{X} = ((X_i)_{i \in I}, (f_{ij})_{(i,j) \in G})$$

* Compact sets are defined in Chapter 9.

and

$$\mathscr{Y} = ((Y_i)_{i \in I}, (g_{ij})_{(i,j) \in G})$$

be two projective systems. Let

$$E = \varprojlim \mathscr{X} \quad \text{and} \quad F = \varprojlim \mathscr{Y}.$$

For each $i \in I$, let f_i be the canonical mapping of E into X_i and g_i the canonical mapping of F into Y_i.

A family $(u_i)_{i \in I}$ is said to be a *projective system of mappings of* $\mathscr{X}$ *into* $\mathscr{Y}$ (or simply a *projective system*) if:

(PSM_1) for each $i \in I, u_i$ is a mapping of X_i into Y_i;

(PSM_2) for each $(i, j) \in G$, $u_i \circ f_{ij} = g_{ij} \circ u_j$.

If $(u_i)_{i \in I}$ is a projective system of mappings of $\mathscr{X}$ into $\mathscr{Y}$, then there is a *unique* mapping $u : E \to F$ satisfying

7.29 $$g_i \circ u = u_i \circ f_i \quad \text{for all} \quad i \in I.$$

In fact, for every $x = (x_i)_{i \in I} \in E$, define

7.30 $$u(x) = (u_i(x_i))_{i \in I}.$$

Clearly, $u : E \to \prod_{i \in I} Y_i$. If $x = (x_i)_{i \in I} \in E$ and $(i, j) \in G$, we have

$$g_{ij}(u_j(x_j)) = g_{ij} \circ u_j(x_j) = u_i \circ f_{ij}(x_j) = u_i(x_i);$$

whence $u(x) = (u_i(x_i))_{i \in I} \in F$. Therefore, u is a mapping of E into F, and clearly u satisfies 7.29. We leave it to the reader to establish the uniqueness of u.

The mapping u defined by 7.30 is called the *projective limit* of $(u_i)_{i \in I}$; we write

$$u = \varprojlim (u_i)_{i \in I}.$$

Let $((X_i, \mathscr{T}_i))_{i \in I}$ be a family of topological spaces. Assume I to be endowed with a preorder relation and let G be the graph of this relation. Assume also that:

7.31 $((X_i)_{i \in I}, (f_{ij})_{(i,j) \in G})$ is a projective system;

7.32 for every $(i, j) \in G$, the mapping $f_{ij} : X_j \to X_i$ is continuous.

In this case, we say that

7.33 $$(((X_i, \mathscr{T}_i))_{i \in I}, (f_{ij})_{(i,j) \in G})$$

is a *projective system of topological spaces.*

Let

$$E = \varprojlim (X_i)_{i \in I}$$

and let $\mathscr{A}$ be the *initial topology* on E associated with the families $((X_i, \mathscr{T}_i))_{i \in I}$ and $(f_i)_{i \in I}$ (for each $i \in I$, f_i is the canonical mapping of E into X_i). The topological space $(E, \mathscr{A})$ is called the *projective limit of the system* 7.33. We write

$$(E, \mathscr{A}) = \varprojlim (((X_i, \mathscr{T}_i))_{i \in I}, (f_{ij})_{(i,j) \in G})$$

or (if there is no ambiguity)

$$(E, \mathscr{A}) = \varprojlim ((X_i, \mathscr{T}_i))_{i \in I}$$

Notice that (see Chapter 6, Exercise 3) if $\mathscr{T}$ is the topology on $X = \prod_{i \in I} X_i$, product of the family $(\mathscr{T}_i)_{i \in I}$, then $\mathscr{A} = \mathscr{T}_E$. It follows that if each $(X_i, \mathscr{T}_i)$, $i \in I$, is separated, then $(E, \mathscr{A})$ *is separated.*

From the definition of E, we see that

$$E = \bigcap_{(i,j) \in G} \{x \mid pr_i(x) = f_{ij}(pr_j(x))\}.$$

Since pr_i and $f_{ij} \circ pr_j$ are continuous, for every $i \in I$ and $(i,j) \in G$, we deduce that (see Theorem 7.21)

7.34 *E is a closed part of X if every $(X_i, \mathscr{T}_i)$, $i \in I$, is separated.*

Now let $(((Y_i, \Sigma_i))_{i \in I}, (g_{ij})_{(i,j) \in G})$ be a second projective system of topological spaces and let

$$(F, \mathscr{B}) = \varprojlim ((Y_i, \Sigma_i))_{i \in I}.$$

Let $(u_i)_{i \in I}$ be a projective system of mappings such that $u_i : X_i \to Y_i$ is continuous for each $i \in I$. Let $u = \varprojlim u_i$. Then

7.35 $$u : E \to F \text{ is continuous.}$$

To establish 7.35, we notice that $g_i \circ u = u_i \circ f_i$ for every $i \in I$, so that $g_i \circ u$ is continuous. The continuity of u follows from 6.4. ▲

Chapter 8

Further Results Concerning Continuous Functions

The functions we discuss in this paragraph have $\mathbf{R}$ for range. To prove their continuity, we shall use 5.5.

8.1 Theorem.—*The mapping* $\alpha: (x, y) \mapsto x + y$ *of* $\mathbf{R}^2$ *into* $\mathbf{R}$ *is continuous on* $\mathbf{R}^2$.

Proof.—Let $(a, b) \in \mathbf{R}^2$ and let $c = \alpha(a, b)$. Let $\varepsilon > 0$ and let

$$U_1 = [a - \varepsilon/2, a + \varepsilon/2] \quad \text{and} \quad U_2 = [b - \varepsilon/2, b + \varepsilon/2].$$

Then $U_1 \in \mathcal{N}_{\mathbf{R}}(a)$ and $U_2 \in \mathcal{N}_{\mathbf{R}}(b)$, whence

$$U = U_1 \times U_2 \in \mathcal{N}_{\mathbf{R}^2}((a,b)).$$

If $(x, y) \in U$, then

$$x \in U_1 \ (\Leftrightarrow |x - a| \leq \varepsilon/2)$$

and

$$y \in U_2 \ (\Leftrightarrow |y - b| \leq \varepsilon/2);$$

hence

$$|\alpha(x, y) - c| = |(x + y) - (a + b)| = |(x - a) + (y - b)|$$
$$\leq |x - a| + |y - b| \leq \varepsilon/2 + \varepsilon/2 = 2(\varepsilon/2) = \varepsilon.$$

Thus, $(x, y) \in U$ implies

$$|\alpha(x, y) - c| \leq \varepsilon.$$

Since $\varepsilon > 0$ was arbitrary, α is continuous at (a, b). Since $(a, b) \in \mathbf{R}^2$ was arbitrary, α is continuous on $\mathbf{R}^2$.

8.2 Theorem.—*The mapping* $\pi:(x,y) \mapsto xy$ *of* $\mathbf{R}^2$ *into* $\mathbf{R}$ *is continuous on* $\mathbf{R}^2$.

Proof.—Let $(a, b) \in \mathbf{R}^2$ and let $c = ab = \pi(a, b)$. Let $\varepsilon > 0$ and let

$$U_1 = [a - \varepsilon/2(1 + |b|), a + \varepsilon/2(1 + |b|)].$$

Notice that if $x \in U_1$, then

$$|x - a| \leq \frac{\varepsilon}{2(1 + |b|)}, \quad \text{hence} \quad |x| \leq |a| + \frac{\varepsilon}{2(1 + |b|)}.$$

Let $\delta = |a| + \dfrac{\varepsilon}{2(1 + |b|)}$, and let

$$U_2 = \left[b - \frac{\varepsilon}{2\delta}, b + \frac{\varepsilon}{2\delta}\right].$$

Thus, $U_1 \in \mathcal{N}_{\mathbf{R}}(a)$ and $U_2 \in \mathcal{N}_{\mathbf{R}}(b)$, whence

$$U = U_1 \times U_2 \in \mathcal{N}_{\mathbf{R}^2}((a, b)).$$

Observe that

$$xy - ab = xy - xb + xb - ab = x(y - b) + (x - a)b;$$

hence

8.3 $$|xy - ab| \leq |x|\,|y - b| + |x - a|\,|b|$$

for all $(x, y) \in \mathbf{R}^2$.

Using 8.3, we deduce, for $(x, y) \in U$,

$$|xy - ab| \leq \delta\frac{\varepsilon}{2\delta} + |b|\frac{\varepsilon}{2(1 + |b|)} \leq \frac{\varepsilon}{2} + \frac{\varepsilon}{2} = 2\left(\frac{\varepsilon}{2}\right) = \varepsilon;$$

that is

$$|\pi(x, y) - c| \leq \varepsilon.$$

Thus, $(x, y) \in U$ implies

$$|\pi(x, y) - c| \leq \varepsilon.$$

Since $\varepsilon > 0$ was arbitrary, π is continuous at (a, b). Since $(a, b) \in \mathbf{R}^2$ was arbitrary, u is continuous on $\mathbf{R}^2$.

If S is a set and $f:S \to \mathbf{R}$, $g:S \to \mathbf{R}$ are two mappings, then we denote by

$f + g$ the mapping $x \mapsto f(x) + g(x)$ of S into $\mathbf{R}$

and by

$$fg \text{ the mapping } x \mapsto f(x) \cdot g(x) \text{ of } S \text{ into } \mathbf{R}.$$

Hence

$$(f + g)(x) = f(x) + g(x) \quad \text{and} \quad (fg)(x) = f(x)g(x)$$

for all $x \in S$.

Observe that

8.4 $$f + g = \alpha(f, g) \quad \text{and} \quad fg = \pi(f, g).$$

8.5 Theorem.—*Let X be a topological space, $S \subset X$, and let f and g be in $C_{\mathbf{R}}(S)$. Then $f + g$ and fg belong to $C_{\mathbf{R}}(S)$.*

Proof.—By 8.4, $f + g = \alpha(f, g)$ and $fg = \pi(f, g)$. Since f, g, α, and π are continuous, we deduce from Theorem 7.23, that $f + g$ and fg belong to $C_{\mathbf{R}}(S)$.

Remark.—If $c \in S$ and if $f:S \to \mathbf{R}$ and $g:S \to \mathbf{R}$ are continuous at c, we deduce by the method of proof used in Theorem 8.5 that $f + g$ and fg are continuous at c (hint: Use 7.22′).

8.6 Theorem.—*The mapping $\gamma: (x, y) \mapsto x/y$ of $\mathbf{R} \times \mathbf{R}^*$ ($\subset \mathbf{R}^2$) into $\mathbf{R}$ is continuous on $\mathbf{R} \times \mathbf{R}^*$.*

Proof.—The mapping $(x, y) \mapsto x$ of $\mathbf{R}^2$ into $\mathbf{R}$ is continuous on $\mathbf{R}^2$. Hence, the mapping $(x, y) \mapsto x$ of $\mathbf{R} \times \mathbf{R}^*$ into $\mathbf{R}$ is continuous on $\mathbf{R} \times \mathbf{R}^*$. By Example 8, Chapter 5, the mapping $y \mapsto 1/y$ of $\mathbf{R}^*$ into $\mathbf{R}$ is continuous on $\mathbf{R}^*$. We deduce that the mapping $(x, y) \mapsto 1/y$ of $\mathbf{R} \times \mathbf{R}^*$ into $\mathbf{R}$ is continuous on $\mathbf{R} \times \mathbf{R}^*$.

Hence the mappings

$$(x, y) \mapsto x \text{ (of } \mathbf{R} \times \mathbf{R}^* \text{ into } \mathbf{R})$$

and

$$(x, y) \mapsto 1/y \text{ (of } \mathbf{R} \times \mathbf{R}^* \text{ into } \mathbf{R})$$

are continuous on $\mathbf{R} \times \mathbf{R}^*$. Denote the first by f and the second by g. By Theorem 8.5 above, fg is continuous on $\mathbf{R} \times \mathbf{R}^*$. Clearly, fg is the mapping γ. Hence, γ is continuous on $\mathbf{R} \times \mathbf{R}^*$.

If S is a set and $f:S \to \mathbf{R}$ and $g:S \to \mathbf{R}$ are two mappings such that

8.7 $g(x) \neq 0$ for $x \in S$,

we denote by f/g $\left(\text{or } \dfrac{f}{g}\right)$ the mapping $x \mapsto f(x)/g(x)$ of S into $\mathbf{R}$. Hence

$$(f/g)(x) = f(x)/g(x) \quad \text{for all} \quad x \in S.$$

Observe that (under Hypothesis 8.7)

8.8 $f/g = \gamma(f, g)$.

8.9 Theorem.—*Let X be a topological space, $S \subset X$, and let f and g be in $C_{\mathbf{R}}(S)$. Suppose $g(x) \neq 0$ for $x \in S$. Then f/g belongs to $C_{\mathbf{R}}(S)$.*

Proof.—By 8.8, $f/g = \gamma(f, g)$. Since f, g, γ are continuous, we deduce from Theorem 7.23, that f/g belongs to $C_{\mathbf{R}}(S)$.

Remark.—If $c \in S$ and if $f: S \to \mathbf{R}$ and $g: S \to \mathbf{R}$ are continuous at c, and if $g(x) \neq 0$ for $x \in S$, then we deduce, by the method of proof used in Theorem 8.9, that f/g is continuous at c (hint: Use 7.22′).

Let S be a set and $f: S \to \mathbf{R}$ and $g: S \to \mathbf{R}$ two mappings. For $\lambda \in \mathbf{R}$ and $\mu \in \mathbf{R}$, we denote by $\lambda f + \mu g$ the mapping $x \mapsto \lambda f(x) + \mu g(x)$ of S into $\mathbf{R}$.

Hence

$$(\lambda f + \mu g)(x) = \lambda f(x) + \mu g(x)$$

for all $x \in S$.

We write $f - g$ instead of $f + (-1)g$; hence $(f - g)(x) = f(x) - g(x)$ for $x \in S$.

To facilitate reference, we also give the following:

8.10 Theorem.—*Let X be a topological space, $S \subset X$, and let f and g be in $C_{\mathbf{R}}(S)$. Then:*

8.11 $\lambda f + \mu g \in C_{\mathbf{R}}(S)$ *for all* $\lambda \in \mathbf{R}$, $\mu \in \mathbf{R}$;

8.12 $fg \in C_{\mathbf{R}}(S)$;

8.13 $f/g \in C_{\mathbf{R}}(S)$ *if* $g(x) \neq 0$ *for* $x \in X$.

Proof.—By Example 10 in Chapter 5, λf and μg belong to $C_{\mathbf{R}}(S)$. By 8.5, $\lambda f + \mu g \in C_{\mathbf{R}}(S)$. Assertion 8.12 was proved in 8.5. Assertion 8.13 was proved in 8.9.

Remark.—Let $c \in S$ and let $f: S \to \mathbf{R}$ and $g: S \to \mathbf{R}$ be two functions continuous at c. We leave it to the reader to establish that $\lambda f + \mu g$

is continuous at c for all $\lambda \in \mathbf{R}$ and $\mu \in \mathbf{R}$. We have already remarked that fg and $f + g$ are continuous at c.

From Theorem 8.10, we deduce then:

8.14 Theorem.—*The set $C_{\mathbf{R}}(S)$ endowed with the addition $(f, g) \mapsto f + g$ and scalar multiplication $(\lambda, f) \mapsto \lambda f$ is a vector space.**

From the remark following 8.10, we deduce that if $c \in S$, then the set of all functions on S to $\mathbf{R}$ continuous at c, endowed with the addition $(f, g) \mapsto f + g$ and scalar multiplication $(\lambda, f) \mapsto \lambda f$, is a *vector space*.

From Theorem 8.10, we also deduce that:

8.15 Theorem.—*The set $C_{\mathbf{R}}(S)$ endowed with the addition $(f, g) \mapsto f + g$, the scalar multiplication $(\lambda, f) \mapsto \lambda f$, and the multiplication $(f, g) \mapsto fg$ is a commutative algebra (over $\mathbf{R}$) having as a unit element the constant function* 1.

From the remark following 8.10, we deduce that, if $c \in S$, then the set of all functions on S to $\mathbf{R}$ continuous, endowed with the addition $(fg) \mapsto f + g$, the scalar multiplication $(x, f) \mapsto \lambda f$, and the multiplication $(f, g) \mapsto fg$, is a commutative algebra having as a unit element the constant function 1.

It follows from the above that

$$\lambda_1 f_1 + \ldots + \lambda_n f_n \quad \text{and} \quad f_1 \ldots f_n$$

are defined and belong to $C_{\mathbf{R}}(S)$ for all $\lambda_1, \ldots, \lambda_n$ in $\mathbf{R}$ and $f_1, \ldots, f_n$ in $C_{\mathbf{R}}(S)$.

Notice that

$$(\lambda_1 f_1 + \ldots + \lambda_n f_n)(x) = \lambda_1 f_1(x) + \ldots + \lambda_n f_n(x)$$

and

$$(f_1, \ldots, f_n)(x) = f_1(x) \ldots f_n(x)$$

for all $x \in S$.

If $S \subset \mathbf{R}$, then we have proved that the mapping $x \mapsto x$ of S into $\mathbf{R}$ is continuous on S. Hence, the mapping $x \mapsto x^n$ (n an integer >1) of S into $\mathbf{R}$ is continuous on S. Hence, the mapping $x \mapsto \lambda x^n$ ($\lambda \in \mathbf{R}$, n an integer >1) of S into $\mathbf{R}$ is continuous on S. Hence, a mapping of the form

$$x \mapsto a_n x^n + \ldots + a_1 x + a_0,$$

where $a_n, \ldots, a_1, a_0$ are real numbers, is *continuous* on S. Such a mapping is called a *polynomial function on* S. If p and q are two

* We assume that the elementary algebraic properties of *vector spaces* and *algebras* are known to the reader.

polynomial functions on S to $\mathbf{R}$ and $q(x) \neq 0$ for $x \in S$, then p/q can be defined and, by 8.13 (of Theorem 8.10), it is *continuous on* S. Such a function is called a *rational function on* S.

Hence, it follows from Theorem 8.10 that polynomials and rational functions are continuous mappings.

Exercises for Chapter 8

1. Let a, b, c be three real numbers such that $(a, b) \neq (0, 0)$. Show that the set

$$L(a, b, c) = \{(x, y) \mid ax + by + c = 0\} \subset \mathbf{R}^2$$

is $\neq \varnothing$, is closed and has a void interior (Hint: Notice that $(x, y) \mapsto ax + by + c$ is a continuous mapping of $\mathbf{R}^2$ into $\mathbf{R}$).

2. Show that $F = \{(x, y) \mid xy = 1\}$ is a closed subset of $\mathbf{R}^2$ with a void interior. Notice that $pr_1(F)$ is not closed.

3. Let $D_1 = \{(x, y) \mid x^2 + y^2 < 1\}$. Find $\bar{D}_1$ and $\mathring{\bar{D}}_1$.

4. Let $D_2 = \{(x, y) : x^2 - y^2 = 1\}$. Is D_2 closed? What is $\mathring{\bar{D}}_2$?

5. Let $h: \mathbf{R}^2 \to \mathbf{R}^2$ be defined by

$$h(x, y) = a(x, y) + (s, t)$$

where $a \in \mathbf{R}^*$ and $(s, t) \in \mathbf{R}^2$. Show that h is a homeomorphism of $\mathbf{R}^2$.

6. Define, for $n \in \mathbf{N}$,

$$D_n = \{(x, y) \mid x^2 + y^2 < 1/n\} \subset \mathbf{R}^2.$$

Show that $\{D_n \mid n \in \mathbf{N}\}$ is a fundamental system of $(0, 0)$. Let h be as in Exercise 5. Show that $\{h(D_n) \mid n \in \mathbf{N}\}$ is a fundamental system of (s, t) (see Exercise 14 at the end of Chapter 5).

* 7. Let G be a group with law of composition $(x, y) \mapsto x \perp y$. If G is endowed with a topology $\mathscr{T}$ such that $(x, y) \mapsto x \perp y$ and $x \mapsto x^{-1}$ are continuous, then $(G, \mathscr{T})$ is called a topological group. Show that $(\mathbf{R}, \mathscr{U})$ (with addition as the law of composition and the inverse of x being $-x$ for each x in $\mathbf{R}$) is a topological group. Show also that $(\mathbf{R}^*, \mathscr{U}_{\mathbf{R}^*})$ (with multiplication) and $(G, \mathscr{P}(G))$ (G any group) are topological groups.

Chapter 9

Compact Spaces

Let X be a set and let $A \subset X$. A family $(U_i)_{i \in I}$ of parts of X is said to be a *covering* of A if

$$A \subset \bigcup_{i \in I} U_i.$$

If I is *finite*, we say that the considered covering is finite.

Example 1.—(i) Let $(X, \mathscr{T})$ be a topological space, $A \subset X$, and for each $x \in A$ let $V_x \in \mathscr{N}(x)$. Then $(V_x)_{x \in X}$ is a covering of A.

(ii) For each $n \in \mathbf{N}$, let $U_n = (-n, +n)$. Then $(U_n)_{n \in \mathbf{N}}$ is a covering of $\mathbf{R}$ (hint: Use Archimedes' property). Notice that there is no finite part $F \subset \mathbf{N}$ such that $(U_n)_{n \in F}$ is still a covering of $\mathbf{R}$.

In fact, suppose that there is such a finite part $F \subset \mathbf{N}$. Let $p = \sup F$. Then

$$U_n \subset U_p = (-p, +p) \text{ for all } n \in \mathbf{N},$$

whence $\mathbf{R} \subset (-p, +p)$. Since this is a contradiction it follows that there is no finite part $F \subset \mathbf{N}$ such that $(U_n)_{n \in F}$ is a covering of $\mathbf{R}$.

(iii) Let $\varepsilon > 0$, and for each $x \in [0, 1]$ let $V_\varepsilon(x) = (x - \varepsilon, x + \varepsilon)$. Then $(V_\varepsilon(x))_{x \in [0,1]}$ is a covering of $[0, 1]$ (notice that this is a particular case of (i), above).

(iv) For each $n \in \mathbf{N}$, let $B_n = (1/n, 1]$. We leave it to the student to verify that $(B_n)_{n \in \mathbf{N}}$ is a covering of $(0, 1]$ and that there is no finite part $F \subset \mathbf{N}$ such that $(B_n)_{n \in F}$ is still a covering of $(0, 1]$.

Below we shall denote by $(X, \mathscr{T})$ a separated topological space.

A covering $(U_i)_{i \in I}$ of A is said to be *open* if $U_i \in \mathscr{T}$ for all $i \in I$.

9.1 Definition.—*A separated* topological space $(X, \mathscr{T})$ is said to be compact if for every open covering $(U_i)_{i,I}$ of X there is a finite set $F \subset I$ such that $(U_i)_{i \in F}$ is still a covering of X.*

---

* A topological space $(Y, \mathscr{S})$, *not necessarily separated*, is said to be *quasi-compact* if, for every open covering $(U_i)_{i \in I}$ of Y, there is a finite set $F \subset I$ such that $(U_i)_{i \in F}$ is still a covering of Y. A set $A \subset Y$ is said to be *quasi-compact* if $(A, \mathscr{T}_A)$ is quasi-compact. We leave it to the reader to determine which of the results given in this section remain valid for quasi-compact spaces.

Thus $(X, \mathscr{T})$ is compact if and only if every open covering of X "contains" a finite covering.

When there is no ambiguity, we shall simply say that X is compact instead of saying that $(X, \mathscr{T})$ is compact.

A set $A \subset X$ *is said to be compact if* $(A, \mathscr{T}_A)$ *is compact.* Since $U \in \mathscr{T}_A$ if and only if $U = V \cap A$ for some $V \in \mathscr{T}$, it follows easily that *a set* $A \subset X$ *is compact if and only if for every given covering* $(U_i)_{i\in I}$ *of* A, *consisting of sets belonging to* $\mathscr{T}$, *there is a finite set* $F \subset I$ *such that* $(U_i)_{i\in F}$ *is still a covering of* A.

Notice that in this text we consider compact sets only in *separated* topological spaces.

We shall give many important examples of compact sets. Let us note first that the results in Example 1 imply immediately that:

Example 2.—(i) The topological space **R** is *not* compact. (ii) The set $(0, 1]$ ($\subset$**R**) is *not* compact.

Example 3.—Let $a \in \mathbf{R}$, $b \in \mathbf{R}$, and $a < b$. Then the set $[a, b]$ is compact.

The proof of this important result is quite long and delicate. However, we shall give it in detail.

Let $(U_i)_{i,I}$ be an open covering of $[a, b]$. Let Y be the set of all $y \in (a, b]$ having the property: there exists a finite set $F \subset I$ such that

$$[a, y] \subset \bigcup_{i\in F} U_i.$$

Then $Y \neq \varnothing$ (why?) We shall now show that $\beta = \sup Y \in Y$. In fact, since $\beta \in [a, b]$, there is $i_0 \in I$ such that $\beta \in U_{i_0}$. Since U_{i_0} is open, U_{i_0} is a neighborhood of β, and hence there exists c and d such that $c < \beta < d$ and $(c, d) \subset U_{i_0}$. Since $\beta = \sup Y$, there exists $y \in Y$ satisfying $c < y \leq \beta$. Since $y \in Y$, it follows that there is a finite set $H \subset I$ such that

$$[a, y] \subset \bigcup_{i\in H} U_i.$$

But then, if $K = H \cup \{i_0\}$, we have

$$[a, \beta] \subset [a, y] \cup U_{i_0} \subset \bigcup_{i\in K} U_i;$$

whence $\beta \in Y$.

We shall now show that $\beta = b$. In fact, suppose $\beta < b$, and let $K \subset I$ be finite, such that

$$[a, \beta] \subset \bigcup_{i\in K} U_i.$$

Let $i_0 \in K$ be such that $U_{i_0} \ni \beta$. Then there exist c and d such that $c < \beta < d$ and $(c, d) \subset U_{i_0}$. If z is such that $\beta < z < d$ and $z \in [a, b]$, then $(U_i)_{i \in K}$ is still a finite covering of $[a, z]$. This implies that $z \in Y$. Since $\beta = \sup Y$ and $z > \beta$, we arrive at a contradiction. Thus, the hypothesis $\beta \neq b$ leads to a contradiction, and hence $\beta = b$.

Hence, there is a finite set $L \subset I$ such that $(U_i)_{i \in L}$ is a covering of $[a, b]$. Since $(U_i)_{i \in I}$ was arbitrary, we deduce that $[a, b]$ is compact.

Exercise 1.—Show that (a, b) is *not* compact.

9.2 Theorem.—*If $A \subset X$ and $B \subset X$ are compact, then $A \cup B$ is compact.*

Proof.—Let $(U_i)_{i \in I}$ be an open covering of $A \cup B$. Then

$$A \subset A \cup B \subset \bigcup_{i \in I} U_i.$$

Since A is compact, there is a finite set, $F_1 \subset I$, such that

$$A \subset \bigcup_{i \in F_1} U_i.$$

In the same way, we see that there is a finite set $F_2 \subset I$ such that

$$B \subset \bigcup_{i \in F_2} U_i.$$

Let $F = F_1 \cup F_2$. Then F is finite and

$$A \cup B \subset \left(\bigcup_{i \in F_1} U_i\right) \cup \left(\bigcup_{i \in F_2} U_i\right) = \bigcup_{i \in F} U_i.$$

Hence, $(U_i)_{i \in F}$ is a finite covering of $A \cup B$. Since $(U_i)_{i \in I}$ was arbitrary we deduce that $A \cup B$ is compact.

9.3 Corollary.—*If $A_1, \ldots, A_n$ are compact parts of X, then $A_1 \cup \ldots \cup A_n$ is also compact.*

Hint: Use 9.2 and reason by induction.

9.4 Theorem.—*If A is compact and $B = \overline{B} \subset A$, then B is compact.*

Proof.—Let $(U_i)_{i \in I}$ be a covering of B. Let $\alpha \notin I$, $U_\alpha = \mathbf{C}B$ and $J = I \cup \{\alpha\}$. Then $(U_j)_{j \in J}$ is an open covering of A. Hence, there exists a finite set $H \subset J$ such that

$$A \subset \bigcup_{i \in H} U_j .$$

If $F = H - \{\alpha\}$, we deduce (notice that $B \cap U_\alpha = \varnothing$) that

$$B \subset \bigcup_{i \in F} U_i.$$

Hence, $(U_i)_{i \in F}$ is a finite covering of B. Since $(U_i)_{i \in I}$ was arbitrary, we deduce that B is compact.

It follows from Theorem 9.4 that if $(A_i)_{i \in I}$ is a family of closed sets such that A_i is compact for some $i_0 \in I$, then $\bigcap_{i \in I} A_i$ is compact. We deduce immediately that if $A_1, \ldots, A_n$ are compact, then their intersection $A_1 \cap \ldots \cap A_n$ is compact.

9.5 Theorem.—*Let $A \subset X$ be a compact set and let $b \notin A$. Then there exist open sets U and V such that*

$$A \subset U, b \in V, U \cap V = \varnothing.$$

Proof.—Since $(X, \mathcal{T})$ is separated, we deduce that, for each $a \in A$, there are open neighborhoods $U_a \in \mathcal{N}(a)$ and $V_a \in \mathcal{N}(b)$ satisfying $U_a \cap V_a = \varnothing$. Clearly, $(U_a)_{a \in A}$ is an open covering of A. Since A is compact, there is a finite set $\{a_1, \ldots, a_n\} \subset A$ such that

$$A \subset U \quad \text{if} \quad U = \bigcup_{j=1}^{n} U_{a_j}.$$

Let $V = \bigcap_{j=1}^{n} V_{a_j}$. Then U and V are open and, clearly, $A \subset U$ and $b \in V$. If $x \in U$, then there is j_0 such that $1 \leq j_0 \leq n$ and such that $x \in U_{a_{j_0}}$. Then $x \notin V_{a_{j_0}}$, whence $x \notin V$. Thus, $U \cap V = \varnothing$, and therefore Theorem 9.5 is proved.

9.6 Corollary.—*A compact space $(X, \mathcal{T})$ is regular.*

Proof.—Let $b \in X$ and let $B \in \mathcal{N}_X(b)$. Since $(X, \mathcal{T})$ is separated, $\{b\}$ is closed. Let W be an open set such that $a \in W \subset B$. By Theorem 9.4, $\mathbf{C}W$ is compact. By Theorem 9.5, there exist open sets U and V such that

$$U \supset \mathbf{C}W, \quad b \in V, \quad U \cap V = \varnothing.$$

Then

$$b \in V \subset \mathbf{C}U \subset W.$$

It follows that $\mathbf{C}U \in \mathcal{N}_X(b)$ and that $\mathbf{C}U \subset B$. Since $\mathbf{C}U$ is closed, and since $b \in X$ and $B \in \mathcal{N}_X(b)$ were arbitrary, we deduce that $(X, \mathcal{T})$ is regular.

9.7 Theorem.—*If A is compact, then $A = \bar{A}$.*

Proof.—Let $b \notin A$. By Theorem 9.5, there is $V \in \mathcal{N}(b)$ such that $V \cap A = \varnothing$. Hence, $b \notin \bar{A}$, and hence $\bar{A} \subset A$. Hence, $A = \bar{A}$.

9.8 Theorem.—*Suppose $(X, \mathscr{T})$ to be regular. Let $K \subset X$ be compact and $F \subset X$ closed and such that $K \cap F = \varnothing$. Then there is an open set $U_K \supset K$ and an open set $U_F \supset F$ such that $\bar{U}_K \cap \bar{U}_F = \varnothing$.*

Proof.—For each $x \in K$, let W_x and V_x be closed neighborhoods of x such that

$$V_x \cap F = \varnothing \quad \text{and} \quad W_x \subset \mathring{V}_x.$$

Then $(\mathring{W}_x)_{x \in K}$ is a covering of K. Hence, there is a finite set $I \subset K$ such that $(\mathring{W}_x)_{x \in I}$ is a covering of K. Let

$$U_K = \bigcup_{x \in I} \mathring{W}_x \quad \text{and} \quad U_F = \complement\left(\bigcup_{x \in I} V_x\right).$$

Then U_K and U_F are open sets, $U_K \supset K$ and $U_F \supset F$. Moreover,

$$\bar{U}_K \subset \bigcup_{x \in I} W_x \subset \bigcup_{x \in I} \mathring{V}_x$$

and

$$\bar{U}_F \subset \complement\left(\bigcup_{x \in I} \mathring{V}_x\right).$$

Hence,

$$\bar{U}_K \cap \bar{U}_F = \varnothing.$$

Remark.—Let K be compact and U be open and containing K. We deduce from 9.8 that there is an open set U_K such that

$$K \subset U_K \subset \bar{U}_K \subset U.$$

Example 4.—Let $A \subset \mathbf{R}$. Then the following assertions are equivalent:

(i) A is bounded and closed;
(ii) A is compact.

Proof of (i) $\Rightarrow$ (ii).—If A is bounded, then there are α and β in $\mathbf{R}$ such that $\alpha \leq x \leq \beta$ for all $x \in A$; hence $A \subset [\alpha, \beta]$. By Example 2, $[\alpha, \beta]$ is compact, and hence, by Theorem 9.4, A is compact.

Proof of (ii) $\Rightarrow$ (i).—If A is compact, then by Theorem 9.7, $A = \bar{A}$. Now let $U_n = (-n, +n)$ for all $n \in \mathbf{N}$. Since $\bigcup_{n \in \mathbf{N}} U_n = \mathbf{R}$ (see Example 1, (b)), $(U_n)_{n \in \mathbf{N}}$ is a covering of A. Since A is compact, there is a finite part $F \subset \mathbf{N}$ such that $(U_n)_{n \in F}$ is a covering of A. Let $p = \sup F$. Then $U_n \subset U_p$ for $n \in F$, whence $A \subset U_p = (-p, +p)$. Hence $-p \leq x \leq p$ for $x \in A$, and hence A is bounded.

An arbitrary closed part of $\mathbf{R}$ *is not necessarily compact. By Example* 4, *to be compact it must be both closed and bounded.*

Let $(X, \mathscr{T})$ and $(Y, \mathscr{S})$ be two *separated* topological spaces.

9.9 Theorem.—*Let $A \subset X$ be a compact set and let $f: A \to Y$ be a continuous mapping. Then $f(A)$ is compact.*

Proof.—Let $(V_j)_{j \in I}$ be an open covering of $f(A)$. Then for every $j \in I$ there is $U_j \in \mathscr{T}$ such that $f^{-1}(V_j) = U_j \cap A$ (use 5.11(b) and the definition of $\mathscr{T}_A$). Hence, $(U_j)_{j \in I}$ is an open covering of A. Since A is compact, there is a finite set $F \subset I$ such that $(U_j)_{j \in F}$ is still an open covering of A. Since we have

$$A = \bigcup_{j \in F} A \cap U_j = \bigcup_{j \in F} f^{-1}(V_j),$$

we deduce that

$$f(A) = \bigcup_{j \in F} f(f^{-1}(V_j)) \subset \bigcup_{j \in F} V_j.$$

Hence, $(V_j)_{j \in F}$ is an open covering of $f(A)$. Since $(V_j)_{j \in I}$ was arbitrary, we deduce that $f(A)$ is compact.

It follows in particular from Theorem 9.9 that $f(A)$ is closed.

9.10 Corollary.—*Let $B \subset X$ and let $f: B \to Y$ be a function continuous on B. If $A \subset B$ is compact, then $f(A)$ is compact.*

Proof.—Note that $h = f \mid A$ is continuous on A.

Example 5.—Let $A \subset X$ be a compact set and let $f: A \to Y$ be a continuous mapping. Then G_f is compact (in $X \times Y$).

For $x \in A$, let $u(x) = (x, f(x))$. Since the mapping $x \mapsto x$ of A into X is continuous on A, it follows from Corollary 7.20 that u is continuous on A. By Theorem 9.9, $G_f = u(A)$ is compact.

Example 6.—Let $B = \{(x, x^2) \mid x \in [0, 1]\}$. Then $B \subset \mathbf{R}^2$ is compact.

9.11 Theorem.—*Let $A \subset X$ be compact and let $f \in C_{\mathbf{R}}(A)$. Then there are c_m and c_M in A such that*

$$f(c_m) \leq f(x) \leq f(c_M)$$

for all $x \in A$.

Proof.—By Theorem 9.9, $f(A)$ is compact and hence bounded and closed (see Example 4). Let

$$\alpha = \inf f(A) \quad \text{and} \quad \beta = \sup f(A).$$

By Example 7, Chapter 3, $\alpha \in f(A)$ and $\beta \in f(A)$. Hence, there are c_m and c_M in A such that $\alpha = f(c_m)$ and $\beta = f(c_M)$. Hence,

$$f(c_m) = \alpha \leq f(x) \leq \beta = f(c_M)$$

for all $x \in A$, and hence Theorem 9.11 is proved.

The hypothesis that A is compact is essential in Theorem 9.11. For instance, let $(0, 1) \subset \mathbf{R}$ and let $f(x) = x$ for $x \in (0, 1)$. Then $f \in C_{\mathbf{R}}((0, 1))$,

$$\inf f((0, 1)) = 0 \quad \text{and} \quad \sup f((0, 1)) = 1.$$

However, there is no $u \in (0, 1)$ such that $f(u) = 0$, and there is no $v \in (0, 1)$ such that $f(v) = 1$.

This argument can be also used to show that $(0, 1)$ is not compact. Denote by $(Y, \mathscr{I})$ a separated topological space.

9.12 Theorem.—*Let $A \subset X$ be a compact set and $f: A \to Y$ an injective continuous function. Let $g: f(A) \to X$ be defined by*

$$g(x) = f^{-1}(x) \quad \textit{for} \quad x \in f(A).$$

Then g is continuous.

Proof.—Let $U \subset X$ be open. Then $A - U = A \cap \mathbf{C}U$ is closed and hence (since it is contained in A) compact. By Theorem 9.9, $f(A - U)$ is compact, and therefore closed. Notice that

$$g^{-1}(A \cap \mathbf{C}U) = f(A - U).$$

In fact, $x \in g^{-1}(A \cap \mathbf{C}U)$ if and only if $g(x) \in A \cap \mathbf{C}U$; that is, if and only if $f^{-1}(x) \in A \cap \mathbf{C}U$; whence $x \in g^{-1}(A \cap \mathbf{C}U)$ if and only if $x \in f(A - U)$.

Now

$$g^{-1}(\mathbf{C}U) = g^{-1}(A \cap \mathbf{C}U),$$

and hence $g^{-1}(\mathbf{C}U)$ is closed. Since

$$g^{-1}(U) = f(A) \cap \mathbf{C}g^{-1}(\mathbf{C}U)$$

(why?), we deduce that g is continuous (see 5.11).

From Theorem 9.12, we deduce the following corollaries.

9.13 Corollary.—*If* $(X, \mathscr{T})$ *is compact and* $f: X \to Y$ *is a continuous bijection, then* f *is a homeomorphism.*

9.14 Corollary.—*Let* f *be as in the statement of Theorem* 9.12. *Then* $f \mid A$ *is a homeomorphism of* A *onto* $f(A)$.

Remarks.—(1) Let $(X, \mathscr{T})$ be compact and let $\mathscr{W}$ be a topology on X. Suppose that:

(i) $\mathscr{W} \subset \mathscr{T}$;
(ii) $(X, \mathscr{W})$ is separated.

Then $\mathscr{W} = \mathscr{T}$. In fact, the mapping $j_X: x \mapsto x$ is a continuous bijection of X, endowed with the topology $\mathscr{T}$, onto X endowed with the topology $\mathscr{W}$. By Theorem 9.13, j_X is a homeomorphism, whence

$$\mathscr{W} = \mathscr{T}.$$

(2) Let $(X, \mathscr{T})$ be compact and let $\mathscr{F} \subset C_{\mathbf{R}}(X)$ be a part such that for any x and y in X, $x \neq y$, there is $f \in \mathscr{F}$ such that

$$f(x) \neq f(y).$$

Then the initial topology associated with $\mathbf{R}$ and $(f)_{f \in \mathscr{F}}$ coincides with $\mathscr{T}$.

In this setting, we shall often say *the initial topology associated with* $\mathscr{F}$ instead of the initial topology associated with (the space) $\mathbf{R}$ and (the family) $(f)_{f \in \mathscr{F}}$.

Let X be a set. A family $(F_i)_{i \in I}$ of parts of X has the *finite intersection property* if

$$\bigcap_{i \in J} F_i \neq \varnothing$$

for every finite set $J \subset I$.

Notice that if $(F_i)_{i \in I}$ has the finite intersection property, then $F_i \neq \varnothing$ for all $i \in I$.

A sequence $(F_n)_{n \in N}$ of non-void parts of X, such that

$$F_1 \supset F_2 \supset \ldots \supset F_n \supset \ldots,$$

is an example of a family having the finite intersection property.

9.15 Theorem.—*Let* $(X, \mathscr{T})$ *be a separated topological space. Then the following assertions are equivalent:*

(a) *the space* $(X, \mathscr{T})$ *is compact;*
(b) *if* $(F_i)_{i \in I}$ *is a family of closed parts of* X *having the finite intersection*

property,

$$\bigcap_{i \in I} F_i \neq \varnothing.$$

Proof of (a) $\Rightarrow$ (b).—Suppose $(X, \mathscr{T})$ is compact. Let $(F_i)_{i \in I}$ be a family of closed parts of X having the finite intersection property. Suppose $\bigcap_{i \in I} F_i = \varnothing$. Then

$$\bigcup_{i \in I} \mathbf{C}F_i = \mathbf{C}(\bigcap_{i \in I} F_i) = \mathbf{C}\varnothing = X.$$

Since F_i is closed for each $i \in I$, it follows that $(\mathbf{C}F_i)_{i \in I}$ is an open covering of X. Since $(X, \mathscr{T})$ is compact, it follows that there is a finite set $J \subset I$ such that $\bigcup_{i \in J} \mathbf{C}F_i = X$. Then

$$\varnothing = \mathbf{C}(\bigcup_{i \in J} \mathbf{C}F_i) = \bigcap_{i \in J} F_i.$$

Since this is a contradiction, it follows that $\bigcap_{i \in I} F_i \neq \varnothing$.

Proof of (b) $\Rightarrow$ (a).—Suppose that (b) is satisfied and that $(X, \mathscr{T})$ is *not* compact. Then there exists an open covering $(U_i)_{i \in I}$ of X such that $\bigcup_{i \in J} U_i \neq X$ for every finite set $J \subset I$. Then

$$\bigcap_{i \in J} \mathbf{C}U_i = \mathbf{C}(\bigcup_{i \in J} U_i) \neq \varnothing$$

for any finite set $J \subset I$; that is, the family $(\mathbf{C}U_i)_{i \in I}$ (consisting of closed parts) has the finite intersection property. By (b), we deduce that $\bigcap_{i \in I} \mathbf{C}U_i \neq \varnothing$; that is,

$$\bigcup_{i \in I} U_i = \mathbf{C}(\bigcap_{i \in I} \mathbf{C}U_i) \neq X.$$

Since this is a contradiction, we conclude that $(X, \mathscr{T})$ is compact.

Theorem 9.15 will be used essentially to show that the product of a family of compact spaces is compact. To facilitate the presentation of the proof of this result, we shall first make several remarks.

Let X be a set. We say that $\mathscr{X} \subset \mathscr{P}(X)$ has the *finite intersection property** if for every finite set $\mathscr{F} \subset \mathscr{X}$ we have

$$\bigcap_{A \in \mathscr{F}} A \neq \varnothing.$$

Notice that if Y is a set, $f: X \to Y$, and $\mathscr{X} \subset \mathscr{P}(X)$ is a set having the finite intersection property, then

$$f(\mathscr{X}) = \{f(A) \mid A \in \mathscr{X}\}$$

* Here we are considering *sets* having the finite intersection property. Previously we have considered families.

also has the finite intersection property.

In fact, for every $\mathscr{F} \subset \mathscr{X}$, we have

$$\bigcap_{A\in\mathscr{F}} f(A) \supset f(\bigcap_{A\in\mathscr{F}} A) \neq \varnothing$$

if $\bigcap_{A\in\mathscr{F}} A \neq \varnothing$.

Let $(X, \mathscr{T})$ be a topological space and denote by $\mathscr{A}$ the set of all parts $\mathscr{X} \subset \mathscr{P}(X)$ having the finite intersection property and consisting of *closed sets*. It follows from Theorem 9.15 *that a separated space* $(X, \mathscr{T})$ *is compact if and only if for every* $\mathscr{X} \in \mathscr{A}$

$$\bigcap_{A\in\mathscr{X}} A \neq \varnothing.$$

In fact, suppose X to be compact and let $\mathscr{X} \in \mathscr{A}$. Then $(X_A)_{A\in\mathscr{X}}$, where $X_A = A$ for $A \in \mathscr{X}$, is a family having the finite intersection property. By Theorem 9.15,

$$\bigcap_{A\in\mathscr{X}} A = \bigcap_{A\in\mathscr{X}} X_A \neq \varnothing.$$

Conversely, suppose now that for every set $\mathscr{X} \in \mathscr{A}$ we have $\bigcap_{A\in\mathscr{X}} A \neq \varnothing$. Let $(F_i)_{i\in I}$ be a family of closed parts of X having the finite intersection property. Then

$$\mathscr{X} = \{F_i \mid i \in I\} \in \mathscr{A}$$

and

$$\bigcap_{i\in I} F_i = \bigcap_{A\in\mathscr{X}} A \neq \varnothing.$$

By Theorem 9.15, the space (X, J) is compact.

For any topological space $(X, \mathscr{T})$, we shall *order* $\mathscr{A}$ by writing

$$\mathscr{X}_1 \leq \mathscr{X}_2 \Leftrightarrow \mathscr{X}_1 \subset \mathscr{X}_2.$$

It is easy to see that $\mathscr{A}$ is inductive when endowed with this order (why?). By Zorn's lemma, every $\mathscr{X} \in \mathscr{A}$ is then contained in a maximal set $\mathscr{M} \in \mathscr{A}$.

We leave it to the reader to show that *a separated topological space* $(X, \mathscr{T})$ *is compact if and only if for every maximal set* $\mathscr{U} \in \mathscr{A}$, *we have*

$$\bigcap_{A\in\mathscr{U}} A \neq \varnothing.$$

Again, denote by $\mathscr{U}$ a *maximal* element of $\mathscr{A}$. We notice that if $F \subset X$ is closed, and if $F \cap A \neq \varnothing$ for all $A \in \mathscr{U}$, then $F \in \mathscr{U}$. In fact, we have

$$\{F\} \cup \mathscr{U} \in \mathscr{A} \quad \text{and} \quad \{F\} \cup \mathscr{U} \supset \mathscr{U};$$

whence $\{F\} \cup \mathscr{U} = \mathscr{U}$; that is, $F \in \mathscr{U}$.

Let us also notice that if $(A_i)_{i \in J}$ is a *finite* family of elements of $\mathscr{U}$, then

$$\bigcap_{i \in J} A_i \in \mathscr{U}.$$

9.16 Theorem (*Tychonoff*).—*If $((X_i, \mathscr{T}_i))_{i \in I}$ is a family of compact spaces, then the topological space $(X, \mathscr{T})$, product of the family $((X_i, \mathscr{T}_i))_{i \in I}$, is compact.*

Proof.—Let $\mathscr{A}$ be as above. To prove the theorem, it is enough to show that for every maximal set $\mathscr{M}$ in $\mathscr{A}$ we have $\bigcap_{A \in \mathscr{M}} A \neq \varnothing$. Hence, let $\mathscr{M}$ be a maximal element of $\mathscr{A}$. For each i in I, let $\mathscr{M}_i = \{pr_i(A) \mid A \in \mathscr{M}\}$; then $\mathscr{M}_i$ is a set of parts of X_i having the finite intersection property. It is clear, then, that for each i in I, we have $\bigcap_{A \in \mathscr{M}_i} \bar{A} \neq \varnothing$. Let $z = (z_i)_{i \in I}$, where $z_i \in \bigcap_{A \in \mathscr{M}_i} \bar{A}$ for each i in I. We shall establish that $z \in \bigcap_{A \in \mathscr{M}} A$; this will show that $\bigcap_{A \in \mathscr{M}} A \neq \varnothing$ and hence will complete the proof.

Let

9.17 $$V = \prod_{i \in I} V_i,$$

where $J = \{i \mid V_i \neq X_i\}$ is finite, $V_i \in \mathscr{N}_{X_i}(z_i)$, and $V_i = \bar{V}_i$ for each i in I. Since $V_i \in \mathscr{N}_{X_i}(z_i)$ for each i in I, we deduce that $V_i \cap pr_i(A) \neq \varnothing$ for all i in I and A in $\mathscr{M}$. By our previous remarks, we conclude that $pr_i^{-1}(V_i) \in \mathscr{M}$ for each i in I. Since $V = \bigcap_{j \in J} pr_j^{-1}(V_j)$, we have also $V \in \mathscr{M}$, and therefore $V \cap A \neq \varnothing$ for every A in $\mathscr{M}$. Since X_i is regular for each i in I (see Corollary 9.7), the set of all parts of the form 9.17 is a fundamental system of z (see 7.10). Thus, we deduce that $z \in A$ for each A in $\mathscr{M}$, that is, $\bigcap_{A \in \mathscr{M}} A \neq \varnothing$. This completes the proof of Theorem 9.16.

Exercise 2.—*If $((X_i, \mathscr{T}_i))_{i \in I}$ is a family of topological spaces such that its product, $(X, \mathscr{T})$, is compact and $X \neq \varnothing$, then, for each i in I, X_i is compact.*

Hint: Notice that X_i is separated for each i in I and use Theorem 9.12.

9.18 Theorem.—*If $((X_i, \mathscr{T}_i))_{i \in I}$ is a family of separated topological spaces, and if, for each i in I, A_i is a compact subset of X_i, then $\prod_{i \in I} A_i$ is compact.*

Proof.—The result follows immediately from 7.28.

Example 7.—A closed, bounded n-dimensional interval

$$[(a_1, \ldots, a_n), (b_1, \ldots, b_n)]$$

(see Example 4, Chapter 7) is compact.

To prove this, note that

$$[(a_1, \ldots, a_n), (b_1, \ldots, b_n)] = \prod_{1 \leq i \leq n} [a_i, b_i].$$

The assertion then follows from the result in Example 3 and Theorem 9.18.

Example 8.—Let $A \subseteq \mathbf{R}^n (n \geq 1)$. Then the following assertions are equivalent:

(i) A is closed and contained in a closed, bounded n-dimensional interval;

(ii) A is compact.

Proof of (i) ⇒ (ii).—This follows from Example 7 and from 9.4.

Proof of (ii) ⇒ (i).—The proof is left to the reader (hint: Use the fact that the projections $(x_i)_{1 \leq i \leq n} \mapsto x_k (1 \leq k \leq n)$ are continuous and use Corollary 9.10).

Example 9.—Let $R^{(i)} = \mathbf{R}$ and $\mathscr{U}^{(i)} = \mathscr{U}$ for each i in some index set I, and let $(\mathbf{R}^I, \mathscr{U}_I)$ be the product of the family $((\mathbf{R}^{(i)}, \mathscr{U}^{(i)}))_{i \in I}$. Then $A = \{(x_i)_{i \in I} \mid |x_i| \leq 1 \text{ for all } i \text{ in } I\}$ is compact.

We have $A = \prod_{i \in I} A_i$ where $A_i = [-1, +1]$ for each i in I. The assertion then follows from Theorem 9.18.

Let $(X, \mathscr{T})$ be a separated topological space.

9.19 Definition.—*A subset A of X is said to be relatively compact if $\bar{A}$ is compact.*

Hence, if A is relatively compact, and $B \subset A$, then B is relatively compact. From Corollary 9.3, it follows that if $B_1, \ldots, B_n$ are relatively compact, then $B_1 \cup \ldots \cup B_n$ is relatively compact. If $(A_i)_{i \in I}$ is any family of relatively compact sets, then $\bigcap_{i \in I} A_i$ is relatively compact.

Example 10.—Let $A \subset \mathbf{R}$. Then the following assertions are equivalent:

(i) A is bounded;

(ii) A is relatively compact.

Proof of (i) ⇒ (*ii*).—If A is bounded, then $\bar{A}$ is bounded (why?). By Example 4, $\bar{A}$ is compact, whence A is relatively compact.

Proof of (ii) ⇒ (i).—If A is relatively compact, then $\bar{A}$ is compact. By Example 4, $\bar{A}$ is bounded, whence A is bounded.

Example 11.—Let $A \subset \mathbf{R}^n$ ($n \geq 1$). Then the following assertions are equivalent:

(i) A is contained in a bounded n-dimensional interval;

(ii) A is relatively compact.

The proof is left to the reader.

9.20 Theorem.—*Let* $((X_i, \mathscr{T}_i))_{i \in I}$ *be a family of separated topological spaces, and for each* i *in* I *let* A_i *be a relatively compact subset of* X_i. *Then* $\prod_{i \in I} A_i$ *is relatively compact.*

Proof.—We have $\prod_{i \in I} A_i \subset \overline{\prod_{i \in I} A_i} = \prod_{i \in I} \bar{A}_i$. Since $\bar{A}_i$ is compact for each i in I, we deduce that $\prod_{i \in I} \bar{A}_i$ is compact. Therefore, $\overline{\prod_{i \in I} A_i}$ is compact and $\prod_{i \in I} A_i$ is relatively compact.

Example 12.—If, for each i in some index set I, A_i is a bounded subset of $\mathbf{R}$, then it follows from Example 10 and Theorem 9.20 that $\prod_{i \in I} A_i$ is relatively compact in $\mathbf{R}^I$ (see Example 9). Conversely, if A is any relatively compact subset of $\mathbf{R}^I$, then $pr_i(A)$ is relatively compact in $\mathbf{R}$ for each i in I. This result follows from a more general fact, which is a corollary of Theorem 9.9: If X and Y are two separated topological spaces, $A \subset X$, A relatively compact and $f: X \to Y$ a continuous function, then $f(A)$ is relatively compact in Y.

Exercises for Chapter 9

1. Let $f: [0, 100] \to \mathbf{R}$ be defined by $f(x) = x^3 + 2x + 1$ for $x \in [0, 100]$. Is f bounded? Is there $t \in [0, 100]$ such that $f(x) \leq f(t)$ for every $x \in [0, 100]$? Answer the same questions in the case of the function $f \mid [0, 100)$.

2. Is the subset $\{(x, 1/x) \mid x \in [\frac{1}{2}, 1]\}$ of $\mathbf{R} \times \mathbf{R}$ compact?

3. Is the subset $\{\sin x \mid x \in \mathbf{R}^*\}$ of $\mathbf{R}$ compact?

4. Let $(X, \mathscr{T})$ be the topological space in Example 2, Chapter 1. If X has at least two points, then there are quasi-compact parts of X that are not closed.

5. Suppose that $f: [0, 1] \to \mathbf{R}$ has the following property: For each $n \in \mathbf{N}$ there is $x_n \in [0, 1]$ such that $f(x_n) > n$. Is f continuous on $[0, 1]$?

6. Consider the topological space $(\mathbf{N}, \mathscr{P}(\mathbf{N}))$. Is there a continuous surjection $f: [0, 1] \to \mathbf{N}$?

7. Consider the subspace $\mathbf{Q}$ of $\mathbf{R}$. Give an example of a sequence $(F_n)_{n\in N}$ of closed subsets of $\mathbf{Q}$ such that $F_n \supset F_{n+1}$ for every $n \in \mathbf{N}$, and $\bigcap \{F_n \mid n \in \mathbf{N}\} = \varnothing$.

8. Let $S = [0, 1]$ and consider the set $B = S \cap \mathbf{Q}$. Then S is a compact subset of $\mathbf{R}$, but B is not compact.

9. Let $f:(0, 1] \to R$ be the function defined by $f(x) = 1/x$. Use Theorem 9.9 and Example 4 to show that $(0, 1]$ is not compact.

10. Let $(X, \mathscr{T})$ be a separated topological space and $(A_i)_{i\in I}$ a locally finite family of subsets of X (see Chapter 2). If K is a compact subset of X, then $\{i \mid A_i \cap K \neq \varnothing\}$ is finite.

11. Let X and Y be two (separated) topological spaces and $f:X \to Y$ a continuous mapping. Let $(K_\alpha)_{\alpha\in A}$ be a directed family of compact parts of X such that $\alpha \leq \beta$ implies $K_\alpha \supset K_\beta$ and let $K = \bigcap_{\alpha\in A} K_\alpha$. Then

$$f(K) = \bigcap_{\alpha\in A} f(K_\alpha).$$

(Hint: For each $y \in \bigcap_{\alpha\in A} f(K_\alpha)$, consider the directed family $(K_\alpha \cap f^{-1}(y))_{\alpha\in A}$. Then

$$K \cap f^{-1}(y) = \bigcap_{\alpha\in A} (K_\alpha \cap f^{-1}(y)) \neq \varnothing,$$

so that $y \in f(K)$).

Chapter 10

Locally Compact Spaces

We start this chapter with the:

10.1 Definition.—*A separated topological space $(X, \mathscr{T})$ is said to be locally compact if for every $x \in X$ there exists a compact neighborhood of x.*

When there is no ambiguity, we shall say simply that X is locally compact, instead of saying that $(X, \mathscr{T})$ is locally compact.

Below, we shall denote by $(X, \mathscr{T})$ a separated topological space.

A set $Z \subset X$ *is said to be locally compact if $(Z, \mathscr{T}_Z)$ is locally compact.* Hence, Z is locally compact if and only if for every $z \in Z$ there is $V \in \mathscr{N}_X(z)$ such that $V \cap Z$ is compact.

For various applications, it is useful to notice that the *following assertions are equivalent:*

10.2 *The space $(X, \mathscr{T})$ is locally compact;*

10.3 *For every $x \in X$, there exists a fundamental system of x consisting of compact neighborhoods;*

10.4 *For every $x \in X$, there exists a relatively compact neighborhood;*

10.5 *For every $x \in X$, there exists a fundamental system of x consisting of relatively compact neighborhoods;*

10.6 *For every $x \in X$, there exists a fundamental system of x consisting of open relatively compact neighborhoods.*

From the equivalence between 10.2 and 10.3, it follows that a locally compact space is *regular.*

10.7 Theorem.—*If $(X, \mathscr{T})$ is a separated topological space and $Z \subset X$ is locally compact, then Z is locally closed.*

Proof.—Since Z is locally compact, for each $z \in Z$, there is $V(z) \in \mathscr{N}_X(z)$ such that $Z \cap V(z)$ is compact and hence closed in $V(z)$. Since z was arbitrary, we deduce that Z is locally closed.

We deduce from Theorem 4.9 that *there is an open set $U \subset X$ and a closed set $F \subset X$ such that $Z = U \cap F$.*

▼ *Remark.*—Let X be locally compact, μ a positive (Radon) measure on X, and $Z \subset X$ locally compact. Since open and closed sets are μ-measurable, we deduce that Z is μ-measurable. ▲

10.8 Theorem.—*Let $(X, \mathscr{T})$ be locally compact and $Z \subset X$. Then the following assertions are equivalent:*

(a) *The set Z is locally compact;*
(b) *the set Z is locally closed.*

Proof of (a) $\Rightarrow$ (b).—This follows from 10.7.

Proof of (b) $\Rightarrow$ (a).—By Theorem 4.9, $Z = U \cap F$ where U is an open part of X and F a closed part of X. Let $z \in Z$ and let $V(z) \in \mathscr{N}_X(z)$ be compact and such that $V(z) \subset U$. Then $Z \cap V(z) = F \cap V(z)$ is compact and belongs to $\mathscr{N}_Z(z)$. Since $z \in Z$ was arbitrary, we deduce that Z is locally compact.

10.9 Corollary.—*Let X be locally compact and $Z \subset X$, either open or closed. Then Z is locally compact.*

Example 1.—The space $\mathbf{R}^n (n \geq 1)$ is locally compact and non-compact.

The fact that $\mathbf{R}^n$ is locally compact follows from 7.16 and from the result in Example 7, Chapter 9. In Example 2(a) of Chapter 9, we observed that $\mathbf{R}^1 = \mathbf{R}$ is not compact. To show that $\mathbf{R}^n$ $(n > 1)$ is not compact, we may recall, for instance, that $(x_1, \ldots, x_n) \mapsto x_1$ is a continuous mapping of $\mathbf{R}^n$ onto $\mathbf{R}$ and use 9.10 (we could also have used the results in Example 11, Chapter 9).

Example 2.—The sets

$$H = \{(x, y) \mid xy = 1\} \quad \text{and} \quad K = \{(x, y) \mid xy \neq 1\}$$

are locally compact and non-compact.

The set H is closed and the set K is open. By 10.9, they are locally compact. The fact that they are not compact follows from the results in Example 11, Chapter 9.

Example 3.—For any set X, the topological space $(X, \mathscr{D})$, where $\mathscr{D} = \mathscr{P}(X)$ (see Example 1, Chapter 1), is locally compact.

Example 4.—Let $(X, \mathscr{T})$ be a topological space and let $(U(i))_{i \in I}$ be a partition of X consisting of sets belonging to $\mathscr{T}$. If $(U(i), \mathscr{T}_{U(i)})$ is locally compact for every $i \in I$, then $(X, \mathscr{T})$ is locally compact.

Example 5.—The space $(\mathbf{Q}, \mathscr{U}_{\mathbf{Q}})$, where $\mathscr{U}$ is the usual topology of $\mathbf{R}$, is *not* locally compact (hint: Use Theorem 9.15).

10.10 Theorem.—*If* $((X_i, \mathscr{T}_i))_{i \in I}$ *is a finite family of locally compact spaces, then the topological space* $(X, \mathscr{T})$ *product of the family* $((X_i, \mathscr{T}_i))_{i \in I}$ *is locally compact.*

Proof.—Let $x = (x_i)_{i \in I} \in X$, and for each $i \in I$ let $V_i \in \mathscr{N}_{X_i}(x)$ be compact. By 9.18, the set $V = \prod_{i \in I} V_i$ is compact in $(X, \mathscr{T})$. But (see, for instance, 7.7) $V \in \mathscr{N}_X(x)$. Since $x \in X$ was arbitrary, we deduce that $(X, \mathscr{N})$ is locally compact.

Exercise 1.—If $((X_i, \mathscr{T}_i))_{i \in I}$ is a family of separated spaces such that the topological space $(X, \mathscr{T})$, product of $((X_i, \mathscr{T}_i))_{i \in I}$, is locally compact and $X \neq \varnothing$, then $(X_i, \mathscr{T}_i)$, $i \in I$, is locally compact.

Note that, although in Theorem 9.18 I is arbitrary, in 10.10 we assume I to be *finite*. In fact, Theorem 10.10 does not hold for I arbitrary. For instance, if $X_i = R$ for all $i \in \mathbf{N}$, then $\prod_{i \in \mathbf{N}} X_i$ is not locally compact (we leave the proof of this assertion to the reader).

We shall now show that a locally compact space can be "imbedded in a compact space by adding a point and that the compact space we so obtain is unique."

10.11 Theorem.—*Let* $(X, \mathscr{T})$ *be locally compact. Then there exists a compact space* $(X', \mathscr{T}')$, *a point* $\omega \in X'$, *and a homeomorphism* h *of* $(X, \mathscr{T})$ *onto* $(X' - \{\omega\}, \mathscr{T}'_{X' - \{\omega\}})$.

Proof.—Let X' be a set such that there is $\omega \in X'$ and a bijection h of X onto $X' - \{\omega\}$.* The set $h(\mathscr{T}) = \{h(U) \mid U \in \mathscr{T}\}$ is a topology on $X' - \{\omega\}$ and h is a *homeomorphism* of $(X, \mathscr{T})$ onto $(X' - \{\omega\}, h(\mathscr{T}))$.

Now let $\mathscr{T}'$ be the set of all parts $V \subset X'$ such that either $V \in h(\mathscr{T})$ or $\mathbf{C}V$ is a compact part of $X' - \{\omega\}$. It is immediate that $\mathscr{T}'$ is a separated topology on X' and that $\mathscr{T}'_{X' - \{\omega\}} = h(\mathscr{T})$. Moreover, $(X', \mathscr{T}')$ is compact. In fact, let $(V_i)_{i \in I}$ be a covering of X' consisting of sets belonging to $\mathscr{T}'$. For some $i_0 \in I$, we have $V_{i_0} \ni \omega$. Then $\mathbf{C}V_{i_0}$ is compact and $(V_i)_{i \in I - \{i_0\}}$ is a covering of $\mathbf{C}V_{i_0}$. Hence, there is a finite set $J \subset I - \{i_0\}$ such that $\bigcup_{i \in J} V_i \supset \mathbf{C}V_{i_0}$. We deduce that $(V_i)_{i \in J \cup \{i_0\}}$ is a finite covering of X'. Since $(V_i)_{i \in I}$ was arbitrary, we conclude that $(X', \mathscr{T}')$ is compact. Hence, the theorem is proved.

Remarks.—(1) If $(X, \mathscr{T})$ is locally compact, and if $(X', \mathscr{T}')$ is compact and such that the conditions of Theorem 10.11 are satisfied,

* Consider the product $X \times \{0, 1\}$. Let $x_0 \in X$, $\omega = (x_0, 1)$, and $X' = (X \times \{0\}) \cup \{\omega\}$. Then $x \mapsto (x, 0)$ is a bijection of X onto $X' - \{\omega\}$.

we say that $(X', \mathcal{T}')$ is a one-point (or Alexandroff) compactification of $(X, \mathcal{T})$. We also say that the point ω, in the statement of Theorem 10.11, is the point at infinity of X'. Theorem 10.13 shows that (in a certain sense) the one-point compactification of a locally compact space is *unique*.

(2) Since h is a homeomorphism of X onto $X' - \{\omega\}$, we may "identify" X with $X' - \{\omega\}$, so that $X \subset X'$ and $X' = X \cup \{\omega\}$.

(3) If X is locally compact and non-compact, then ω belongs to the adherence of $X' - \{\omega\}$. In fact, otherwise $X' - \{\omega\}$ would be closed and hence compact; hence X would be compact. Conversely, if ω belongs to the closure of $X' - \{\omega\}$, then X is not compact.

(4) By 4.7, a subspace of a regular space is regular. We deduce from Corollary 9.6 and Theorem 10.11 that a *locally compact space is regular.*

▼ **10.12 Theorem.**—*Let X' and X'' be two compact spaces and $\omega' \in X'$, $\omega'' \in X''$. Let $u: X' \to X''$ be a bijection. Suppose that $u(\omega') = \omega''$ and that $u_1 = u \mid (X' - \{\omega'\})$ is a homeomorphism of $X' - \{\omega'\}$ onto $X'' - \{\omega''\}$. Then u is a homeomorphism of X' onto X''.*

Proof.—Let K be a closed, and hence a compact part of X''. If $\omega'' \notin K$, then $u^{-1}(K) = u_1^{-1}(K)$ is a compact part of $X' - \{\omega'\}$, and hence a closed part of X'. If $\omega'' \in K$, then

$$u^{-1}(K) = u_1^{-1}(K) \cup \{\omega'\}.$$

If $y \in \overline{u_1^{-1}(K)}$ and $y \notin u_1^{-1}(K)$, then $y = \omega'$, since $u_1^{-1}(K)$ is a closed part of $X' - \{\omega'\}$. We deduce that $u^{-1}(K)$ is a closed part of Y. Hence, u is continuous. In the same way, we show that $v = u^{-1}$ is continuous. Hence, u is a homeomorphism of X' onto X''.

10.13 Theorem.—*Let X be a locally compact space. Let X' and X'' be compact spaces and let $\omega' \in X'$ and $\omega'' \in X''$. Suppose that h' is a homeomorphism of X onto $X' - \{\omega'\}$ and h'' is a homeomorphism of X onto $X'' - \{\omega''\}$. Then there is a unique homeomorphism f of X' onto X'' such that $f \circ h' = h''$.*

Proof.—Define $f: X' \to X''$ by

$$f(x) = \begin{cases} h'' \circ (h')^{-1}(x) & \text{for } x \in X' - \{\omega'\} \\ \omega'' & \text{for } x = \omega'. \end{cases}$$

Clearly, f is a bijection of X' onto X'' such that $f(\omega') = \omega''$; moreover, it follows from the definition that $f \mid (X' - \{\omega'\})$ is a homeomorphism of $X' - \{\omega'\}$ onto $X'' - \{\omega''\}$. By Theorem 10.12, f is a homeomorphism of X' onto X''. The relation $f \circ h' = h''$ follows immediately from the definition of f. ▲

Example 6.—Consider the locally compact space $\mathbf{R}$ (see Example 1). We shall show that

$$\mathbf{T}^1 = \{(x, y) \mid x^2 + y^2 = 1\} \subset \mathbf{R}^2$$

is the one-point compactification of $\mathbf{R}$ and that $\omega = (0, 1)$ is the "point at the infinity" of the compactification.

If $f(x, y) = x^2 + y^2 - 1$ for $(x, y) \in \mathbf{R}^2$, then f is continuous and $\mathbf{T}^1 = f^{-1}(\{0\})$. Hence $\mathbf{T}^1$ is closed. Furthermore, $\mathbf{T}^1 \subset [(-1, -1), (1, 1)]$. By the result in Example 7, Chapter 9, we deduce that $\mathbf{T}^1$ is compact.

Now let

$$h(x) = \left(\frac{2x}{x^2 + 1}, \frac{x^2 - 1}{x^2 + 1}\right) \quad \text{for} \quad x \in \mathbf{R}$$

and

$$u(x, y) = \frac{x}{1 - y} \quad \text{for} \quad (x, y) \in \mathbf{T}^1 - \{\omega\}.$$

Clearly, h is a continuous mapping of $\mathbf{R}$ into $\mathbf{T}^1 - \{\omega\}$ and u is a continuous mapping of $\mathbf{T}^1 - \{\omega\}$ into $\mathbf{R}$. Moreover,

$$h(u(x, y)) = (x, y) \quad \text{if} \quad (x, y) \in \mathbf{T}^1 - \{\omega\}$$

and

$$u(h(x)) = x \quad \text{if} \quad x \in \mathbf{R}.$$

We deduce that h is a homeomorphism of $\mathbf{R}$ onto $\mathbf{T}^1 - \{\omega\}$. Hence, $\mathbf{T}^1$ is the one-point compactification of $\mathbf{R}$ and $(1, 0)$ is the point at infinity of the compactification.

Example 7.—Consider the locally compact space $\mathbf{R}^2$ (see Example 1). Then

$$\mathbf{T}^2 = \{(x, y, z) \mid x^2 + y^2 + z^2 = 1\} \subset \mathbf{R}^3$$

is the one-point compactification of $\mathbf{R}^2$ and $\omega = (0, 0, 1)$ is the point at the infinity of this compactification. We leave the proof to the reader. (Hint: Use the mappings

$$h(x, y) = \left(\frac{2x}{x^2 + y^2 + 1}, \frac{2y}{x^2 + y^2 + 1}, \frac{x^2 + y^2 - 1}{x^2 + y^2 + 1}\right) \quad \text{for} \quad (x, y) \in \mathbf{R}^2$$

and

$$u(x, y, z) = \left(\frac{x}{1 - z}, \frac{y}{1 - z}\right) \quad \text{for} \quad (x, y, z) \in \mathbf{T}^2 - \{\omega\}.)$$

10.14 Theorem.—*Let* $(X, \mathcal{T})$ *be locally compact, let* $K \subset X$ *be compact, and let* $F \subset X$ *be closed and such that* $K \cap F = \varnothing$. *Then there is an open, relatively compact set* $U_K \supset K$ *and an open set* $U_F \supset F$ *such that* $\bar{U}_K \cap \bar{U}_F = \varnothing$.

Proof.—For each $x \in K$, let W_x and V_x be compact neighborhoods of x such that

$$V_x \cap F = \varnothing \quad \text{and} \quad W_x \subset \mathring{V}_x.$$

Then $(\mathring{W}_x)_{x \in K}$ is a covering of K. Hence, there is a finite set $I \subset K$ such that $(\mathring{W}_x)_{x \in I}$ is a covering of K. Let

$$U_K = \bigcup_{x \in I} \mathring{W}_x \quad \text{and} \quad U_F = \complement\left(\bigcup_{x \in I} V_x\right).$$

Then U_K and U_F are open sets $U_K \supset K$, U_K is relatively compact, and $U_F \supset F$. Moreover,

$$\bar{U}_K \subset \bigcup_{x \in I} W_x \subset \bigcup_{x \in I} \mathring{V}_x$$

and

$$\bar{U}_F \subset \complement\left(\bigcup_{x \in I} \mathring{V}_x\right).$$

Hence,

$$\bar{U}_K \cap \bar{U}_F = \varnothing.$$

10.15 Corollary.—*Let* K *be compact and* U *open and containing* K. *Then there is an open, relatively compact set* U_K *such that*

$$K \subset U_K \subset \bar{U}_K \subset U.$$

Remark.—The proof of 10.14 is similar to that of Theorem 9.8. The only difference is that U_K is relatively compact in Theorem 10.14.

A topological space $(X, \mathcal{T})$ is *σ-compact* if X is the union of a sequence of compact parts of X. The σ-compact locally compact spaces are important in applications.

Example 8.—The locally compact space $\mathbf{R}^n$ $(n \geq 1)$ is σ-compact. In fact, for each $p \in \mathbf{N}$, let

$$X_p = [(-p, \ldots, -p), (p, \ldots, p)] \subset \mathbf{R}^n;$$

by Example 7, Chapter 9, the set X_p is compact. Since, obviously, $\bigcup_{p \in \mathbf{N}} X_p = \mathbf{R}^n$, we deduce that $\mathbf{R}^n$ is σ-compact.

Example 9.—The locally compact space $(X, \mathscr{D})$ (see Example 3) is σ-compact if and only if X is countable.

Example 10.—The space $(\mathbf{Q}, \mathscr{U}_{\mathbf{Q}})$ (see Example 5) is σ-compact but *not* locally compact.

10.16 Theorem.—*Let $(X, \mathscr{T})$ be a σ-compact, locally compact space. Then there is a sequence $(U_n)_{n\in\mathbf{N}}$ of open, relatively compact sets satisfying:*

(i) $\bigcup_{n\in\mathbf{N}} U_n = X$;
(ii) $\bar{U}_n \subset U_{n+1}$ *for all* $n \in \mathbf{N}$.

Proof.—Let $(X_i)_{i\in\mathbf{N}}$ be a sequence of compacts such that $\bigcup_{i\in\mathbf{N}} X_i = X$. Choose an open, relatively compact set U containing X_1. Suppose that $n \in \mathbf{N}$, $n > 1$, and that we have chosen the open, relatively compact sets $U_1, \ldots, U_n$ such that $U_i \supset X_i$ for $i = 1, \ldots, n$ and $\bar{U}_{i-1} \subset U_i$ for $i = 2, \ldots, n$. Choose then an open, relatively compact set U_{n+1} containing $\bar{U}_n \cup X_{n+1}$. Proceeding this way, we define a sequence $(U_n)_{n\in\mathbf{N}}$ of open, relatively compact sets satisfying (i) and (ii).

Before the statement of Theorem 2.9, we introduced the notion of a locally finite family of parts of a topological space $(X, \mathscr{T})$. Recall that a family $(A_i)_{i\in I}$ of parts of X is said to be *locally finite* if for every $x \in X$ there is $V \in \mathscr{N}(x)$ such that

$$\{i \mid V \cap A_i \neq \varnothing\}$$

is finite.

A finite family is obviously locally finite. Let $X = \mathbf{R}$ and let

$$A_n = \{x \mid n < x < n + 2\} \quad \text{and} \quad B_n = \{x \mid 0 < x < n\}$$

for all $n \in \mathbf{N}$. Then $(A_n)_{n\in\mathbf{N}}$ is a locally finite covering of $\mathbf{R}$ and $(B_n)_{n\in\mathbf{N}}$ is a covering of $\mathbf{R}_+^* = \{x \mid x > 0\}$ that is *not* locally finite.

10.17 Definition.—*Let X be a set and $(A_i)_{i\in I}$ and $(B_j)_{j\in J}$ two families of parts of X. We say that $(B_j)_{j\in J}$ is finer than $(A_i)_{i\in I}$ if for every $j \in J$ there is $i(j) \in I$ such that $B_j \subset A_{i(j)}$.*

With these notations, we may now state and prove the following.

10.18 Theorem.—*Let $(X, \mathscr{T})$ be a σ-compact, locally compact space. Then for every open covering $\mathscr{C} = (C_i)_{i\in I}$ of X, there exists a locally finite open covering $\mathscr{W} = (W_j)_{j\in J}$ of X, finer than $\mathscr{C}$.*

Proof.—By Theorem 10.16, there exists a sequence $(U_n)_{n\in\mathbf{N}}$ of open, relatively compact parts of X such that $\bigcup_{n\in\mathbf{N}} U_n = X$ and $\bar{U}_n \subset U_{n+1}$ for all $n \in \mathbf{N}$.

For each $n \in \mathbf{N}$, let (we define $U_n = \varnothing$ if $n \leq 1$)

$$A_n = \overline{U}_n \cap \mathbf{C}U_{n-1} \quad \text{and} \quad B_n = U_{n+1} \cap \mathbf{C}\overline{U}_{n-2};$$

then A_n is compact and B_n is open. Moreover, for each $n \in \mathbf{N}$, $U_{n-1} \supset \overline{U}_{n-2}$, so that $\mathbf{C}U_{n-1} \subset \mathbf{C}\overline{U}_{n-2}$; hence $A_n \subset B_n$. Let us notice that

$$|n - p| > 2 \Rightarrow B_n \cap B_p = \varnothing.$$

In fact, suppose $n - p > 2$. Then $n - 2 \geq p + 1$, so that $\overline{U}_{n-2} \supset U_{p+1}$, and hence $U_{p+1} \cap \mathbf{C}\overline{U}_{n-2} = \varnothing$. Hence,

$$B_n \cap B_p = (U_{n+1} \cap \mathbf{C}\overline{U}_{n-2}) \cap (U_{p+1} \cap \mathbf{C}\overline{U}_{p-2}) = \varnothing.$$

In the same way, we show that $B_p \cap B_n = \varnothing$ if $p - n > 2$.

Now let $n \in \mathbf{N}$ and $x \in A_n$. Let $i \in I$ such that $x \in C_i$ and let W_x be an open neighborhood of x contained in $C_i \cap B_n$. Since A_n is compact, there is a finite set $x(n, 1), \ldots, x(n, j(n))$ of elements of A_n such that

$$A_n \subset W_{x(n,1)} \cup \ldots \cup W_{x(n,j(n))}.$$

For each $n \in \mathbf{N}$ and $1 \leq j \leq j(n)$, let $W_{(n,j)} = W_{x(n,j)}$; let

$$J = \bigcup_{n \in \mathbf{N}} \{(n, j) \mid 1 \leq j \leq j(n)\}.$$

Then $\mathscr{W} = (W_{(n,j)})_{(n,j) \in J}$ is an open covering of X *finer* than $\mathscr{C}$. Moreover, for (n, i) and (p, j) in J, we have

$$W_{(n,i)} \cap W_{(p,j)} = \varnothing \quad \text{if} \quad |n - p| > 2.$$

Hence, $\mathscr{W}$ is locally finite, and hence the theorem is proved.

The proof of Theorem 10.18 shows that the following stronger result holds:

10.19 Theorem.—*Let $(X, \mathscr{T})$ be a σ-compact, locally compact space and let $\mathscr{B} \subset \mathscr{T}$ be a basis of $\mathscr{T}$. Then for every open covering $\mathscr{C} = (C_i)_{i \in I}$ of X, there exists a locally finite open covering $\mathscr{W} = (W_j)_{j \in J}$ finer than $\mathscr{C}$ and consisting of sets belonging to the basis $\mathscr{B}$.*

Remark.—Notice that the sets in the covering $\mathscr{W}$ can be chosen relatively compact.

10.20 Definition.—*A topological space $(X, \mathscr{T})$ is paracompact if it is separated and if for every open covering $\mathscr{C}$ of X there is a locally finite open covering $\mathscr{W}$ of X finer than $\mathscr{C}$.*

From Theorem 10.17, it follows that every σ-compact, locally compact space is paracompact. In particular, a compact space is paracompact.

Exercise 2.—Let $(X, \mathscr{T})$ be a locally compact space. If there exists a partition $(U_i)_{i\in I}$ of X consisting of σ-compact *open* sets, then $(X, \mathscr{T})$ is paracompact.*

Exercises for Chapter 10

1. Let $(X, \mathscr{T})$ be a locally compact topological space and $(Y, \mathscr{S})$ a separated space. If $f: X \to Y$ is continuous and open ($V \in \mathscr{T} \Rightarrow f(V) \in \mathscr{S}$), then $(f(X), \mathscr{S}_{f(X)})$ is locally compact.

2. Let $((X_i, \mathscr{T}_i))_{i\in I}$ be a family of locally compact topological spaces such that $\{i \mid X_i \text{ is not compact}\}$ is finite. Then $\prod_{i\in I} X_i$ is locally compact.

3. Let $(X, \mathscr{T})$ be a paracompact topological space and F a closed subset of X. Then $(F, \mathscr{T}_F)$ is paracompact.

4. Show that a locally compact space is regular, using 9.8 and 10.3.

* It can be shown that if the locally compact space $(X, \mathscr{T})$ is paracompact, then there exists a partition $(U_i)_{i\in I}$ of X consisting of σ-compact open sets (see, for instance, [2], Chapter 1 (fourth edition), pp. 109–110).

Chapter 11

Connected Spaces

We start with the:

11.1 Definition.—*A topological space* $(X, \mathscr{T})$ *is connected if, for any non-void open sets* U *and* V *such that* $U \cup V = X$, *we have* $U \cap V \neq \varnothing$.

When there is no ambiguity, we shall say simply that X is connected, instead of saying that $(X, \mathscr{T})$ is connected.

If $(X, \mathscr{T})$ is a topological space, we deduce from Definition 11.1 that *the following assertions are equivalent:*

11.2 *the space* $(X, \mathscr{T})$ *is connected;*

11.3 *for any non-void closed sets* F *and* G *such that* $F \cup G = X$, *we have* $F \cap G \neq \varnothing$.

Proof of 11.2 $\Rightarrow$ 11.3.—Let F and G be non-void closed sets such that $F \cup G = X$. Suppose $F \cap G = \varnothing$. Then $G = \mathbf{C}F$ and $F = \mathbf{C}G$. Hence, F and G are non-void open sets and $F \cup G = X$. Thus we have arrived at a contradiction. Hence, $F \cap G \neq \varnothing$.

Proof of 11.3 $\Rightarrow$ 11.2.—The proof is similar to that of 11.2 $\Rightarrow$ 11.3.

Note that *a topological space* $(X, \mathscr{T})$ *is connected if and only if the only parts of* X *that are both open and closed are* $\varnothing$ *and* X.

Exercise 1.—Let $(X, \mathscr{T})$ be a topological space. Then the following assertions are equivalent:

(i) the space $(X, \mathscr{T})$ is connected;

(ii) there is no partition $(U_i)_{i \in I}$ of X consisting of open sets and such that I contains at least two elements;

(iii) there is no *finite* partition $(F_i)_{i \in I}$ of X consisting of closed sets and such that I contains at least two elements.

Let $(X, \mathcal{T})$ be a topological space. A set $A \subset X$ *is said to be connected if* $(A, \mathcal{T}_A)$ *is connected.* Since

$$\mathcal{T}_A = \{A \cap U \mid U \in \mathcal{T}\},$$

and since $F \subset A$ is closed (in A) if and only if there is $F' \subset X$ closed (in X) such that $F = A \cap F'$, we deduce that *the following assertions are equivalent:*

11.4 *The set A is connected;*

11.5 *For any sets U_1 and U_2 open (in X), the union of which contains A, and such that $A \cap U_1 \neq \varnothing$ and $A \cap U_2 \neq \varnothing$, we have $A \cap U_1 \cap U_2 \neq \varnothing$;*

11.6 *For any sets F_1 and F_2 closed (in X), the union of which contains A and such that $A \cap F_1 \neq \varnothing$ and $A \cap F_2 \neq \varnothing$, we have $A \cap F_1 \cap F_2 \neq \varnothing$.*

11.7 Theorem.—*Let X and Y be two topological spaces, $A \subset X$ a connected set, and $f: A \to Y$ a continuous mapping. Then $f(A)$ is connected.*

Proof.—Let $V_1 \subset Y$ and $V_2 \subset Y$ be two open sets such that

$$f(A) \cap V_1 \neq \varnothing, \qquad f(A) \cap V_2 \neq \varnothing \quad \text{and} \quad f(A) \subset V_1 \cup V_2.$$

Then $B_1 = f^{-1}(V_1)$ and $B_2 = f^{-1}(V_2)$ are open parts of A and $B_1 \cup B_2 = A$. Since $f(A) \cap V_1 \neq \varnothing$, and since

$$B_1 = f^{-1}(V_1) = f^{-1}(f(A) \cap V_1),$$

we deduce that $B_1 \neq \varnothing$; in the same way we obtain $B_2 \neq \varnothing$. Since A is connected, we must have $B_1 \cap B_2 \neq \varnothing$. If $x \in B_1 \cap B_2$, then $f(x) \in f(A) \cap V_1 \cap V_2$. Hence, $f(A) \cap V_1 \cap V_2 \neq \varnothing$. Since V_1 and V_2 were arbitrary, it follows that $f(A)$ is connected.

In the following theorem, we assume that $\{0, 1\}$ is endowed with the *discrete topology*. Clearly, then, $A \subset \{0, 1\}$ is connected if and only if $A = \varnothing$, $A = \{0\}$, or $A = \{1\}$.

11.8 Theorem.—*Let $(X, \mathcal{T})$ be a topological space. Then the following assertions are equivalent:*

(a) *the space $(X, \mathcal{T})$ is not connected;*
(b) *there is a continuous surjection $f: X \to \{0, 1\}$.*

Proof of (a) $\Rightarrow$ (b).—If $(X, \mathcal{T})$ is not connected, then there are two non-void disjoint open sets U and V, the union of which is X.

Let $f: X \rightarrow \{0, 1\}$ be defined by $f(x) = 0$ if $x \in U$ and $f(x) = 1$ if $x \in V$. Then f is a continuous surjection of X onto $\{0, 1\}$.

Proof of (b) $\Rightarrow$ (a).—Let $f: X \rightarrow \{0, 1\}$ be a continuous surjection and let $U = f^{-1}(\{0\})$ and $V = f^{-1}(\{1\})$. Then U and V are non-void disjoint open sets, the union of which is X. Hence, X is not connected.

11.9 Theorem.—*Let $(X, \mathscr{T})$ be a topological space and $A \subset X$ a connected set. If $A \subset B \subset \bar{A}$, then B is connected.*

Proof.—Let U and V be two *open* sets such that

$$U \cap B \neq \varnothing, \qquad V \cap B \neq \varnothing, \qquad U \cup V \supset B.$$

If $x \in U \cap B$, then $x \in \bar{A}$, and hence (since U is open) $U \cap A \neq \varnothing$. In the same way, we see that $V \cap A \neq \varnothing$. Since A is connected, we have $A \cap U \cap V \neq \varnothing$, so that $B \cap U \cap V \neq \varnothing$. Since the sets U and V were arbitrary, we deduce that B is connected.

11.10 Theorem.—*Let $(X, \mathscr{T})$ be a topological space and let $Z \subset X$. Let $A \subset X$ be connected and such that*

$$A \cap Z \neq \varnothing \quad \text{and} \quad A \cap \mathbf{C}Z \neq \varnothing.$$

Then $A \cap Fr(Z) \neq \varnothing$.

Proof.—The sets $\mathring{Z}$, $Fr(Z)$, and $\widehat{\mathbf{C}Z}^{\circ}$ are disjoint and

$$X = \mathring{Z} \cup Fr(Z) \cup \mathring{\mathbf{C}}Z.$$

Assume that $A \cap Fr(Z) = \varnothing$. Then $A \subset \mathring{Z} \cup \widehat{\mathbf{C}Z}^{\circ}$, $A \cap \mathring{Z} \neq \varnothing$, and $A \cap \widehat{\mathbf{C}Z}^{\circ} \neq \varnothing$. Since A is connected, this is a contradiction. Hence, $A \cap Fr(Z) \neq \varnothing$.

Exercise 2.—If X is connected, and if Z is a non-void set contained in X and distinct from X, then $Fr(Z) \neq \varnothing$.

11.11 Definition.—*Let $(X, \mathscr{T})$ be a topological space. For any two sets $A \subset X$ and $B \subset X$, we write*

$$A \odot B = (A \cap \bar{B}) \cup (\bar{A} \cap B).$$

Notice that $A \odot B = B \odot A$ and that

$$A \cap B \subset A \odot B,$$

so that $A \cap B \neq \varnothing$ implies $A \odot B \neq \varnothing$. Notice also that $A' \subset A$ and $B' \subset B$ implies

$$A' \odot B' \subset A \odot B.$$

11.12 Theorem.—*Let* $(X, \mathscr{T})$ *be a topological space and* $A \subset X$ *and* $B \subset X$ *be connected sets such that* $A \odot B \neq \varnothing$. *Then* $C = A \cup B$ *is connected.*

Proof.—Let U and V be open sets such that

$$U \cap C \neq \varnothing, \qquad V \cap C \neq \varnothing, \quad \text{and} \quad C \subset U \cup V.$$

Now let $y \in A \odot B$. Then either $y \in A \cap \bar{B}$ or $y \in \bar{A} \cap B$. Suppose that $y \in A \cap \bar{B}$ (the case in which $y \in \bar{A} \cap B$ would be treated the same way). The point y belongs to at least one of the sets U and V; assume, for instance, that $y \in U$. Then $U \cap A \neq \varnothing$ and $U \cap B \neq \varnothing$ (since U is open). Since $V \cap C \neq \varnothing$, we have either $V \cap A \neq \varnothing$ or $V \cap B \neq \varnothing$. If $V \cap A \neq \varnothing$, we deduce that $U \cap V \cap A \neq \varnothing$, since A is connected. If $V \cap B \neq \varnothing$, we deduce that $U \cap V \cap B \neq \varnothing$, since B is connected. Hence, in either case, $U \cap V \cap C \neq \varnothing$. Since U and V were arbitrary, C is connected.

11.13 Theorem.—*Let* $(X, \mathscr{T})$ *be a topological space and* $(A_i)_{i \in I}$ *a family of connected parts of* X. *Suppose that* $A_i \odot A_j \neq \varnothing$ *for every* $(i, j) \in I \times I$. *Then* $A = \bigcup_{i \in I} A_i$ *is connected.*

Proof.—Assume first that

(I) $\bigcap_{i \in I} A_i \neq \varnothing$.

Let U_1 and U_2 be open sets such that

$$U_1 \cap A \neq \varnothing, \qquad U_2 \cap A \neq \varnothing \quad \text{and} \quad A \subset U_1 \cap U_2.$$

Clearly, then $A_i \subset U_1 \cup U_2$ for all $i \in I$. Now let $\alpha \in \bigcap_{i \in I} A_i$. Then α belongs either to U_1 or to U_2. Suppose, for instance, that $\alpha \in U_1$ (the case in which $\alpha \in U_2$ can be treated in the same way); then $U_1 \cap A_i \neq \varnothing$ for all $i \in I$. Since $U_2 \cap A \neq \varnothing$, it follows that there is $i_0 \in I$ such that $U_2 \cap A_{i_0} \neq \varnothing$. Hence,

$$U_1 \cap A_{i_0} \neq \varnothing, \qquad U_2 \cap A_{i_0} \neq \varnothing \quad \text{and} \quad A_{i_0} \subset U_1 \cup U_2.$$

Since A_{i_0} is connected, we deduce that $A_{i_0} \cap U_1 \cap U_2 \neq \varnothing$. Since $A_{i_0} \subset A$, we obtain $A \cap U_1 \cap U_2 \neq \varnothing$. Since U_1 and U_2 were arbitrary, we conclude that A is a connected set.

If condition (I) is not satisfied, we reason as follows: Let $i_1 \in I$

and let $A'_i = A_{i_0} \cup A_i$ for all $i \in I$. By Theorem 11.12, A'_i is connected for each $i \in I$; moreover,

$$\bigcap_{i \in I} A'_i = A_{i_0} \neq \varnothing.$$

Hence $(A'_i)_{i \in I}$ satisfies (I), and hence $A = \bigcup_{i \in I} A'_i$ is connected.

11.14 Corollary.—*Let $(A_n)_{n \in N}$ be a sequence of connected parts of X such that $A_n \cap A_{n+1} \neq \varnothing$ for all $n \in \mathbf{N}$. Then $A = \bigcup_{n \in \mathbf{N}} A_n$ is connected.*

Proof.—By Theorem 11.12, the set $A(p) = \bigcup_{n=1}^{p} A_n$ is connected for each $p \in \mathbf{N}$ (reason by induction on p). Since $\bigcap_{p \in \mathbf{N}} A(p) = A_1 \neq \varnothing$ and $A = \bigcup_{p \in \mathbf{N}} A(p)$, the assertion in Corollary 11.14 is proved.

Example 1.—The set $\mathbf{Q} \subset \mathbf{R}$ is not connected.
Since $\mathbf{Q} \neq \mathbf{R}$, there is $z \in \mathbf{R} - \mathbf{Q}$. Let

$$U_1 = \{x \mid x \in \mathbf{R}, x < z\} \quad \text{and} \quad U_2 = \{x \mid x \in \mathbf{R}, x > z\}.$$

Then U_1 and U_2 are open (see Example 6, Chapter 1), $\mathbf{Q} \subset U_1 \cup U_2$ and $U_1 \cap U_2 = \varnothing$. Since there are rational numbers r_1 and r_2 such that $r_1 < z < r_2$, we deduce that $\mathbf{Q} \cap U_1 \neq \varnothing$ and that $\mathbf{Q} \cap U_2 \neq \varnothing$. Now, $U_1 \cap U_2 = \varnothing$ implies that $\mathbf{Q} \cap U_1 \cap U_2 = \varnothing$. Hence, $\mathbf{Q}$ is not connected.

Example 2.—Let $A \subset \mathbf{R}$ be a bounded and closed set. Then the following two assertions are equivalent:

(i) A is connected;
(ii) $A = [a, b]$ for some $a \in \mathbf{R}$, $b \in \mathbf{R}$, $a < b$.

Proof of (i) $\Rightarrow$ (ii).—Since A is bounded and closed, we have (see Example 7, Chapter 3)

$$a = \inf A \in A \quad \text{and} \quad b = \sup A \in A.$$

Suppose that there is $z \in [a, b] - A$; then $a < z < b$. Let

$$U_1 = \{x \mid x \in \mathbf{R}, x < z\} \quad \text{and} \quad U_2 = \{x \mid x \in \mathbf{R}, x > z\}.$$

Then U_1 and U_2 are open, $A \subset U_1 \cup U_2$, and $U_1 \cap U_2 = \varnothing$. Since $a \in A$ and $b \in A$, we have $A \cap U_1 \neq \varnothing$ and $A \cap U_2 \neq \varnothing$. Since $U_1 \cap U_2 = \varnothing$ implies that $A \cap U_1 \cap U_2 = \varnothing$ (recall that we supposed A to be connected), we arrive at a contradiction. Hence $[a, b] - A = \varnothing$, and hence $A = [a, b]$.

Proof of (ii) $\Rightarrow$ (i).—Suppose that $[a, b]$ is not connected and let $U_1 \subset \mathbf{R}$ and $U_2 \subset \mathbf{R}$ be open sets such that $A \cap U_1 \neq \varnothing$, $A \cap U_2 \neq \varnothing$, $A \subset U_1 \cup U_2$, and

$$A \cap U_1 \cap U_2 = \varnothing.$$

Notice that

$$U_1 \cap A = A \cap \mathbf{C}U_2,$$

and hence that $U_1 \cap A$ is *closed.*

Now let $\alpha \in A \cap U_1$ and $\beta \in A \cap U_2$; then $\alpha \neq \beta$. Suppose, for instance, that $\alpha < \beta$ (the case in which $\alpha > \beta$ is treated in the same way). Now let H be the set of all $x \in [a, b]$ such that $x > \alpha$ and $[\alpha, x] \subset U_1$. Let $r = \sup H$. Since $U_1 \cap A$ is closed, it follows that $r \in U_1 \cap A$. Also, $r < \beta \leq b$. But U_1 is open, and hence there is r' such that $r < r' < \beta$ and $[r, r'] \subset U_1$. Then $[\alpha, r'] \subset U_1$, and hence $r' \in H$. This leads to a contradiction. Hence, there are no sets U_1 and U_2 with the properties indicated above, and hence A is connected.

Example 3.—Let $A \subset \mathbf{R}$ be a non-void set. Then the following assertions are equivalent:

(j) A is connected;
(jj) if s and t belong to A, and $s < t$, then $[s, t] \subset A$.

Proof of (j) $\Rightarrow$ (jj).—Suppose that s and t belong to A, $s < t$, and $[s, t] \not\subset A$. There is then $x \in (s, t)$ that does not belong to A. Let

$$U_1 = \{x \mid x \in \mathbf{R}, x < z\} \quad \text{and} \quad U_2 = \{x \mid x \in \mathbf{R}, x > z\}.$$

Then U_1 and U_2 are open, $A \subset U_1 \cup U_2$, and $U_1 \cap U_2 = \varnothing$. Moreover, $s \in A \cap U_1 \Rightarrow A \cap U_1 \neq \varnothing$ and $t \in A \cap U_2 \Rightarrow A \cap U_2 \neq \varnothing$. Hence, if there are s and t in A such that $s < t$ and $[s, t] \not\subset A$, then A is not connected. Thus (j) $\Rightarrow$ (jj).

Proof of (jj) $\Rightarrow$ (j).—Let $a \in A$ and let

$$A_1 = \bigcup_{a<x,x\in A} [a, x] \quad \text{and} \quad A_2 = \bigcup_{x<a,x\in A} [x, a]$$

Then A_1 and A_2 are connected. Since $A = A_1 \cup A_2$ and $A_1 \cap A_2 = \{a\} \neq \varnothing$, it follows that A is connected.

For any real numbers $a < b$, let:

$$[a, b] = \{x \in \mathbf{R} \mid a \leq x \leq b\};$$
$$(a, b) = \{x \in \mathbf{R} \mid a < x < b\};$$
$$[a, b) = \{x \in \mathbf{R} \mid a \leq x < b\};$$
$$(a, b] = \{x \in \mathbf{R} \mid a < x \leq b\}.$$

Sets of the form $[a, b]$, (a, b), $[a, b)$, and $(a, b]$ are called *bounded intervals* (see also Chapter 1). For any $a \in \mathbf{R}$, let:

$$\begin{aligned}
[a, +\infty) &= \{x \in \mathbf{R} \mid x \geq a\}; \\
(a, +\infty) &= \{x \in \mathbf{R} \mid x > a\}; \\
(-\infty, a] &= \{x \in \mathbf{R} \mid x \leq a\}; \\
(-\infty, a) &= \{x \in \mathbf{R} \mid x < a\}.*
\end{aligned}$$

Sets of the form $[a, +\infty)$, $(a, +\infty)$, $(-\infty, a]$, $(-\infty, a)$, and $\mathbf{R} = (-\infty, +\infty)$ are called *unbounded intervals* (see also Chapter 1). By an *interval of* $\mathbf{R}$, we mean a part of $\mathbf{R}$ that is either a bounded interval or an unbounded interval.

It is immediate that if $A \subset R$ is an interval, then A satisfies condition (jj) of Example 3. Hence, *intervals are connected parts of* $\mathbf{R}$. Conversely, it can be shown that *every set* $A \subset \mathbf{R}$ *satisfying* (jj) *is an interval.*† We shall prove this assertion only in the case in which A has a minorant‡ but does not have a majorant.

Let $\alpha = \inf A$. We shall distinguish the cases $\alpha \in A$ *and* $\alpha \notin A$. *Suppose first that* $\alpha \in A$. Let $x \in \mathbf{R}$, $x > \alpha$. Since A does not have a majorant, there is $\beta \in A$ such that $\beta \geq x$. Then $[\alpha, \beta] \subset A$, and hence $x \in A$. Hence $[\alpha, +\infty) \subset A$; since, clearly, $A \subset [\alpha, +\infty)$, we deduce that $A = [\alpha, +\infty)$. *Suppose now that* $\alpha \notin A$. If $y > \alpha$, then there is $x \in A$ such that $\alpha < x \leq y$ (otherwise y would be a minorant of A). Since A has no majorant, there is $z \in A$, $z > y$. Since x and z belong to A, it follows that $[x, z] \subset A \Rightarrow y \in A$. Hence $(\alpha, +\infty) \subset A$. Since, clearly, $A \subset (\alpha, +\infty)$, we deduce that $A = (\alpha, +\infty)$.

Remark.—Let $(X, \mathscr{T})$ be a topological space, $A \subset X$ a connected set, and $f: A \to \mathbf{R}$ continuous. By Theorem 11.7, the set $f(A)$ is connected, and hence (by the above results) $f(A)$ is an interval.

Example 4.—Let

$$A = \{(0, y) \mid -1 \leq y \leq 1\} \quad \text{and} \quad B = \{(x, \sin(2\pi/x)) \mid x \in (0, 1]\}.$$

Then $C = A \cup B$ is connected (see the figure on the front cover).

It follows from the result in Example 2 that A is connected (why?). Furthermore, $x \mapsto (x, \sin(2\pi/x))$ is a continuous mapping of $(0, 1]$ into $\mathbf{R}^2$ and $(0, 1]$ is connected. By Theorem 11.7, the set B is connected. For $x = 1/n$ $(n \in \mathbf{N})$, we have $\sin(2\pi/(1/n)) = \sin(2\pi n) = 0$. Hence, $(1/n, 0) \in B$ for all $n \in \mathbf{N}$. We deduce easily that $A \cap \bar{B} \neq \varnothing$, so that $A \odot B \neq \varnothing$. By Theorem 11.12, the set $C = A \cup B$ is connected.

---

* Notice that $-\infty$ and $+\infty$ are only notational symbols. We do not give them any other significance. In particular, $-\infty$ and $+\infty$ are *not* real numbers.

† We have already proved this assertion when A is bounded and closed (see Example 2).

‡ Sometimes we shall say minorant instead of lower bound and majorant instead of upper bound.

Remark.—It can be shown that $C = \bar{B}$ (we leave the proof of this assertion to the reader).

Let $((X_i, \mathscr{T}_i))_{i\in I}$ be a family of connected spaces and let $(X, \mathscr{T})$ be the topological space product of the family $((X_i, \mathscr{T}_i))_{i\in I}$. Let $a' = (a'_i)_{i\in I}$ and $a'' = (a''_i)_{i\in I}$ be two elements of X and $f: X \to \{0, 1\}$ a continuous mapping. If*

$$c(\{i \mid a'_i \neq a''_i\}) = 1,$$

then

$$f(a') = f(a'').$$

In fact, let $t \in I$ be such that $a'_t \neq a''_t$ and let $u = (u^{(i)})_{i\in I}$ be the mapping of X_t into X, where

$$u^{(i)}(x) = \begin{cases} a'_i = a''_i & \text{if } i \neq t \\ x & \text{if } i = t \end{cases}$$

for all $i \in I$ and $x \in X_t$. Then u is continuous, whence $f \circ u(X_t)$ is a connected part of $\{0, 1\}$ so that

$$f(a') = f \circ u(a'_t) = f \circ u(a''_t) = f(a'').$$

11.15 Theorem.—*If $((X_i, \mathscr{T}_i))_{i\in I}$ is a family of connected spaces, then the topological space $(X, \mathscr{T})$ product of the family $((X_i, \mathscr{T}_i))_{i\in I}$ is connected.*

Proof.—Suppose that $(X, \mathscr{T})$ is not connected. By Theorem 11.8, there is a continuous surjection $f: X \to \{0, 1\}$. We shall now show that this leads to a contradiction.

Let $a = (a_i)_{i\in I}$ be an element of X, for every $p \in \mathbf{N}$ let $S(p)$ be the set of all $x \in X$ such that $c(\{i \mid x_i \neq a_i\}) = p$, and let $S = \bigcup_{p\in\mathbf{N}} S(p)$. We shall show (by induction) that $f(x) = f(a)$ for $x \in S$. In fact, if $x \in S(1)$, this follows from the remark made before the statement of the theorem. Suppose now that $p \in \mathbf{N}$ is *given* and that $f(x) = f(a)$ for $x \in S(p)$. Let $x \in S(p + 1)$ and let $s \in \{i \mid x_i \neq a_i\}$. Define $y = (y_i)_{i\in I}$ by

$$y_i = \begin{cases} x_i & \text{if } i \neq s \\ a_i & \text{if } i = s. \end{cases}$$

Then $y \in S(p)$, so that $f(y) = f(a)$. But $c(\{i \mid x_i \neq y_i\}) = 1$, and hence $f(x) = f(y)$. Hence $f(x) = f(a)$ for $x \in S(p + 1)$. We deduce that $f(x) = f(a)$ for all $n \in \mathbf{N}$ and $x \in S(n)$, that is, for all $x \in S$.

* If A is a set, $c(A)$ is its cardinal number.

Since $\bar{S} = X$ (use the fact that the elementary parts of X form a basis of the product topology), and since f is continuous, we deduce that $f(x) = f(a)$ for $x \in X$. Since we assumed f to be a surjection, we arrive at a contradiction. Hence $(X, \mathscr{T})$ is connected.

Exercise 3.—If $((X_i, \mathscr{T}_i))_{i \in I}$ is a family of topological spaces such that the topological space $(X, \mathscr{T})$, product of the family $((X_i, \mathscr{T}_i))_{i \in I}$, is connected, and if $X \neq \varnothing$, then each $(X_i, \mathscr{T}_i)$, $i \in I$ is connected.

11.16 Theorem.—*Let* $((X_i, \mathscr{T}_i))_{i \in I}$ *be a family of topological spaces and* $(X, \mathscr{T})$ *the topological space product of the family* $((X_i, \mathscr{T}_i))_{i \in I}$. *If* $A_i \subset X_i$, $i \in I$ *is connected, then* $\prod_{i \in I} A_i$ *is connected.*

Proof.—This follows from Theorems 7.28 and 11.15.

Example 5.—The space $\mathbf{R}^n$ $(n \geq 1)$, endowed with the usual topology, is connected.

By the result in Example 3, $\mathbf{R}$ is connected. The fact that $\mathbf{R}^n$ $(n \geq 1)$ is connected follows from Theorem 11.16.

Example 6.—If I_j, $1 \leq j \leq n$ is an interval of $\mathbf{R}$, then $\prod_{1 \leq j \leq n} I_j$ is a connected part of $\mathbf{R}^n$. In particular, any open or closed n-dimensional interval is a connected part of $\mathbf{R}^n$.

Let $(X, \mathscr{T})$ be a topological space and let C be the set of all $(x, y) \in X \times X$ such that $\{x, y\} \subset A$ for *some* connected set of $A \subset X$. It is easy to see that C is an *equivalence relation* in X (to prove that C is transitive, use Theorem 11.12). For each $x \in X$, denote by $C(x)$ the equivalence class of x corresponding to the equivalence relation C.

11.17 Definition.—*For each* $x \in X$, *the set* $C(x)$ *is called the connected component of* x.

Using Theorem 11.13, we see that for each $x \in X$, the set $C(x)$ is the "largest" connected set containing x. By Theorem 11.9, $\overline{C(x)}$ is also connected; we deduce that the connected component of a point $x \in X$ is *closed*.

Clearly, the space $(X, \mathscr{T})$ is connected if and only if $C(x) = X$ for some $x \in X$ (or $C(x) = X$ for all $x \in X$).

A topological space $(X, \mathscr{T})$ is said to be *totally discontinuous* if $C(x) = \{x\}$ for each $x \in X$. A set $A \subset X$ is said to be *totally discontinuous* if $(A, \mathscr{T}_A)$ is totally discontinuous.

Example 6.—The set $\mathbf{Q} \subset \mathbf{R}$ is totally discontinuous. In fact, let $x \in \mathbf{Q}$ and $y \in C(x)$. If $y > x$ and if $z \in \mathbf{R} - \mathbf{Q}$ and $x < z < y$, then

$$(-\infty, z) \cap C(x) \quad \text{and} \quad C(x) \cap (z, +\infty)$$

are open and non-void and have for union $C(x)$; since $C(x)$ is connected this is a contradiction. Hence, there is no $y \in C(x), y > x$. In the same way, we show that there is no $y \in C(x), y < x$. Therefore, $C(x) = \{x\}$. Since $x \in X$ was arbitrary, we conclude that $\mathbf{Q}$ is *totally discontinuous.*

11.17 Theorem.—*Let* $((X_i, \mathcal{T}_i))_{i \in I}$ *be a family of topological spaces and let* $(X, \mathcal{T})$ *be the topological space product of the family* $((X_i, \mathcal{T}_i))_{i \in I}$. *Let* $a = (a_i)_{i \in I} \in X$. *Then*

$$C(a) = \prod_{i \in I} C(a_i).$$

Proof.—For each $i \in I$, $pr_i(C(a)) \ni a_i$. By Theorem 11.7, this set is connected, whence $pr_i(C(a)) \subset C(a_i)$. Using Theorem 11.16, we obtain

$$\prod_{i \in I} C(a_i) \subset C(a) \subset \prod_{i \in I} pr_i(C(a)) \subset \prod_{i \in I} C(a_i),$$

so that Theorem 11.17 is proved.

11.18 Definition.—*A topological space* $(X, \mathcal{T})$ *is said to be locally connected if, for each* $x \in X$, *there is a fundamental system of* x *consisting of connected sets.*

Example 7.—The space $\mathbf{R}^n$ $(n \geq 1)$ is locally connected (see Example 6 and 7.15 or 7.16).

Example 8.—A set X endowed with a discrete topology is locally connected. However, if X contains more than one point, it is not connected. Hence a space can be locally connected without being connected.

Example 9.—Let $Y = \{(x, 0) \mid x \in \mathbf{R}\}$, and for each $r \in \mathbf{Q}$, let $Y_r = \{(r, y) \mid y \in \mathbf{R}\}$. Let

$$X = Y \cup \bigcup_{r \in \mathbf{Q}} Y_r$$

and let $\mathcal{U}_2$ be the usual topology of $\mathbf{R}^2$. Then $(X, (\mathcal{U}_2)_X)$ is connected but not locally connected (we leave the proof to the reader). Hence, a space can be connected without being locally connected.

Arcwise Connected Spaces

We shall introduce here the definition of *arcwise connected spaces.* We shall show that an arcwise connected space is connected and that the converse is not necessarily true.

11.19 Definition.—*A curve in a topological space* $(X, \mathscr{T})$ *is a continuous mapping* $\varphi: [\alpha, \beta] \to X$ where $[\alpha, \beta] \subset \mathbf{R}$.

The *trace* of a curve $\varphi: [\alpha, \beta] \to X$ is the set

$$\varphi([\alpha, \beta]) = \{\varphi(t) \mid t \in [\alpha, \beta]\}.$$

By Theorem 11.7, *the trace of a curve is a connected set.*

11.20 Definition.—*A topological space* $(X, \mathscr{T})$ *is arcwise connected if for every* $a \in X$ *and* $b \in X$ *there exists a curve* $\varphi: [\alpha, \beta] \to X$ *such that* $\varphi(\alpha) = a$ *and* $\varphi(\beta) = b$.

A set $A \subset X$ is arcwise connected if $(A, \mathscr{T}_A)$ is arcwise connected. Clearly, A is arcwise connected if for every $a \in A$ and $b \in A$ there exists a curve $\varphi: [\alpha, \beta] \to A$ such that $\varphi(\alpha) = a$ and $\varphi(\beta) = b$.

Example 10.—Let $n \geq 1$. For every

$$a = (a_1, \ldots, a_n) \in \mathbf{R}^n, \qquad b = (b_1, \ldots, b_n) \in \mathbf{R}^n, \qquad t \in [0, 1],$$

let

$$(1 - t)a + tb = ((1 - t)a_1 + tb_1, \ldots, (1 - t)a_n + tb_n).$$

Clearly, $t \mapsto (1 - t)a + tb$ is a continuous mapping of $\mathbf{R}$ into $\mathbf{R}^n$.

A set $A \subset \mathbf{R}^n$ is *convex* if, for any $a \in A$, $b \in B$, and $t \in [0, 1]$, we have

$$(1 - t)a + tb \in A.$$

Since the mapping $\varphi: t \mapsto (1 - t)a + tb$ of $[0, 1]$ into A is a curve in A, and $\varphi(0) = a$, $\varphi(1) = b$, we deduce that a *convex subset of* $\mathbf{R}^n$ *is arcwise connected.* In particular, an n-dimensional interval and $\mathbf{R}^n$ are arcwise connected.

▼ In the same way, it can be shown that any convex part of a topological vector space is arcwise connected. ▲

11.21 Theorem.—*An arcwise connected topological space is connected.*

Proof.—Let $(X, \mathscr{T})$ be an arcwise connected space and let $a \in X$. For every $y \in X$, there exists a curve $\varphi:[\alpha, \beta] \to X$ such that $\varphi(\alpha) = a$ and $\varphi(\beta) = y$. Hence, both a and y belong to $\varphi([\alpha, \beta])$. Since $\varphi([\alpha, \beta])$ is connected, we deduce that $y \in C(a)$. Hence $X = C(a)$, and hence $(X, \mathscr{T})$ is connected.

Example 11.—We shall show here that the set $C = A \cup B$ defined in Example 4 is *not* arcwise connected (recall that we have shown in Example 4 that this set is connected). We shall proceed as follows:

I. We shall show first that there is *no* curve $\varphi:[\alpha, \beta] \to C$ such that $\varphi(\alpha) \in A$ and $\varphi(x) \in B$ if $\alpha < x \leq \beta$. In fact, suppose that there were such a φ. Then $\varphi = (\varphi_1, \varphi_2)$ where φ_1, φ_2 are continuous mappings of $[\alpha, \beta]$ into $\mathbf{R}$ (see 7.18). Since φ is continuous, there is $\gamma > 0$ (which we may and shall assume to be inferior to $\beta - \alpha$) such that

$$|\varphi_2(t) - \varphi_2(\alpha)| \leq \tfrac{1}{2} \quad \text{for} \quad t \in [\alpha, \alpha + \gamma]$$

(see 5.6(b)). Now, $\varphi(\alpha + \gamma) \in B$, and hence there is $z \in (0, 1]$ such that $\varphi(\alpha + \gamma) = (z, \sin(2\pi/z))$. Assume now that $\varphi_2(\alpha) \geq 0$, and let $s \in (0, z)$ such that $\sin(2\pi/s) = -1$. Let $F = \{(x, y) \mid x \leq s\}$. Then $\varphi([\alpha, \alpha + \gamma])$ is connected and its intersections with F and $\mathbf{C}F$ are non-void. By 11.10, there is $t_0 \in [\alpha, \alpha + \gamma]$ such that

$$\varphi(t_0) \in F_r(F) = \{(x, y) \mid x = s\}.$$

But $\varphi(t_0) \in B$ also; that is,

$$\varphi(t_0) \in B \cap \{(x, y) \mid x = s\} = \{(s, \sin(2\pi/s))\} = \{(s, -1)\}.$$

We conclude that $\varphi_2(t_0) = -1$. Since $t_0 \in [\alpha, \alpha + \gamma]$, we must have

$$|(-1) - \varphi_2(\alpha)| = |\varphi_2(t_0) - \varphi_2(\alpha)| \leq \tfrac{1}{2}.$$

But this is a contradiction, since $(-1) - \varphi_2(\alpha) \leq -1$. In the same way, we arrive at a contradiction if we *assume* $\varphi_2(\alpha) \leq 0$. Hence, there is no curve $\varphi:[\alpha, \beta] \to C$ such that $\varphi(\alpha) \in A$ and $\varphi(x) \in B$ if $\alpha < x \leq \beta$.

II. We shall show now that there is *no* curve $\varphi:[\alpha, \beta] \to C$ such that $\varphi(\alpha) \in A$ and $\varphi(\beta) \in B$. Suppose that there were such a φ. Notice that if $t \in [\alpha, \beta]$ and $\varphi(t) \in B$, then there is $\delta > 0$ such that $\varphi(x) \in B$

if $x \in [\alpha, \beta]$ and $|x - t| < \delta$. If

$$D = \{x \mid x \in (\alpha, \beta], \varphi(y) \in B \text{ if } x \leq y \leq \beta\},$$

then $D \neq \phi$. Let $d = \inf D$. It is easy to see that $\varphi(d) \in A$ and $\varphi(y) \in B$ if $y \in (d, \beta]$. If $\psi = \varphi \mid [d, \beta]$, then ψ is a curve in C such that $\varphi(d) \in A$ and $\varphi(x) \in B$ if $x \in (d, \beta]$. But we have shown in (I) that there is no such curve. Hence, there is no curve $\varphi:[\alpha, \beta] \to C$ such that $\varphi(\alpha) \in A$ and $\varphi(\beta) \in B$.

We conclude that C is not arcwise connected (although it is connected).

Exercises for Chapter 11

1. Let $(X_1, \mathscr{T}_1)$ and $(X_2, \mathscr{T}_2)$ be topological spaces and $(X, \mathscr{T})$ their product. Suppose A is a connected subset of X_1 and $f: A \to X_2$ is continuous. Then $\{(x, f(x)) \mid x \in A\}$ (the graph of f) is a connected subset of X.

2. Let $(X, \mathscr{T})$ be a topological space. Then X is connected if and only if $A \subset X$, $A \neq X$, and $A \neq \varnothing \Rightarrow Fr(A) \neq \varnothing$ (see Exercise 2 earlier in the chapter).

3. (The Intermediate Value Theorem)* Let $S \subset \mathbf{R}$ be a set and let $f \in C_{\mathbf{R}}(S)$. Suppose that a and b are two real numbers such that $a < b$ and $[a, b] \subset S$. (i) If $f(a) \leq f(b)$ and $f(a) \leq k \leq f(b)$, then there exists at least one $c \in [a, b]$ such that $f(c) = k$. (ii) If $f(a) \geq f(b)$ and $f(a) \geq k \geq f(b)$, then there exists at least one $c \in [a, b]$ such that $f(c) = k$ (hint: Use Theorem 11.7 and Example 3).

4. Using the notations of Exercise 3, if $f(a) < 0$ and $f(b) > 0$, or if $f(a) > 0$ and $f(b) < 0$ (that is, if $f(a)$ and $f(b)$ have opposite signs), then there is at least one $c \in [a, b]$ such that $f(c) = 0$.

5. Let S be a set, $g: S \to \mathbf{R}$, and $J \subset S$. We say that g has a *constant sign* on J if either $g(x) > 0$ for every $x \in J$ or $g(x) < 0$ for every $x \in J$. Let $S \subset R$ be a set and let $g \in C_{\mathbf{R}}(S)$. Let $J \subset S$ be an interval and suppose that $g(x) \neq 0$ for $x \in J$. Then g has a constant sign on J.

6. Is the subset

$$\{(x, y) \mid (x + 1)^2 + y^2 < 1\} \cup \{(x, y) \mid (x - 1)^2 + y^2 \leq 1\}$$

a connected part of $\mathbf{R}^2$? Is it arcwise connected?

---

* The theorem asserts that for every k contained between $f(a)$ and $f(b)$ there exists at least one $c \in [a, b]$ such that $f(c) = k$.

7. Is $\mathbf{R}^*$ connected?

8. Is $\mathbf{R}^2 - \{(0, 0)\}$ connected?

9. Let $I \subset \mathbf{R}$ be an interval. Show that any continuous injection $f: I \to \mathbf{R}$ is strictly monotone.

10. Give an example of two connected parts of $\mathbf{R}^2$, the intersection of which is not connected.

11. (The Customs Passage Theorem) Let $(X, \mathscr{T})$ be a topological space, $A \subset X$, $u \in \mathring{A}$, and $v \in \mathbf{C}A$. Let $f: [0, 1] \to X$ be a continuous mapping such that $f(0) = u$ and $f(1) = v$. Then $f([0, 1]) \cap Fr(A) \neq \varnothing$ (see Theorem 11.10).

12. Let $p: \mathbf{R} \to \mathbf{R}$ be a polynomial of odd degree. Show that there is at least one $t \in \mathbf{R}$ such that $p(t) = 0$.

13. Let $(X, \mathscr{T})$ be a topological space and $\mathscr{C}$ a set of connected subsets of X such that $A \in \mathscr{C}$ and $B \in \mathscr{C} \Rightarrow$ there is a sequence $A = A_0, A_1, \ldots, A_n = B$ of elements of $\mathscr{C}$ with $A_{i-1} \circ A_i \neq \varnothing$ for all i in $\{1, \ldots, n\}$. Then $\bigcup \mathscr{C}$ is connected.

Chapter 12

Final Topologies. Quotient Spaces

Let Y be a set, $((X_i, \mathcal{T}_i))_{i\in I}$ a family of topological spaces, and for each $i \in I$ let f_i be a mapping of X_i into Y.

Denote by $\mathcal{I}$ the set of all parts U of Y such that

$$f_i^{-1}(U) \in \mathcal{T}_i$$

for *all* $i \in I$. It is easy to verify that $\mathcal{I}$ *is a topology* on Y.

In fact,

$$f_i^{-1}(\varnothing) = \varnothing \in \mathcal{T}_i \quad \text{and} \quad f_i^{-1}(X) = X_i \in \mathcal{T}_i$$

for all $i \in I$, so that $\mathcal{I}$ *satisfies* (C_1) of Chapter 1. If U and V belong to $\mathcal{I}$, then

$$f_i^{-1}(U \cap V) = f_i^{-1}(U) \cap f_i^{-1}(V) \in \mathcal{T}_i$$

for all $i \in I$, so that $U \cap V \in \mathcal{I}$. Since U and V were arbitrary in $\mathcal{I}$, we deduce that $\mathcal{I}$ *satisfies* (C_2) of Chapter 1. If $(U_j)_{j\in J}$ is a family of elements of $\mathcal{I}$, then

$$f_i^{-1}(\bigcup_{j\in J} U_j) = \bigcup_{j\in J} f_i^{-1}(U_j) \in \mathcal{T}_i$$

for all $i \in I$, so that $\bigcup_{j\in J} U_j \in \mathcal{I}$. Since $(U_j)_{j\in J}$ was arbitrary, we deduce that $\mathcal{I}$ *satisfies* (C_3) of Chapter 1.

12.1 Definition.—*The topology $\mathcal{I}$ is called the final topology on Y associated with the families $((X_i, \mathcal{T}_i))_{i\in I}$ and $(f_i)_{i\in I}$.*

If $I = \{i_0\}$, then we shall usually say that $\mathcal{I}$ is the initial topology associated with $(Y_{i_0}, \mathcal{T}_{i_0})$ and f_{i_0}, instead of the initial topology associated with the families $((X_i, \mathcal{I}_i))_{i\in I}$ and $(f_i)_{i\in I}$.

If $X_i = X$ and $\mathcal{T}_i = \mathcal{T}$ for all $i \in I$, we shall often say that $\mathcal{I}$ is the final topology associated with $(X, \mathcal{T})$ and $(f_i)_{i\in I}$.

When there is no ambiguity, we shall say simply that $\mathscr{I}$ is the final topology on Y, instead of saying that $\mathscr{I}$ is the final topology on Y associated with the families $((Y_i, \mathscr{T}_i))_{i\in I}$ and $(f_i)_{i\in I}$.

Let Y be a set, $((X_i, \mathscr{T}_i))_{i\in I}$ a family of topological spaces, and for each $i \in I$ let f_i be a mapping of X_i into Y. We denote by $\mathscr{I}$ the final topology on Y associated with the families $((X_i, \mathscr{T}_i))_{i\in I}$ and $(f_i)_{i\in I}$.

12.2 Theorem.—(a) *If Y is endowed with the topology $\mathscr{I}$, then $f_i:X_i \to Y$ is continuous for each $i \in I$.* (b) *If $\mathscr{I}'$ is a topology on Y such that $f_i:X \to Y$ is continuous for each $i \in I$ (when Y is endowed with $\mathscr{I}'$), then $\mathscr{I}' \subset \mathscr{I}$.*

Proof of (a).—If $U \in \mathscr{I}$, then by the definition of $\mathscr{I}$, $f_i^{-1}(U) \in \mathscr{T}_i$ for all $i \in I$. Hence, f_i is continuous for all $i \in I$.

Proof of (b).—Let $\mathscr{I}'$ be a topology on Y such that $f_i:X_i \to Y$ is continuous for each $i \in I$ (when Y is endowed with $\mathscr{I}'$). Hence, $f_i^{-1}(U) \in \mathscr{T}_i$ if $U \in \mathscr{I}'$ and $i \in I$. We deduce that $U \in \mathscr{I}$. Since U was arbitrary, we conclude that $\mathscr{I}' \subset \mathscr{I}$.

Hence, the final topology $\mathscr{I}$ is the *largest* topology on Y such that $f_i:X \to Y$ is continuous for all $i \in I$.

12.3 Theorem.—*Let $(Z, \mathscr{W})$ be a topological space and $g: Y \to Z$ a mapping. Then g is continuous if and only if $g \circ f_i$ is continuous for all $i \in I$.*

Proof.—If g is continuous, then $g \circ f_i$ is continuous for each $i \in I$. Conversely, suppose that $g \circ f_i$ is continuous for each $i \in I$ and let $W \in \mathscr{W}$. Then

$$f_i^{-1}(g^{-1}(W)) = (g \circ f_i)^{-1}(W) \in \mathscr{T}_i$$

for each $i \in I$. Hence, $g^{-1}(W) \in \mathscr{I}$. Since $W \in \mathscr{W}$ was arbitrary, we deduce that g is continuous.

12.4 Theorem.—(a) *A set $U \subset Y$ is open if and only if $f_i^{-1}(U) \subset X_i$ is open for each $i \in I$.* (b) *A set $F \subset Y$ is closed if and only if $f_i^{-1}(F) \subset X_i$ is closed for each $i \in I$.*

Proof.—The fact that (a) holds follows immediately from the definition of $\mathscr{I}$. To show that (b) holds, we noticed that $\mathbf{C}f_i^{-1}(F) = f_i^{-1}(\mathbf{C}F)$ for all $i \in I$.

Example 1.—Let X be a set and $(\mathscr{T}_i)_{i\in I}$ a family of topologies on X. Let j_X be the identity mapping $x \mapsto x$ of X into X and $f_i = j_X$

for $i \in I$. Let $\mathscr{I}$ be the final topology on X associated with the families $((X, \mathscr{T}_i))_{i \in I}$ and $(f_i)_{i \in I}$. It is easy to see that

$$\mathscr{I} = \bigcap_{i \in I} \mathscr{T}_i.$$

Observe* that

$$\bigcap_{i \in I} \mathscr{T}_i = \inf_{i \in I} \mathscr{T}_i.$$

From Example 3, Chapter 6, and from the above, it follows that for every family $(\mathscr{T}_i)_{i \in I}$ of topologies on X, the *infimum* and the *supremum* of this family exist. The infimum is a final topology; the supremum is an initial topology.

Quotient Spaces.—Let $(X, \mathscr{T})$ be a topological space, R an equivalence relation in X, $Y = X/R$, and φ the canonical mapping† of X onto Y. In this case, the final topology associated with $(X, \mathscr{T})$ and φ is called the *quotient topology* of X by R. It will usually be denoted $\mathscr{T}^{(\varphi)}$. The space $(X/R, \mathscr{T}^{(\varphi)})$ is called the *topological space quotient of* $(X, \mathscr{T})$ *by* R.

From the above, we deduce that:

12.5 *A set* $U \subset Y$ *belongs to* $\mathscr{T}^{(\varphi)}$ *if and only if* $\varphi^{-1}(U) \in \mathscr{T}$ (see the definition of the final topology);

12.6 *The mapping* $\varphi : X \to Y$ *is continuous* (see 12.2);

12.7 *If* $\mathscr{T}'$ *is a topology on* Y *such that* $\varphi : X \to Y$ *is continuous* (*when* Y *is endowed with* $\mathscr{T}'$), *then* $\mathscr{T}' \subset \mathscr{T}^{(\varphi)}$ (see 12.2);

12.8 *Let* $(Z, \mathscr{W})$ *be a topological space and* $g : Y \to Z$. *Then* g *is continuous if and only if* $g \circ \varphi$ *is continuous* (see 12.3);

12.9 *A set* $F \subset Y$ *is closed if and only if* $\varphi^{-1}(F) \subset X$ *is closed* (see 12.4).

Let $(X, \mathscr{T})$ be a topological space and R an equivalence relation in X. We say that R is *separated* if $(Y, \mathscr{T}^{(\varphi)})$ is separated. We say that R is *open*‡ if φ is an open mapping (see Chapter 5).

There are examples of separated topological spaces $(X, \mathscr{T})$ and of equivalence relations R in X that are neither separated nor open (see Exercise 1). Hence, a topological space quotient of a separated topological space is not necessarily separated.

---

* The set of all topologies on X is ordered by inclusion (see Example 3, Chapter 6).

† We shall (usually) denote by $\dot{x}$ the equivalence class of $x \in X$. The canonical mapping is the mapping $x \mapsto \dot{x}$ of X onto Y.

‡ Recall that φ is open if $\varphi(A)$ is open for every open set $A \subset X$.

For applications, it is important to have certain conditions implying that $(Y, \mathscr{T}^{(\varphi)})$ is separated. Such conditions are given in the following theorems.

12.10 Theorem.—*Let $(X, \mathscr{T})$ be a topological space and R an equivalence relation in X that is open and such that the set*

$$G = \{(x, y) \mid x \equiv y(R)\}$$

is closed. Then $(Y, \mathscr{T}^{(\varphi)})$ is separated.

Proof.—Let $\dot{x}$ and $\dot{y}$ be two distinct elements in $Y = X/R$. Then $(x, y) \notin G$. Since G is closed, there are open neighborhoods $U \in \mathscr{N}_X(x)$ and $V \in \mathscr{N}_X(y)$ such that $U \times V \subset \complement G$. Since R is open, $\varphi(U)$ and $\varphi(V)$ are open neighborhoods of $\dot{x}$ and $\dot{y}$ respectively.

If $\dot{z} \in \varphi(U) \cap \varphi(V)$, then $\dot{z} = \varphi(u)$ with $u \in U$ and $\dot{z} = \varphi(v)$ with $v \in V$. Since $\varphi(u) = \varphi(v) = \dot{z}$, we deduce that $u \equiv v$, that is $(u, v) \in G$. But $(u, v) \in U \times V$, so that $G \cap (U \times V) \neq \varnothing$. Since this leads to a contradiction, it follows that $\varphi(U) \cap \varphi(V) = \varnothing$. Since $\dot{x}$ and $\dot{y}$ were arbitrary, we deduce that $(X, \mathscr{T}^{(\varphi)})$ is separated.

12.11 Theorem.—*Let $(X, \mathscr{T})$ and $(X', \mathscr{T}')$ be two topological spaces and $f: X \to X$ be a continuous mapping. Suppose $(X', \mathscr{T}')$ is separated and let*

$$R = \{(x, y) \mid f(x) = f(y)\}.$$

Then R is a separated equivalence relation.

Proof.—The set $Y = X/R$ can be identified with $f(X)$. Since f is continuous, it is obvious that $\mathscr{T}^{(\varphi)} \supset \mathscr{T}_Y$. Since $(Y, \mathscr{T}_Y)$ is separated, we deduce that $(Y, \mathscr{T}^{(\varphi)})$ is separated.

Let $(X, \mathscr{T})$ be a topological space and R an equivalence relation in R. Then

12.12 Theorem.—*If $(X, \mathscr{T})$ is connected, then $(Y, \mathscr{T}^{(\varphi)})$ is connected.*

Proof.—This follows from Theorem 11.7.

12.13 Theorem.—*If $(X, \mathscr{T})$ is compact and $(Y, \mathscr{T}^{(\varphi)})$ is separated, then $(Y, \mathscr{T}^{(\varphi)})$ is compact.*

Proof.—This follows from Theorem 9.9.

Remark.—To deduce that $(Y, \mathscr{T}^{(\varphi)})$ is compact, it is enough to assume that $(X, \mathscr{T})$ is separated and that there is a set $A \subset X$ such that $(A, \mathscr{T}_A)$ is compact and

$$A \cap \varphi^{-1}(\dot{x}) \neq \varnothing$$

for *every* $\dot{x} \in Y$.

Example 2.—Consider the topological space $(\mathbf{R}, \mathscr{U})$, and let

$$R_1 = \{(x, y) \mid x - y \in \mathbf{Z}\}.$$

Then R_1 is an *equivalence relation* in $\mathbf{R}$ (we denote by φ the canonical mapping of $\mathbf{R}$ onto $\mathbf{R}/R_1$), which is *open* and *separated.*

To show that R_1 is *open*, we reason as follows: We notice first that for every $A \subset \mathbf{R}$, we have

12.14 $\varphi^{-1}(\varphi(A)) = \bigcup_{n \in \mathbf{Z}} \{x + n \mid x \in A\}.$

In fact, if $y \in \varphi^{-1}(\varphi(A))$, then $\varphi(y) = \varphi(x)$ for some $x \in A$, whence $y - x \in \mathbf{Z}$. Hence $y = x + n$ for some $n \in \mathbf{Z}$, so that

$$y \in \bigcup_{n \in \mathbf{Z}} \{x + n \mid x \in A\}.$$

Conversely, if

$$y \in \bigcup_{n \in \mathbf{Z}} \{x + n \mid x \in A\},$$

Then $y = x + n$ for some $n \in \mathbf{N}$ and $x \in A$. Hence $\varphi(y) = \varphi(x)$ for some $x \in A$, so that $y \in \varphi^{-1}(\varphi(A))$. Hence 12.14 holds.

If A is open, then for each $n \in \mathbf{N}$, $\{x + n \mid x \in A\}$ is open (notice that $\{n + n \mid x \in A\}$ is the image of A by the homeomorphism $x \mapsto x + n$ of $\mathbf{R}$ onto $\mathbf{R}$); so that $\varphi^{-1}(\varphi(A))$ is open. We deduce that $\varphi(A)$ is open. Since A was arbitrary, φ is *open.*

Notice now that $(x, y) \mapsto x - y$ is a continuous mapping of $\mathbf{R} \times \mathbf{R}$ into $\mathbf{R}$. Since $\mathbf{Z}$ is a closed part of $\mathbf{R}$, we deduce that *the set* R_1 is closed.

Using Theorem 12.10, we conclude that the equivalence relation R_1 is *separated.*

Hence, the topological space $(\mathbf{R}/R_1, \mathscr{T}^{(\varphi)})$, quotient of $(\mathbf{R}, \mathscr{T})$ by R_1, is separated. The space $(\mathbf{R}/R_1, \mathscr{T}^{(\varphi)})$ is often denoted by $\mathbf{T}^1$ and called the *one-dimensional Torus.*

From Theorem 12.12, it follows that $\mathbf{T}^1$ *is connected.* From the remark following Theorem 12.13 (notice that $[0, 1] \cap \varphi^{-1}(\dot{x}) \neq \varnothing$ for every $\dot{x} \in \mathbf{R}/R_1$), $\mathbf{T}^1$ *is compact.*

Exercises for Chapter 12

1. Let S be an equivalence relation on $\mathbf{R}$ defined by $(x, y) \in S \Leftrightarrow x$ and y belong to $(n, n + 1]$ for some n in $\mathbf{Z}$. Show that the canonical mapping $\varphi: \mathbf{R} \to \mathbf{R}/S$ is neither open nor closed.

2. Let $(Y, \mathscr{I})$ be a topological space and denote by D the set $\{y \mid y \in Y, \{y\} \in \mathscr{I}\}$. Let $E = Y - D$. Then $\mathscr{I}$ is the final topology on Y associated with the topological space $(E, \mathscr{I}_E)$ and the mapping $y \mapsto y$ of E into Y if and only if $E \in \mathscr{I}$.

3. Let Y be a set and $((X_i, \mathscr{T}_i))_{i \in I}$ a family of topological spaces, and for each i in I let f_i be a mapping of X_i into Y. For each i in I, denote by $\mathscr{I}_i$ the topology $\{V \mid f_i^{-1}(V) \in \mathscr{T}_i\}$. Then the final topology on Y associated with the families

$$((X_i, \mathscr{T}_i))_{i \in I} \quad \text{and} \quad (f_i)_{i \in I} \quad \text{is} \quad \inf \{\mathscr{I}_i \mid i \in I\} = \bigcap \{\mathscr{I}_i \mid i \in I\}.$$

4. Let $(Y, \mathscr{I})$ be a topological space and $((X_i, \mathscr{T}_i))_{i \in I}$ a family of topological spaces, and for each i in I let f_i be a mapping of X_i into Y. Suppose that for each topological space $(Z, \mathscr{W})$ and mapping f of Y into Z we have f continuous if and only if $f \circ f_i$ is continuous for every i in I. Then $\mathscr{I}$ is the final topology on Y associated with the families $((X_i, \mathscr{T}_i))_{i \in I}$ and $(f_i)_{i \in I}$.

Chapter 13

Metrics. Metric Spaces*

Let X be a set. A *metric* on X is a mapping $d:(x,y) \mapsto d(x,y)$ of $X \times X$ into $\mathbf{R}$ having the following properties:

13.1 $d(x,y) = 0$ if and only if $x = y$;

13.2 $d(x,y) = d(y,x)$ for all $x \in X, y \in X$;

13.3 $d(x,y) \leq d(x,z) + d(z,y)$ for all $x \in X, y \in X, z \in X$.

If d is a metric on X, we have

13.4 $d(x,y) \geq 0$ for all $x \in X, y \in X$. In fact, using 13.1, 13.3, and 13.2, we deduce that

$$0 = d(x,x) \leq d(x,y) + d(y,x) = 2d(x,y).$$

Now let X be a set, d a metric on X, $a \in X$, and A a *non-void* part of X. Define

13.5 $d(a, A) = \inf_{x \in A} d(a, x)$.

The number $d(x, A)$ is called *the distance from a to A*. Notice that $d(a, \{x\}) = d(a, x)$ for every $x \in X$. We shall show now that for every non-void $A \subset X$ and every x in X and y in X, we have:

13.6 $|d(x, A) - d(y, A)| \leq d(x,y)$.

Let $\varepsilon > 0$ and let $a \in A$ be such that $d(y, a) \leq d(y, A) + \varepsilon$. Then

$$d(x, a) \leq d(x,y) + d(y, a) \leq d(x,y) + d(y, A) + \varepsilon.$$

* Some of the definitions and results that we give concerning *metric spaces* will be presented in a more general setting in the chapters on uniform spaces.

Since $\varepsilon > 0$ was arbitrary,

$$d(x, A) \leq d(x, y) + d(y, A);$$

that is,

$$d(x, A) - d(y, A) \leq d(x, y).$$

In this same way, we obtain

$$-(d(x, A) - d(y, A)) = d(y, A) - d(x, A) \leq d(y, x) = d(x, y).$$

Hence

$$|d(x, A) - d(y, A)| \leq d(x, y).$$

If A is a non-void part of X, we call

13.7 $$\delta(A) = \sup_{x \in A, y \in A} d(x, y)$$

the *diameter* of A (notice that $0 \leq \delta(A) \leq +\infty$). By definition, the diameter of $\varnothing$, $\delta(\varnothing)$, is zero. A set $A \subset X$ is *bounded* if $\delta(A) < +\infty$. Clearly, every finite set is bounded.

Let X be a set, d a metric on X, and $a \in X$. We shall write

$$V_r(a) = \{x \mid d(a, x) < r\} \quad \text{for} \quad r > 0$$

and

$$W_r(a) = \{x \mid d(a, x) \leq r\} \quad \text{for} \quad r \geq 0.$$

We call $V_r(a)$ the *open ball* of center a and radius $r > 0$; we call $W_r(a)$ the *closed ball* of center a and radius $r \geq 0$.

Remark.—Notice that a set $A \subset X$ is bounded if and only if A is contained in some ball (open and closed). Notice also that if $A \subset X$ is bounded and $a \in X$, then there is a ball of center a containing A.

We shall give now several examples of metrics.

Example 1.—Define $d: \mathbf{R} \times \mathbf{R} \to \mathbf{R}$ by

$$d(x, y) = |x - y|$$

for $(x, y) \in \mathbf{R} \times \mathbf{R}$. It is immediate that d is a metric on $\mathbf{R}$.

Let $a \in \mathbf{R}$, $x \in \mathbf{R}$, and $r > 0$. Then

$$d(a, x) < r \Leftrightarrow |a - x| < r \Leftrightarrow |x - a| < r \Leftrightarrow -r < x - a < r;$$

hence

$$d(a, x) < r \Leftrightarrow a - r < x < a + r.$$

We deduce that

13.8 $V_r(a) = (a - r, a + r).$

In the same way, we show that

13.9 $W_r(a) = [a - r, a + r]$ *if** $r > 0.$

Example 2.—Define $\eta:\mathbf{Q} \times \mathbf{Q} \to \mathbf{R}$ by

$$\eta(x, y) = |x - y|$$

for $(x, y) \in \mathbf{Q} \times \mathbf{Q}$. It is immediate that η is a metric on $\mathbf{Q}$.

Notice that if d is the metric on $\mathbf{R}$ defined in Example 1, then

$$\eta = d \mid \mathbf{Q} \times \mathbf{Q}.$$

Example 3.—Let $n \in \mathbf{N}$ and define $d_n:\mathbf{R}^n \times \mathbf{R}^n \to \mathbf{R}$ by†

$$d_n(x, y) = \left(\textstyle\sum_{i=1}^{n} (x_i - y_i)^2\right)^{1/2}$$

for $x = (x_1, \ldots, x_n) \in \mathbf{R}^n$ and $y = (y_1, \ldots, y_n) \in \mathbf{R}^n$. It is immediate that d_n satisfies 13.1 and 13.2. We show in the first appendix to this chapter that d_n also satisfies 13.3. Therefore, d_n is a metric on $\mathbf{R}^n$, which is usually called the Euclidean metric. Notice that the metric on $\mathbf{R}$ introduced in Example 1 is d_1. When there is no ambiguity, we shall write d instead of d_n.

Remark.—In $(\mathbf{R}^n, d_n)$, a set is bounded if and only if it is contained in some bounded n-dimensional interval (open or closed).

Example 4.—Let X be a set and $f:X \to R$ an *injection.* Define $d:X \times X \to \mathbf{R}$ by

$$d(x, y) = |f(x) - f(y)|$$

for $(x, y) \in X \times X$. Then

$$d(x, y) = 0 \Leftrightarrow |f(x) - f(y)| = 0 \Leftrightarrow f(x) = f(y) \Leftrightarrow x = y.$$

Also,

$$d(x, y) = |f(x) - f(y)| = |f(y) - f(x)| = d(y, x).$$

---

* If $r = 0$, then $W_r(a) = \{a\}$.

† For each $x \in \mathbf{R}_+$, we denote by $\sqrt{x}$ or $(x)^{1/2}$ the unique $y \in \mathbf{R}_+$ satisfying $y^2 = x$. The mapping $x \mapsto (x)^{1/2}$ of $\mathbf{R}_+$ into $\mathbf{R}_+$ is strictly increasing and continuous.

Moreover,

$$\begin{aligned} d(x,y) &= |f(x) - f(y)| = |f(x) - f(z) + f(z) - f(y)| \\ &\le |f(x) - f(z)| + |f(z) - f(y)| \\ &= d(x, z) + d(z,y) \end{aligned}$$

for all x, y, z in X. Hence, d is a metric on X.

Example 5.—Define $\delta:\mathbf{R} \times \mathbf{R} \to \mathbf{R}$ by

$$\delta(x,y) = \left| \frac{x}{1 + |x|} - \frac{y}{1 + |y|} \right|$$

for $(x,y) \in \mathbf{R} \times \mathbf{R}$. Then δ is a metric on $\mathbf{R}$. This follows from Example 5 if we notice that $x \mapsto x/(1 + |x|)$ is an injection of $\mathbf{R}$ into $\mathbf{R}$.

Example 6.—Let X be a set and let

$$d(x,y) = \begin{cases} 1 & \text{if} \quad x \ne y \\ 0 & \text{if} \quad x = y \end{cases}$$

for $(x,y) \in X \times X$. Then d is a metric on X.

Example 7.—Let X be a set and let d be a metric on X. Let

$$\delta(x,y) = \inf\{d(x,y), 1\}$$

for $(x,y) \in X \times X$. Then δ is a metric on X.

If $\delta(x,y) = 0$, then $d(x,y) = 0$, whence $x = y$. Since $d(x,y) = d(y, x)$, we deduce that

$$\delta(x,y) = \inf(d(x,y), 1) = \inf(d(y, x), 1) = \delta(y, x).$$

If $d(x, z) \le 1$ and $d(z,y) \le 1$, then $d(x,y) \le 1$, whence

$$\delta(x,y) = d(x,y) \le d(x, z) + d(z,y) = \delta(x, z) + \delta(z,y).$$

If $d(x, z) > 1$, then

$$\delta(x,y) \le 1 + \delta(z,y) = \delta(x, z) + \delta(z,y);$$

If $d(z,y) > 1$, then

$$\delta(x,y) \le \delta(x, z) + 1 = \delta(x, z) + \delta(z,y).$$

Example 8.—Let I be a countable set, and for each $i \in I$ let X be a set, d_i a metric on X and $c_i > 0$ such that $\sum_{i \in I} c_i < +\infty$. Define $d: X \times X \to \mathbf{R}$ by

$$d(x, y) = \sum_{i \in I} c_i \inf(1, d_i(x_i, y_i))$$

for $x = (x_i)_{i \in I} \in X$ and $y = (y_i)_{i \in I} \in X$. Then d is a metric on X.

Let X be a set and let d be a metric on X. Let $a \in X$, $r_1 > 0$ and $b \in X$, $r_2 > 0$.

13.10 Theorem.—*If* $c \in V_{r_1}(a) \cap V_{r_2}(b)$, *then there exists* $r > 0$ such that

$$V_r(c) \subset V_{r_1}(a) \cap V_{r_2}(b).$$

Proof.—Since $c \in V_{r_1}(a)$, we have $d(a, c) < r_1$, whence $r_1 - d(a, c) > 0$. In the same way, we see that $r_2 - d(b, c) > 0$. Let

$$r = \inf(r_1 - d(a, c), r_2 - d(b, c)).$$

Then $V_r(c) \subset V_{r_1}(a) \cap V_{r_2}(b)$. In fact, let $x \in V_r(c)$. Then

$$d(x, a) \leq d(x, c) + d(c, a) < r + d(c, a)$$
$$\leq (r_1 - d(c, a)) + d(c, a) = r_1,$$

and thus

$$d(x, a) < r_1; \quad \text{that is,} \quad x \in V_{r_1}(a).$$

In the same way, we see that

$$d(x, b) \leq d(x, c) + d(c, b) < r + d(c, b)$$
$$\leq (r_2 - d(c, b)) + d(c, b) = r_2,$$

and thus

$$d(x, b) < r_2; \quad \text{that is,} \quad x \in V_{r_2}(b).$$

Hence, $x \in V_r(c)$ implies $x \in V_{r_1}(a)$ *and* $x \in V_{r_2}(b)$, and hence Theorem 13.10 is proved.

Let X be a set and d a metric on X. For each $x \in X$, denote by $\mathscr{B}(x)$ the set of all parts of X that contain some open ball of center x (in particular, $V_r(x) \in \mathscr{B}(x)$ for all $x > 0$).

It is obvious that (a), (b), and (c) of Theorem 2.6 are satisfied. Now let $x \in X$ and $V \in \mathscr{B}(x)$ and let $r > 0$ such that $V_r(x) \subset V$. By Theorem 13.10, for each $c \in V_r(x)$, there is $s > 0$ such that $V_s(c) \subset V_r(x) \subset V$. Hence, (*d*) of Theorem 2.6 is also satisfied. By Theorem 2.6 there is a *unique* topology $\mathscr{T}$ on X such that $\mathscr{N}_{(X,\mathscr{T})}(x) = \mathscr{B}(x)$ for every $x \in X$.

13.11 Definition.—*The topology* $\mathcal{T}$ *is denoted* $\mathcal{T}_d$ *and is said to be the topology defined by d.*

From Theorem 13.10, it follows that $V_r(x) \in \mathcal{T}_d$ for every $x \in X$ and $r > 0$. Notice also that $U \in \mathcal{T}_d$ if and only if for every $x \in U$ there is $r > 0$ such that $V_r(x) \subset U$.

Let X be a set and d a metric on X. Consider the metric δ on X defined by:

13.12 $\delta(x, y) = \inf\{d(x, y), 1\}$ for $(x, y) \in X \times X$ (see Example 6). Then:

13.13 Theorem.—*We have* $\mathcal{T}_d = \mathcal{T}_\delta$.

Proof.—It is easy to see that for every $x \in X$,

$$\mathcal{N}_{(X,\mathcal{T}_d)}(a) = \mathcal{N}_{(X,\mathcal{T}_\delta)}(a).$$

The conclusion follows then from Theorem 2.5.

Theorem 13.13 shows that two metrics may be quite different (see also Example 6, Chapter 15) and yet define the same topology.

Exercise 1.—Let X be a set and d and δ two metrics on X such that

$$A\delta(x, y) \leq d(x, y) \leq B\delta(x, y) \qquad (A > 0,\ B > 0)$$

for all $(x, y) \in X \times X$. Then $\mathcal{T}_d = \mathcal{T}_\delta$.

We shall introduce now the following definition.

13.14 Definition.—*A couple* (X, d), *where X is a set and d a distance on X, is called a metric space.*

If (X, d) is a metric space, we say that d is the metric of (X, d). When there is no ambiguity as to what metric we consider, we shall often say *the metric space* X instead of *the metric space* (X, d).

If (X, d) is a metric space, then we always suppose that X is endowed with the topology $\mathcal{T}_d$. We shall often say that $\mathcal{T}_d$ is the topology of the metric space (X, d).

Example 9.—Consider the metric space $(\mathbf{R}^n, d_n)$ $(n \geq 1)$, where d_n is the metric introduced in Example 3. Then $\mathcal{T}_{d_n}$ is the usual topology $\mathcal{U}^n$ of $\mathbf{R}^n$.

Let $a = (a_j)_{1 \leq j \leq n} \in \mathbf{R}^n$. By the definition of $\mathscr{T}_{d_n}$, the set $\{V_r(a) \mid r > 0\}$ is a fundamental system of a in the topological space $(\mathbf{R}^n, \mathscr{T}_{d_n})$.

For each $r > 0$, let

$$I_r(a) = \{(x_j)_{1 \leq j \leq n} \mid a_j - r \leq x_j \leq a_j + r, \ 1 \leq j \leq n\}.$$

Then $\{I_r(a) \mid r > 0\}$ is a fundamental system of a (see Example 2, Chapter 2, and Example 4, Chapter 7). Notice now that for every $r > 0$, we have

$$I_{r/\sqrt{n}}(a) \subset V_r(a) \subset I_r(a).$$

Now let $V \in \mathscr{N}_{(\mathbf{R}^n, \mathscr{U}^n)}(a)$. Then there is $r > 0$ such that $V \supset I_r(a)$; hence, $V \supset V_r(a)$, so that $V \in \mathscr{N}_{(\mathbf{R}^n, \mathscr{T}_{d_n})}(a)$. Since V was arbitrary,

$$\mathscr{N}_{(\mathbf{R}^n, \mathscr{U}^n)}(a) \subset \mathscr{N}_{(\mathbf{R}^n, \mathscr{T}_{d_n})}(a).$$

In the same way, we show that

$$\mathscr{N}_{(\mathbf{R}^n, \mathscr{T}_{d_n})}(a) \subset \mathscr{N}_{(\mathbf{R}^n, \mathscr{U}^n)}(a),$$

so that

$$\mathscr{N}_{(\mathbf{R}^n, \mathscr{U}^n)}(a) = \mathscr{N}_{(\mathbf{R}^n, \mathscr{T}_{d_n})}(a).$$

Since $a \in \mathbf{R}^n$ was arbitrary, we deduce from Theorem 2.5 that $\mathscr{U}^n = \mathscr{T}_{d_n}$.

13.15 Definition.—(i) *Let X be a set and $\mathscr{T}$ a topology on X. We say that $\mathscr{T}$ is metrizable if there is a metric d on X such that $\mathscr{T} = \mathscr{T}_d$.* (ii) *We say that a topological space $(X, \mathscr{T})$ is metrizable if $\mathscr{T}$ is metrizable.*

13.16 Theorem.—*A metrizable topological space $(X, \mathscr{T})$ is separated.*

Proof.—Let $a \in X$ and $b \in X$ such that $a \neq b$. Let d be a metric on X satisfying $\mathscr{T} = \mathscr{T}_d$. Since $a \neq b$, we have $r = d(a, b) > 0$. Then $V_{r/2}(a)$ and $V_{r/2}(b)$ are neighborhoods of a and b respectively. Now,

$$V_{r/2}(a) \cap V_{r/2}(b) = \varnothing.$$

In fact, if $x \in V_{r/2}(a) \cap V_{r/2}(b)$, then

$$r = d(a, b) \leq d(a, x) + d(x, b) < r/2 + r/2 = r;$$

that is, $r < r$; since this is a contradiction, Theorem 13.16 is proved.

13.17 Theorem.—*Let $(X, \mathscr{T})$ be a metrizable topological space and let d be a metric on X such that $\mathscr{T} = \mathscr{T}_d$. Let $a \in X$. Then:*

(a) $\{V_r(a) \mid r > 0\}$ *is a fundamental system of* a;
(b) $\{V_{1/n}(a) \mid n \in \mathbf{N}\}$ *is a fundamental system of* a;
(c) $\{W_r(a) \mid r > 0\}$ *is a fundamental system of* a;
(d) $\{W_{1/n}(a) \mid n \in \mathbf{N}\}$ *is a fundamental system of* a.

Proof.—The proof is left to the reader.

Let (X_1, d_1) and (X_2, d_2) be two metric spaces and $S \subset X_1$. It follows from Theorems 5.4 and 13.17 that:

13.18 *A function $f: S \to X$ is continuous at $c \in S$ if and only if for every $\varepsilon > 0$ there is $\delta > 0$ such that*

$$d_2(f(x), f(c)) < \varepsilon \quad \textit{whenever} \quad x \in S \quad \textit{and} \quad d_1(x, c) < \delta$$

(*or* $d_2(f(x), f(c)) \leq \varepsilon$ *whenever* $x \in S$ *and* $d_1(x, c) \leq \delta$).

Denote by (X, d) a metric space. Then:

13.19 Theorem.—*For each non-void set $A \subset X$ the mapping $x \mapsto d(x, A)$, of X into $\mathbf{R}$, is continuous on X.*

Proof.—Denote by f the considered mapping. Let $c \in X$ and let $\varepsilon > 0$. If $x \in V_\varepsilon(c)$, we have (see 13.6)

$$|f(x) - f(c)| \leq d(x, c) < \varepsilon.$$

Since $\varepsilon > 0$ was arbitrary, we deduce that f is continuous at c (see 5.5(b)). Since $c \in X$ was arbitrary, we deduce that f is continuous on X.

We have shown previously that for every $a \in X$ and $r > 0$, the set $V_r(a)$ is open. Since

$$V_r(a) = \{x \mid d(a, x) < r\} = f^{-1}(-r, +r)$$

(if $f(x) = d(a, x)$ for $x \in X$) and since $(-r, +r)$ is open, this follows also from Theorem 5.11. We also see that

$$W_r(a) = \{x \mid d(a, x) \leq r\} = f^{-1}([-r, +r])$$

is *closed* for every $r \geq 0$.

13.20 Corollary.—*If A is a non-void part of X, then $a \in \bar{A}$ if and only if $d(a, A) = 0$.*

Proof.—Since $f: x \mapsto d(x, A)$ is continuous, the set $\{x \mid f(x) = 0\}$ is closed and contains A. Hence, this set contains $\bar{A}$. Hence, $a \in \bar{A}$ implies that $d(a, A) = 0$. Now let $a \in X$ be such that $d(a, A) = 0$. Let $r > 0$. Then there is $x \in A$ such that $d(x, a) < r$. Hence, $V_r(a) \cap A \neq \varnothing$. Since $r > 0$ was arbitrary, we deduce that $a \in \bar{A}$.

Exercise 2.—Let $A \subset X$ be non-void and let $a \in X$. Then $d(a, A) = d(a, \bar{A})$.

Exercise 3.—Let $((X_i, d_i))_{i \in I}$ be a *countable* family of metric spaces and let d be the metric on $X = \prod_{i \in I} X_i$ defined in Example 8. Show that $\mathscr{T}_d$ is the product of the family of topologies $(\mathscr{T}_{d_i})_{i \in I}$.

13.21 Theorem.—*Let (X, d) be a metric space. Then the mapping $(x, y) \mapsto d(x, y)$ of $X \times X$ into $\mathbf{R}$ is continuous on $X \times X$.*

Proof.—Let $(a, b) \in X \times X$. For every $(x, y) \in X \times X$, we have

$$d(x, y) \leq d(x, a) + d(a, b) + d(b, y)$$

and

$$d(a, b) \leq d(a, x) + d(x, y) + d(y, b);$$

hence

$$|d(a, b) - d(x, y)| \leq d(a, x) + d(b, y).$$

Now let $\varepsilon > 0$ and let $(x, y) \in V_{\varepsilon/2}(a) \times V_{\varepsilon/2}(b)$. Then

$$|d(a, b) - d(x, y)| < 2(\varepsilon/2) = \varepsilon.$$

Since $\varepsilon > 0$ was arbitrary, and since $V_{\varepsilon/2}(a) \times V_{\varepsilon/2}(a)$ is a neighborhood of (a, b) (see, for instance, 7.6′), we deduce that $(x, y) \mapsto d(x, y)$ is continuous at (a, b). Since $(a, b) \in X \times X$ was arbitrary, we deduce that $(x, y) \mapsto d(x, y)$ is continuous on $X \times X$.

▼ *Remark.*—In fact, the mapping $(x, y) \mapsto d(x, y)$ of $X \times X$ into $\mathbf{R}$ is *uniformly* continuous (see 17.1, 17.2, and Exercise 3 above). ▲

The Metric d_n on $\mathbf{R}^n$ $(n \in \mathbf{N})$

In Example 3 in this chapter, we defined a mapping $d_n:\mathbf{R}^n \times \mathbf{R}^n \to \mathbf{R}$ and we asserted that this mapping is a *metric* on $\mathbf{R}^n$. We shall prove this assertion here.

First, we shall introduce several definitions and prove several results. These definitions and results have their own interest, and this is why they are discussed in detail.

The Scalar Product on $\mathbf{R}^n$

For any $x = (x_1, \ldots, x_n)$ and $y = (y_1, \ldots, y_n)$, elements of $\mathbf{R}^n$, let

$$\varphi(x,y) = x_1y_1 + \ldots + x_ny_n.$$

We define this way a mapping $\varphi:\mathbf{R}^n \times \mathbf{R}^n \to \mathbf{R}$. We call φ the *scalar product* on $\mathbf{R}^n$. For x and y in $\mathbf{R}^n$, $\varphi(x,y)$ is called the scalar product of x and y (it is often denoted $x \cdot y$ and read "x times y"). It is easy to see that:

13.22(i) *if $b \in \mathbf{R}^n$, then $y \mapsto \varphi(x, b)$ is a linear* mapping of $\mathbf{R}^n$ into $\mathbf{R}$;*

13.22(ii) *if $a \in \mathbf{R}^n$, then $y \mapsto \varphi(a,y)$ is a linear mapping of $\mathbf{R}^n$ into $\mathbf{R}$;*

13.23 $\varphi(x,y) = \varphi(y, x)$ *for any x and y in $\mathbf{R}^n$;*

13.24 $\varphi(0,y) = \varphi(x, 0) = 0$ *for any x and y in $\mathbf{R}^n$;*

13.25 $\varphi(x, x) \geq 0$ *for every $x \in \mathbf{R}^n$;*

13.26 *If $x \in \mathbf{R}^n$, then $\varphi(x, x) = 0$ if and only if $x = 0$.*

The properties 13.22–13.26 of φ follow immediately from the definition of φ.

* We assume that the definition and elementary properties of linear mappings are known to the reader.

13.27 *If $\lambda \in \mathbf{R}$ and x and y belong to $\mathbf{R}^n$, then*

$$\varphi(\lambda x + y, \lambda x + y) = \lambda^2 \varphi(x, x) + 2\lambda\varphi(x, y) + \varphi(y, y).$$

In fact,

$$\begin{aligned} \varphi(\lambda x + y, \lambda x + y) &= \lambda\varphi(x, \lambda x + y) + \varphi(y, \lambda x + y) \\ &= \lambda^2\varphi(x, x) + \lambda\varphi(x, y) + \lambda\varphi(y, x) + \varphi(y, y) \\ &= \lambda^2\varphi(x, x) + 2\lambda\varphi(x, y) + \varphi(y, y), \end{aligned}$$

whence 13.27 is proved.

13.28 Theorem (Cauchy's Inequality).—*If a and b belong to $\mathbf{R}^n$, then*

$$\varphi(a, b)^2 \leq \varphi(a, a)\varphi(b, b).$$

Proof.—Let

$$\alpha = \varphi(a, a), \qquad \beta = \varphi(b, b), \quad \text{and} \quad \Gamma = \varphi(a, b).$$

If $\alpha = 0$, then $a = 0$ (see 13.26); hence $\varphi(a, b) = 0$ (see 13.24) and hence Theorem 13.28 holds.

Now suppose $\alpha \neq 0$. By 13.27,

$$\lambda^2\alpha + 2\lambda\Gamma + \beta \geq 0$$

for all $\lambda \in R$. In particular, for $\lambda = -\Gamma/\alpha$,

$$(-\Gamma/\alpha)^2\alpha + 2(-\Gamma/\alpha)\Gamma + \beta \geq 0.$$

Hence,

$$\Gamma^2/\alpha - 2\Gamma^2/\alpha + \beta \geq 0 \Rightarrow \beta - \Gamma^2/\alpha \geq 0,$$

whence

$$(\alpha\beta - \Gamma^2)/\alpha \geq 0 \Rightarrow \Gamma^2 \leq \alpha\beta.$$

Hence, Theorem 13.28 holds.

If $a = (a_1, \ldots, a_n)$, $b = (b_1, \ldots, b_n)$, and if we recall the definition of φ, we see that

$$|\textstyle\sum_{i=1}^n a_i b_i|^2 \leq (\sum_{i=1}^n a_i^2)(\sum_{i=1}^n b_i^2),$$

or equivalently,*

$$|\sum_{i=1}^n a_i b_i| \le (\sum_{i=1}^n a_i)^{1/2} (\sum_{i=1}^n b_i^2)^{1/2}.$$

The Euclidean Norm on $\mathbf{R}^n$

For any $x = (x_1, \ldots, x_n)$, let

$$\|x\| = (x_1^2 + \ldots + x_n^2)^{1/2};$$

hence $\|x\| = \varphi(x, x)^{1/2}$. This way we define a mapping $x \mapsto \|x\|$ of $\mathbf{R}^n$ into $\mathbf{R}$. We call this mapping the Euclidean norm on $\mathbf{R}^n$. For $x \in \mathbf{R}^n$, $\|x\|$ is called the norm of x and is read "norm x."

13.29 Theorem.—*We have:*

(i) $\|0\| = 0$ *and* $\|x\| > 0$ *if* $x \in \mathbf{R}^n$, $x \neq 0$;

(ii) $\|\lambda x\| = |\lambda| \, \|x\|$ *if* $\lambda \in \mathbf{R}$ *and* $x \in \mathbf{R}^n$;

(iii) $\|x + y\| \le \|x\| + \|y\|$ *if* $x \in \mathbf{R}^n$ *and* $y \in \mathbf{R}^n$.

Proof.—The proof of (i) is immediate. To prove (ii), we note that

$$\|\lambda x\|^2 = \varphi(\lambda x, \lambda x) = \lambda^2 \varphi(x, x) = (|\lambda| \, \|x\|)^2.$$

To prove (iii), we note that

$$\begin{aligned}\|x + y\|^2 &= \varphi(x + y, x + y) = \varphi(x, x) + \varphi(x, y) + \varphi(y, x) + \varphi(y, y) \\ &= \varphi(x, x) + 2\varphi(x, y) + \varphi(y, y) = \|x\|^2 + 2\varphi(x, y) + \|y\|^2;\end{aligned}$$

by Cauchy's inequality,

$$\varphi(x, y) \le |\varphi(x, y)| \le \varphi(x, x)^{1/2} \varphi(y, y)^{1/2} = \|x\| \, \|y\|,$$

whence

$$\|x + y\|^2 \le \|x\|^2 + 2 \, \|x\| \, \|y\| + \|y\|^2 = (\|x\| + \|y\|)^2.$$

We conclude that (iii) holds.

The Metric d_n

If $x = (x_1, \ldots, x_n) \in \mathbf{R}^n$ and $y = (y_1, \ldots, y_n)$ (see Example 3), then

$$d_n(x, y) = (\sum_{i=1}^n (x_i - y_i)^2)^{1/2} = \|x - y\|.$$

---

* The mapping $x \mapsto \sqrt{x}$ of $\mathbf{R}_+$ into $\mathbf{R}$ is *strictly increasing*; hence $x < y \Leftrightarrow \sqrt{x} < \sqrt{y}$. If $x \in \mathbf{R}_+$ and $y \in \mathbf{R}_+$, then $(\sqrt{x}\sqrt{y})^2 = xy$; that is, $\sqrt{x}\sqrt{y} = \sqrt{xy}$.

We have already noted (in Example 3) that d_n satisfies the conditions of a metric given in 13.1 and 13.2. To prove 13.3, we observe that

$$\begin{aligned} d_n(x,y) &= \|x - y\| = \|x - z + z - y\| \\ &= \|(x - z) + (z - y)\| \leq \|x - z\| + \|z - y\| \\ &= d_n(x, z) + d_n(z,y). \end{aligned}$$

Hence, d_n is a *metric* on $\mathbf{R}^n$.

A Non-Metrizable Topological Space

Denote by X the set of all functions on $\mathbf{R}$ to $\mathbf{R}$. For every $A \subset \mathbf{R}$ finite, $\varepsilon > 0$, and $f \in X$, we define*

$$V(f; A, c) = \{g \mid g \in X, |f(x) - g(x)| < \varepsilon \text{ for all } x \in A\}.$$

Now let $f \in X$, $A \subset \mathbf{R}$ finite, $\varepsilon > 0$, $g \in X$, $B \subset \mathbf{R}$ finite, and $\eta > 0$. If

$$h \in V(f; A, \varepsilon) \cap V(g; B, \eta),$$

there exists $C \subset \mathbf{R}$ finite and $r > 0$ such that

13.30 $V(h; C, r) \subset V(f; A, \varepsilon) \cap V(g; B, \eta).$

In fact, let

$$\alpha = \sup \{|h(x) - f(x)| \mid x \in A\} \quad \text{and} \quad \beta = \sup \{|h(x) - g(x)| \mid x \in B\};$$

obviously $\alpha < \varepsilon$ and $\beta < \eta$. Let

$$r = \inf \{\varepsilon - \alpha, \eta - \beta\}.$$

Then 13.30 is satisfied if $C = A \cup B$. In fact, if $u \in V(h; A \cup B, r)$, then for $x \in A$ $(\subset A \cup B)$ we have

$$|u(x) - f(x)| \leq |u(x) - h(x)| + |h(x) - f(x)| < r + \alpha \leq \varepsilon - \alpha + \alpha = \varepsilon$$

for all $x \in A$. Hence, $u \in V(f; A, \varepsilon)$. In a similar way, we show that $u \in V(g; B, \eta)$. Hence 13.30 is proved.

For each $f \in X$, let

$$\mathscr{F}_f = \{V(f; A, \varepsilon) \mid A \subset \mathbf{R} \text{ finite}, \varepsilon > 0\}$$

and let $\mathscr{B}(f)$ be the set of all parts of X that contain some set belonging to $\mathscr{F}_f$. It is easy to verify that (a), (b), (c), and (d) of Theorem 2.6

* $V(f; \varnothing, \varepsilon) = X$ for all $f \in X$ and $\varepsilon > 0$.

are satisfied. Hence, there exists a unique topology $\mathscr{T}$ on X such that $\mathscr{N}_{(X,\mathscr{T})}(f) = \mathscr{B}(f)$ for every $f \in X$.

We shall now show that:

13.31 *$\mathscr{F}_f$ is a fundamental system of f;*

13.32 *$(X, \mathscr{T})$ is separated;*

13.33 *$(X, \mathscr{T})$ is not metrizable.*

Proof of 13.31.—The assertion is obvious.

Proof of 13.32.—Let $f \in X$ and $g \in X$, $f \neq g$. Then there is $x_0 \in \mathbf{R}$ such that $f(x_0) \neq g(x_0)$. Let $\varepsilon = |f(x_0) - g(x_0)|/3$. Then

13.34 $V(f; \{x_0\}, \varepsilon) \cap V(g; \{x_0\}, \varepsilon) = \varnothing$.

In fact, if

$$h \in V(f; \{x_0\}, \varepsilon) \cap V(g; \{x_0\}, \varepsilon),$$

then

$$|h(x_0) - f(x_0)| < \varepsilon \quad \text{and} \quad |h(x_0) - g(x_0)| < \varepsilon,$$

and thus

$$|f(x_0) - g(x_0)| \leq |f(x_0) - h(x_0)| + |h(x_0) - g(x_0)| < 2\varepsilon$$
$$= (\tfrac{2}{3})\, |f(x_0) - g(x_0)|.$$

Since this is a contradiction, 13.34 holds. Since f and g were arbitrary, we conclude that X is separated.

Proof of 13.33.—Suppose $(X, \mathscr{T})$ is metrizable and let d be a metric on X such that $\mathscr{T}_d = \mathscr{T}$. Let $f \in X$. By Theorem 13.17, the set $\{V_{1/n}(f) \mid n \in \mathbf{N}\}$ is a fundamental system of f. Hence, for every $n \in \mathbf{N}$, there is a finite set $A_n \subset X$ and $\varepsilon_n > 0$ such that

$$V(f; A_n, \varepsilon_n) \subset V_{1/n}(f).$$

Since $\{V_{1/n}(f) \mid n \in \mathbf{N}\}$ is a fundamental system of f, $\{V(f; A_n, \varepsilon_n) \mid n \in \mathbf{N}\}$ is a fundamental system of f. Since $\bigcup_{n\in\mathbf{N}} A_n$ is countable, there exists $x_0 \in \mathbf{R} - \bigcup_{n\in\mathbf{N}} A_n$. Let $\varphi \in X$ be the function defined by

$$\varphi(x) = \begin{cases} f(x_0) + 1 & \text{for } x = x_0 \\ f(x) & \text{for } x \neq x_0. \end{cases}$$

Then $\varphi \neq f$. Since $(X, \mathscr{T})$ is separated, there is $V \in \mathscr{N}(f)$ such that $\varphi \notin V$. Hence, there is $p \in \mathbf{N}$ such that $\varphi \notin V(f; A_p, \varepsilon_p)$. But $\varphi(x) = f(x)$ for all $x \neq x_0$, and hence $\varphi \in V(f; A_p, \varepsilon_p)$ $(x_0 \notin \bigcup_{n\in\mathbf{N}} A_n \Rightarrow x_0 \notin A_p)$. Thus we arrive at a contradiction. Hence 13.33 is proved.

Exercise 4.—Let X_0 be the set of all $f \in X$ such that

$$\{x \mid f(x) \neq 0\}$$

is finite. Show that $\overline{X}_0 = X$.

▼ *Exercise* 5.—Let X_1 be the set of all $f \in X$ such that

$$\{x \mid f(x) \neq 0\}$$

is countable. Show that for every sequence $(f_n)_{n\in\mathbf{N}}$ of elements of X_1 there exists a subsequence $(f_{\varphi(n)})_{n\in\mathbf{N}}$ converging to a $f \in X$. Show that X_1 is not compact.

Exercise 6.—For each $i \in \mathbf{R}$, let $X_i = \mathbf{R}$ and $\mathscr{T}_i = \mathscr{U}$ (= the usual topology of $\mathbf{R}$). Let $(X_\infty, \mathscr{T}_\infty)$ be the topological space product of the family $((X_i, \mathscr{T}_i))_{i\in\mathbf{R}}$. Let u be the bijection $f \mapsto (f(i))_{i\in R}$ of X onto X_∞. Show that u is a homeomorphism of $(X, \mathscr{T})$ onto $(X_\infty, \mathscr{T}_\infty)$.

Exercise 7.—If $u: \mathbf{R} \to \mathbf{R}_+$ is a function, then

$$\{f \mid f \in X, |f(x)| \leq u(x) \text{ for every } x \in \mathbf{R}\}$$

is compact. ▲

Exercises for Chapter 13

1. Suppose $f: \mathbf{R}_+ \to \mathbf{R}_+$ is such that (i) $f(x) = 0 \Leftrightarrow x = 0$; (ii) $s \leq t \Rightarrow f(s) \leq f(t)$ for all s and t in $\mathbf{R}_+$; (iii) $f(s + t) \leq f(s) + f(t)$ for all s and t in $\mathbf{R}_+$.

(a) If (X, d) is a metric space, then $f \circ d$ is a metric on X. Moreover, if f is continuous at 0, then $\mathscr{T}_d = \mathscr{T}_{f\circ d}$.

(b) The mappings $t \mapsto \inf(1, t)$, $t \mapsto t/(1 + t)$, and $t \mapsto ct$ $(c > 0)$ all satisfy the conditions (i), (ii), and (iii). If f_1 and f_2 satisfy all three conditions, then so does $f_1 \circ f_2$.

2. On $\mathbf{R}_+^n$ $(n \in \mathbf{N})$, define "$\leq$" by $s \leq t \Leftrightarrow s_i \leq t_i$ for every i, $1 \leq i \leq n$ (here $s = (s_1, \ldots, s_n)$ and $t = (t_1, \ldots, t_n)$). Suppose that

$g:\mathbf{R}_+^n \to \mathbf{R}_+$ is such that (i) $g(x) = 0 \Leftrightarrow x = 0$; (ii) $s \leq t \Rightarrow g(s) \leq g(t)$ for all s and t in R_+^n; (iii) $g(s + t) \leq g(s) + g(t)$ for all s and t in $\boldsymbol{R}_+^n$.

If X is a set and $d_1, \ldots, d_n$ are metrics on X, then

$$g \circ (d_1, \ldots, d_n) : (x, y) \mapsto g(d_1(x, y), \ldots, d_n(x, y))$$

is a metric on X. If g is continuous at 0, then

$$\sup \{\mathscr{T}_{d_i} \mid 1 \leq i \leq n\} = \mathscr{T}_{g \circ (d_1, \ldots, d_n)}.$$

3. Let $X = \{0\} \cup \{x \mid |x| \geq 1\}$ and let d be the distance on X defined by $d(x, y) = |x - y|$ for $(x, y) \in X \times X$. Show that $\mathscr{T}_d = \mathscr{U}_X$ and that $\overline{V_1(0)} \neq W_1(0)$.

4. A topological space $(X, \mathscr{T})$ is said to be completely normal (or a T_5 space) if and only if $(X, \mathscr{T})$ is separated and whenever A and B are subsets of X such that $A \odot B = \varnothing$ (see Chapter 11), there are U and V in $\mathscr{T}$ such that $A \subset U$, $B \subset V$, and $U \cap V = \varnothing$. Show that every metrizable space is a T_5 space.

5. Let (X, d) be a metric space. Suppose there is a countable dense subset of X. Then the topology $\mathscr{T}_d$ has a *countable* base.

6. Let (X, d) be a metric space and $A \neq \varnothing$, $A \subset X$. Then $x \in \bar{A} \Leftrightarrow d(x, A) = 0$; furthermore, if A is compact, then for each x in X there is a a_x in A such that $d(x, a_x) = d(x, A)$.

Chapter 14

Sequences

Denote by $(X, \mathscr{T})$ a separated* topological space. We shall introduce the following:

14.1 Definition.—*A sequence*† $(x_n)_{n \in \mathbf{N}}$ *of elements of* X *converges to* $a \in X$ *if for every* $V \in \mathscr{N}(a)$ *there exists some* $A \in \mathbf{N}$ *(depending on* V *and on the sequence) such that*

$$x_n \in V \quad \textit{whenever} \quad n \geq A.$$

The element $a \in X$ is called the *limit of the sequence* $(x_n)_{n \in \mathbf{N}}$ (see Theorem 14.3 below). If $a \in X$, the notations

14.2 $\quad a = \lim_{n \in \mathbf{N}} x_n \quad$ or $\quad a = \lim_{n \to +\infty} x_n$

mean that $(x_n)_{n \in \mathbf{N}}$ *converges to* a. The first notation is read, "a equals the limit, n in $\mathbf{N}$, of x_n"; the second, "a equals the limit of x_n, as n tends to plus infinity." Instead of saying that $(x_n)_{n \in \mathbf{N}}$ converges to $a \in X$, we shall often say that $(x_n)_{n \in \mathbf{N}}$ tends to a. Whenever we say that $(x_n)_{n \in \mathbf{N}}$ is *convergent in* X, we mean that the considered sequence converges to some $a \in X$ in the sense of Definition 14.1.

Since A depends on V, we shall often use some notation to indicate this. For instance, instead of writing only A, we shall use a symbol such as $A(V)$, A_V, $n(V)$, n_V. Usually we shall not indicate by an additional symbol the dependence of A on the sequence.

14.3 Theorem.—*Suppose that* $(x_n)_{n \in \mathbf{N}}$ *converges to* a' *and to* a''. *Then* $a' = a''$.

* Definition 14.1 can be formulated for non-separated spaces. Also, some of the results that follow are valid in non-separated spaces.

† A sequence of elements of X is a mapping having for domain a part of $\mathbf{Z}$, and for range, X. If $f: A \to X$ is a sequence, we often write $f = (f(i))_{i \in A}$. In this book, we consider *mostly* sequences having for domain, $\mathbf{N}$.

Proof.—If $a' \neq a''$, then (since $(X, \mathcal{T})$ is separated) there are $V' \in \mathcal{N}(a')$ and $V'' \in \mathcal{N}(a'')$ such that

$$V' \cap V'' = \varnothing.$$

Since $(x_n)_{n \in \mathbf{N}}$ converges to a', there exists $n_{V'} \in \mathbf{N}$ such that $x_n \in V'$ whenever $n \geq n_{V'}$. Since $(x_n)_{n \in \mathbf{N}}$ converges to a'', there exists $n_{V''} \in \mathbf{N}$ such that $x_n \in V''$ whenever $n \geq n_{V''}$. Let $p = \sup \{n_{V'}, n_{V''}\}$. Then $p \geq n_{V'}$ and $p \geq n_{V''}$; hence

$$x_p \in V' \quad \textit{and} \quad x_p \in V''.$$

This implies that $V' \cap V'' \neq \varnothing$, and hence leads to a contradiction. Hence $a' = a''$.

14.4 Theorem.—*Let $(x_n)_{n \in \mathbf{N}}$ be a sequence of elements of X that converges to $a \in X$. If $E \subset X$, and if there is some $p \in \mathbf{N}$ such that $x_n \in E$ for $n \geq p$, then $a \in \bar{E}$.*

Proof.—Let $V \in \mathcal{N}_X(a)$. Then there is some $q \in \mathbf{N}$ such that $x_n \in V$ if $n \geq q$. Hence, $x_n \in E$ if $n \geq \sup \{q, p\}$, so that $V \cap E \neq \varnothing$. Since $V \in \mathcal{N}_X(a)$ was arbitrary, we deduce that $a \in \bar{E}$.

Example 1.—Let $a \in X$ and let $x_n = a$ for all $n \in \mathbf{N}$. Then $(x_n)_{n \in \mathbf{N}}$ converges to a.

Let $V \in \mathcal{N}(a)$. Then $x_n \in V$ for all $n \in \mathbf{N}$. Hence, the condition of Definition 14.1 is satisfied, it does not matter how we choose A (for instance, in this case we may take $A = 1$). Hence $\lim_{n \in \mathbf{N}} x_n = a$.

Example 2.—In the topological space $\mathbf{R}$, the sequence $(1/n)_{n \in \mathbf{N}}$ converges to zero.

In fact, let $V \in \mathcal{N}(0)$. Then there is $\varepsilon > 0$ such that $[-\varepsilon, +\varepsilon] \subset V$. By Archimedes' property, there is $p \in \mathbf{N}$ such that $p \geq 1/\varepsilon$; that is, $1/p \leq \varepsilon$. Hence, if $n \geq p$, then $1/n \leq 1/p \leq \varepsilon$. Since $0 \leq 1/n$, we deduce that

$$1/n \in [-\varepsilon, +\varepsilon] \quad \text{whenever} \quad n \geq p.$$

Example 3.—Let X be a separated topological space, a and b two distinct elements of X, and $(x_n)_{n \in \mathbf{N}}$ a sequence of elements of X such that $x_{2n} = a$ and $x_{2n+1} = b$ for all $n \in \mathbf{N}$. Then $(x_n)_{n \in \mathbf{N}}$ is not convergent.

In fact, suppose that $(x_n)_{n \in \mathbf{N}}$ converges to some $c \in X$. Clearly, either $c \neq a$ or $c \neq b$. We shall discuss here the case in which $c \neq a$ (the case in which $c \neq b$ can be treated in the same way). Let $V \in \mathcal{N}(c)$ be such that $V \not\ni a$. Since $(x_n)_{n \in \mathbf{N}}$ converges to c, there is $p \in \mathbf{N}$ such that $x_n \in V$ if $n \geq p$. However, $2p \geq p$, and $x_{2p} = a \notin V$. Hence, we arrive at a contradiction if we suppose that $(x_n)_{n\in\mathbf{N}}$ is convergent. Hence, $(x_n)_{n\in\mathbf{N}}$ is not convergent in X.

Example 4.—The sequence $((-1)^n)_{n\in\mathbf{N}}$ is not convergent in **R**.

Hint: Use the result in Example 3.

Example 5.—Let $(x'_n)_{n\in\mathbf{N}}$ and $(x''_n)_{n\in\mathbf{N}}$ be two sequences of elements of a separated topological space X. Suppose that there is $q \in \mathbf{N}$ such that

$$x'_n = x''_n \quad \text{for all} \quad n \geq q.$$

Then

$$\lim_{n\in\mathbf{N}} x'_n = a \quad \text{if and only if} \quad \lim_{n\in\mathbf{N}} x''_n = a.$$

We leave the proof of this assertion to the reader.

Example 6.—Let $\varphi:\mathbf{N} \to \mathbf{N}$ be a strictly increasing mapping, that is,

$$\varphi(n) < \varphi(m)$$

whenever n and m belong to **N** and $n < m$. Let $(x_n)_{n\in\mathbf{N}}$ be a sequence of elements of a separated topological space X and let $y_n = x_{\varphi(n)}$ for all $n \in \mathbf{N}$. Then

$$\lim_{n\in\mathbf{N}} x_n = a \Rightarrow \lim_{n\in\mathbf{N}} y_n = a.$$

Hint: Show first that $\varphi(n) \geq n$ for all $n \in \mathbf{N}$. For this, note that $\varphi(n) \geq 1$ and use induction.

If $(x_n)_{n\in\mathbf{N}}$ is a sequence, then any sequence of the form $(x_{\varphi(n)})_{n\in\mathbf{N}}$ with $\varphi:\mathbf{N} \to \mathbf{N}$ a strictly increasing mapping is called a *subsequence* of $(x_n)_{n\in\mathbf{N}}$. Whenever we say that $(x_{f(n)})_{n\in\mathbf{N}}$ is a subsequence of $(x_n)_{n\in\mathbf{N}}$, we mean, implicitly or explicitly, that f is a strictly increasing mapping of **N** into **N**.

With this terminology, the above result says that *if a sequence converges to a, then any subsequence of the considered sequence converges to a.*

Let $(X, \mathscr{T})$ be a separated topological space, $a \in X$, and $\mathscr{F}$ a fundamental system of $a \in X$. Then we get the following:

14.5 Theorem.—*Let $(x_n)_{n\in\mathbf{N}}$ be a sequence of elements of X. Then the following assertions are equivalent:*

(a) *the sequence $(x_n)_{n\in\mathbf{N}}$ converges to a;*

(b) *for every $V \in \mathscr{F}$, there exists some $A \in \mathbf{N}$ such that $x_n \in V$ whenever $n \geq A$.*

Proof.—The proof is obvious and is left to the reader.

Theorem 14.5 shows that, to prove that a sequence $(x_n)_{n\in\mathbf{N}}$ of elements of a separated topological space X converges to $a \in X$, it is enough to verify the condition of Definition 14.1 for all V in some fundamental system of a.

14.6 Corollary.—*Suppose $(X, \mathscr{T})$ to be metrizable and let d be a metric on X such that $\mathscr{T} = \mathscr{T}_d$. Let $a \in X$ and let $(x_n)_{n\in\mathbf{N}}$ be a sequence of elements of X. The following assertions are then equivalent:*

(a) *the sequence $(x_n)_{n\in\mathbf{N}}$ converges to a;*
(b) *for every $\varepsilon > 0$, there is $n_\varepsilon \in \mathbf{N}$ such that $d(a, x_n) < \varepsilon$ whenever $n \geq n_\varepsilon$.*

Proof.—Note that $d(a,y) < \varepsilon \Leftrightarrow y \in V_\varepsilon(a)$. Hence, (b) is the same as

(b′) for every $\varepsilon > 0$, there is $n_\varepsilon \in \mathbf{N}$ such that $x_n \in V_\varepsilon(a)$ whenever $n \geq n_\varepsilon$.

Since $\mathscr{F} = \{V_\varepsilon(a) \mid \varepsilon > 0\}$ is a fundamental system of a, we deduce (using 14.5) that (a) and (b) are equivalent.

14.7 Corollary.—*A sequence $(x_n)_{n\in\mathbf{N}}$ of real numbers converges to $a \in R$ if and only if for every $\varepsilon > 0$ there is $n_\varepsilon \in \mathbf{N}$ such that*

$$|x_n - a| < \varepsilon \quad \textit{whenever} \quad n \geq n_\varepsilon.$$

The following result is useful in applications.

14.8 Corollary.—*Let $(X, \mathscr{T})$ be metrizable and let d be a metric on X such that $\mathscr{T} = \mathscr{T}_d$. Let $a \in X$ and let $(x_n)_{n\in\mathbf{N}}$ be a sequence of elements of X. Suppose that there is a sequence $(a_n)_{n\in\mathbf{N}}$ of elements of $\mathbf{R}_+$ such that:*

(a) $\lim_{n\in\mathbf{N}} a_n = 0$;
(b) *$d(a, x_n) < a_n$ for all $n \in \mathbf{N}$.*

Then

$$\lim_{n\in\mathbf{N}} x_n = a.$$

Proof.—Let $\varepsilon > 0$. Since $(a_n)_{n\in\mathbf{N}}$ converges to zero, there is $n_\varepsilon \in N$ such that

$$a_n = |a_n - 0| < \varepsilon \quad \text{whenever} \quad n \geq n_\varepsilon.$$

Hence

$$d(a, x_n) < \varepsilon \quad \text{whenever} \quad n \geq n_\varepsilon.$$

Since $\varepsilon > 0$ was arbitrary, we deduce that

$$\lim_{n \in \mathbf{N}} x_n = a.$$

Example 7.—For each $a \in \mathbf{R}$, there is a sequence $(r_n)_{n \in \mathbf{N}}$ of rational numbers that converges to a.

In fact, let $n \in \mathbf{N}$; there is then a rational number $r_n \in (a - 1/n, a + 1/n)$. Hence, for each $n \in \mathbf{N}$, there is $r_n \in \mathbf{Q}$ such that

$$|a - r_n| < 1/n.$$

By 14.8, the sequence $(r_n)_{n \in \mathbf{N}}$ converges to a.

14.9 Theorem.—*Let* $(X, \mathscr{T})$ *be a metrizable space and let* $A \subset X$ *and* $a \in X$. *Then the following are equivalent:*

(a) $a \in \bar{A}$;
(b) *there is a sequence of elements of* A *that converges to* a.

Proof of (a) $\Rightarrow$ (b).—Let d be a metric on S such that $\mathscr{T} = \mathscr{T}_d$. Since $a \in \bar{A}$, we have

$$V_{1/n}(a) \cap A \neq \varnothing$$

for all $n \in \mathbf{N}$. Let $x_n \in V_{1/n}(a) \cap A$ for $n \in \mathbf{N}$. Then $(x_n)_{n \in \mathbf{N}}$ is a sequence of elements of A, and (since $x_n \in V_{1/n}(a)$)

$$d(x_n, a) < 1/n$$

for all $n \in \mathbf{N}$. By Corollary 14.8, $(x_n)_{n \in \mathbf{N}}$ converges to a.

Proof of (b) $\Rightarrow$ (a).—Let $V \in \mathscr{N}(a)$. Since $(x_n)_{n \in \mathbf{N}}$ converges to a, there is $p \in \mathbf{N}$ such that $x_n \in V$ whenever $n \geq p$. In particular, $x_p \in V$. But $x_p \in A$ (since $(x_n)_{n \in \mathbf{N}}$ is a sequence of elements of A). Hence, $V \cap A \neq \varnothing$. Since $V \in \mathscr{N}(a)$ was arbitrary, we deduce that $a \in \bar{A}$.

While it is obvious that the implication (b) $\Rightarrow$ (a) holds in any separated topological space, it can be shown that (a) $\Rightarrow$ (b) does not necessarily hold in non-metrizable spaces. For instance, (a) $\Rightarrow$ (b) does not hold in the space discussed in the Appendix II to Chapter 13.

In the following theorem, we denote by $(X, \mathscr{T})$ a metrizable topological space and by $(Y, \mathscr{S})$ a separated topological space.

14.10 Theorem.—*Let* $A \subset X$, $a \in A$, *and let* $f: A \to Y$ *be a continuous mapping. Then the following assertions are equivalent:*

(a) *the function* f *is continuous at* a;
(b) *for every sequence* $(x_n)_{n \in \mathbf{N}}$ *of elements of* A *that converges to* a, *the sequence* $(f(x_n))_{n \in \mathbf{N}}$ *converges to* $f(a)$.

Proof of (a) $\Rightarrow$ (b).—Let $(x_n)_{n\in\mathbf{N}}$ be a sequence of elements of A that converges to a. Let $V \in \mathcal{N}(f(a))$. Since f is continuous at a, there exists $U \in \mathcal{N}(a)$ such that $x \in U \cap A \Rightarrow f(x) \in V$. Since $(x_n)_{n\in\mathbf{N}}$ converges to a, there is $n_U \in \mathbf{N}$ such that $x_n \in U$ whenever $n \geq n_U$. Hence,

$$n \geq n_U \Rightarrow x_n \in U \Rightarrow f(x_n) \in V.$$

Since $V \in \mathcal{N}(f(a))$ was arbitrary, we deduce that $(f(x_n))_{n\in\mathbf{N}}$ converges to $f(a)$.

Proof of (b) $\Rightarrow$ (a).—We assume that (b) holds. Let d be a metric on X such that $\mathcal{T} = \mathcal{T}_d$. *Suppose that f is not continuous at a.* From Definition 5.1, it follows that there is $V \in \mathcal{N}(f(a))$ such that for *no* $U \in \mathcal{N}(a)$ do we have $f(U \cap A) \subset V$. Hence, for every $U \in \mathcal{N}(a)$, there is $x_U \in U \cap A$ such that $f(x_U) \notin V$. In particular, for every $n \in \mathbf{N}$ there is $x_n \in V_{1/n}(a) \cap A$ such that $f(x_n) \notin V$. This shows that $(f(x_n))_{n\in\mathbf{N}}$ does not converge to $f(a)$. On the other hand, $d(x_n, a) < 1/n$ for all $n \in \mathbf{N}$, whence $(x_n)_{n\in\mathbf{N}}$ converges to a. Since (b) is supposed to hold, this is a contradiction. Hence f is continuous, and hence (b) $\Rightarrow$ (a) holds.

While it is obvious that (a) $\Rightarrow$ (b) holds without supposing that $(X, \mathcal{T})$ is metrizable, there are examples that show that (b) $\Rightarrow$ (a) does not necessarily hold if $(X, \mathcal{T})$ is arbitrary.

▼ Let $((X_i, \mathcal{T}_i))_{i\in I}$ be a family of separated topological spaces, $(X, \mathcal{T})$ the topological space product of this family, and $(Y, \mathcal{S})$ a separated topological space. Let $\varphi: X \to Y$ be a continuous mapping. For each $i \in I$, let $(x_n(i))_{n\in\mathbf{N}}$ be a sequence of elements of X_i that converges to a_i. For each $n \in \mathbf{N}$, let

$$x_n = (x_n(i))_{i\in I}.$$

The sequence $(x_n)_{n\in\mathbf{N}}$ converges to $a = (a_i)_{i\in I}$. The sequence $(\varphi(x_n))_{n\in\mathbf{N}}$ converges to $\varphi(a)$. ▲

Exercises for Chapter 14

1. Prove the result in Example 6.

2. If $(x_n)_{n\in\mathbf{N}}$ is a convergent sequence of elements of a metric space (X, d), then $\{x_n \mid n \in \mathbf{N}\}$ is bounded.

3. A sequence of real numbers $(x_n)_{n\in\mathbf{N}}$ is said to be *eventually increasing* if there is n_0 in $\mathbf{N}$ such that $n \geq m \geq n_0 \Rightarrow x_n \geq x_m$. Show that a bounded eventually increasing sequence of real numbers is convergent.

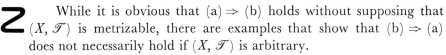

4. Show that $(1/n + 1)_{n \in \mathbf{N}}$ is convergent in **R**.

5. Show that $(\sum_{k=1}^{n} 1/k)_{n \in \mathbf{N}}$ is not convergent.

* 6. Prove the statements made at the end of this chapter.

7. Let (X, d) be a metric space, $(x_n)_{n \in \mathbf{N}}$ a sequence of elements of X, and $a \in X$. Then $(x_n)_{n \in \mathbf{N}}$ converges to a if and only if $(d(a, x_n))_{n \in \mathbf{N}}$ converges to zero.

8. Let (X, d) be a metric space, $(x_n)_{n \in \mathbf{N}}$ a sequence of elements of X, and suppose $a \in X$. Then $(x_n)_{n \in \mathbf{N}}$ does not converge to a if and only if there is $\varepsilon > 0$ and a subsequence $(x_{\varphi(n)})_{n \in \mathbf{N}}$ such that $d(x_{\varphi(n)}, a) > \varepsilon$ for all n in **N**.

9. Let $(X, \mathscr{T})$ be a separated topological space and $(x_n)_{n \in \mathbf{N}}$ a sequence of elements of X. Suppose $x \in X$ and

$$\lim_{n \in \mathbf{N}} x_n = x.$$

Show that $A = \{x_n \mid n \in \mathbf{N}\} \cup \{x\}$ is compact.

10. Let $(x_n)_{n \in \mathbf{N}}$ and $(y_n)_{n \in \mathbf{N}}$ be sequences of real numbers such that $\lim_{n \in \mathbf{N}} x_n = x$ and $\lim_{n \in \mathbf{N}} y_n = y$. Show that:

(i) $\lim_{n \in \mathbf{N}} (x_n + y_n) = x + y$;

(ii) $\lim_{n \in \mathbf{N}} (x_n \cdot y_n) = x \cdot y$;

(iii) $\lim_{n \in \mathbf{N}} (x_n / y_n) = x/y$ if $y \neq 0$ and $y_n \neq 0$ for every $n \in \mathbf{N}$.

Chapter 15

Complete Metric Spaces

Let (X, d) be a metric space. Then:

15.1 Definition.—*A sequence* $(x_n)_{n \in \mathbf{N}}$ *of elements of* X *is said to be a Cauchy sequence in* (X, d) *if for every* $\varepsilon > 0$ *there is some* $A \in \mathbf{N}$ *such that*

$$d(x_n, x_m) < \varepsilon \quad \textit{whenever} \quad n \geq A \quad \textit{and} \quad m \geq A.$$

Since A depends on ε, we shall often use some notation to indicate this. For instance, instead of writing only A, we shall use a symbol such as $A(\varepsilon)$, A_ε, $n(\varepsilon)$, or n_ε.

An immediate result is the following theorem.

15.2 Theorem.—*Let* (X, d) *be a metric space and* $(x_n)_{n \in \mathbf{N}}$ *a convergent sequence* of elements of* X. *Then* $(x_n)_{n \in \mathbf{N}}$ *is a Cauchy sequence.*

Proof.—Let $\varepsilon > 0$ and let a be the limit of $(x_n)_{n \in \mathbf{N}}$. By Corollary 14.6, there is $A \in \mathbf{N}$ such that

$$d(x_p, a) < \varepsilon/2 \quad \text{whenever} \quad p \geq A.$$

Hence, if $n \geq A$ and $m \geq A$, we have

$$d(x_n, x_m) \leq d(x_n, a) + d(a, x_m) < \varepsilon/2 + \varepsilon/2 = \varepsilon.$$

Therefore,

$$d(x_n, x_m) < \varepsilon \quad \text{whenever} \quad n \geq A \quad \text{and} \quad m \geq A,$$

whence $(x_n)_{n \in \mathbf{N}}$ is a Cauchy sequence.

---

* We mean, of course, convergent for the topology $\mathcal{T}_d$ (that is, in the topological space $(X, \mathcal{T}_d)$).

As we shall see in Example 1, a Cauchy sequence in an arbitrary metric space is not necessarily convergent.

Example 1.—Let $(r_n)_{n \in \mathbf{N}}$ be a sequence of rational numbers. Then (see Examples 1 and 2, Chapter 13):

(1) *$(r_n)_{n \in \mathbf{N}}$ is a Cauchy sequence in $(\mathbf{Q}, \eta)$ if and only if $(r_n)_{n \in \mathbf{N}}$ is a Cauchy sequence in $(\mathbf{R}, d)$;*

(2) *$(r_n)_{n \in \mathbf{N}}$ converges to $x \in \mathbf{Q}$ in the space $(\mathbf{Q}, \eta)$ if and only if $(r_n)_{n \in \mathbf{N}}$ converges to x in the space $(\mathbf{R}, d)$.*

Since $\mathbf{Q} \neq \mathbf{R}$, there is $x \in \mathbf{R} - \mathbf{Q}$. Let $(r_n)_{n \in \mathbf{N}}$ be a sequence of rational numbers that converges to x. By Theorem 15.2, $(r_n)_{n \in \mathbf{N}}$ is a Cauchy sequence in $(\mathbf{R}, d)$. By remark (1), $(r_n)_{n \in \mathbf{N}}$ is a Cauchy sequence in $(\mathbf{Q}, \eta)$. If $(r_n)_{n \in \mathbf{N}}$ converges in $(\mathbf{Q}, \eta)$ to some $y \in \mathbf{Q}$, then $(r_n)_{n \in \mathbf{N}}$ converges in $(\mathbf{R}, d)$ to y, and hence $y = x$. But this is not possible, since x is *not* rational.

Hence, *there are Cauchy sequences in the metric space $(\mathbf{Q}, \eta)$ that are not convergent.*

Now let (X, d) be a metric space. We shall introduce the following:

15.3 Definition.—*We say that (X, d) is complete if every Cauchy sequence of elements of X converges to some element (belonging to X).*

When there is no ambiguity, instead of saying that (X, d) is complete, we say that X is complete.

Example 2.—The metric space $(\mathbf{Q}, \eta)$ is not complete. This follows from the discussion in Example 1.

Let $A \subset X$ and let $d_A = d \mid A \times A$. Then (A, d_A) is a metric space. We say that *A is complete if (A, d_A) is complete.* It is obvious that A is complete if and only if every Cauchy sequence consisting of elements of A converges to some element *belonging to A.*

Example 3.—Let (X, d) be a metric space such that every bounded closed part is complete. Then (X, d) is complete.

Let $(x_n)_{n \in \mathbf{N}}$ be a Cauchy sequence of element of X. Let $p \in \mathbf{N}$ be such that

$$d(x_n, x_m) \leq 1$$

if $n \geq p$ and $m \geq p$. Let $a \in X$ and let

$$r = \sup \{d(a, x_1), \ldots, d(a_1, x_p)\}.$$

Then $x_j \in W_r(a) \subset W_{r+1}(a)$ if $1 \leq j \leq p$. Furthermore, if $j \geq p$,

$$d(a, x_j) \leq d(a, x_p) + d(x_p, x_j) \leq r + 1.$$

Hence $x_j \in W_{r+1}(a)$ for $j \geq p$, and hence

$$\{x_n \mid n \in \mathbf{N}\} \subset W_{r+1}(a).$$

Since $W_{r+1}(a)$ is closed and obviously bounded, it is *complete*. We deduce that $(x_n)_{n \in \mathbf{N}}$ converges to some $x \in W_{r+1}(a) \subset X$. Since $(x_n)_{n \in \mathbf{N}}$ was arbitrary, we deduce that (X, d) is complete.

Example 4.—The metric space $(\mathbf{R}^n, d_n)$ $(n \geq 1)$ is complete.

This will be established in Example 2, Chapter 16.

15.4 Theorem.—*Let (X, a) be a complete metric space and let $A \subset X$. Then the following assertions are equivalent:*

(a) $A = \bar{A}$;
(b) A *is complete.*

Proof of (a) $\Rightarrow$ (b).—Let $(x_n)_{n \in \mathbf{N}}$ be a Cauchy sequence of elements belonging to A. Then, $(x_n)_{n \in \mathbf{N}}$ is a Cauchy sequence of elements of X. Since (X, d) is complete, $(x_n)_{n \in \mathbf{N}}$ converges to some $x \in X$. By Theorem 14.4, $x \in \bar{A} = A$. Hence, A is complete.

Proof of (b) $\Rightarrow$ (a).—Suppose A is complete and let $x \in \bar{A}$. By Theorem 14.9, there is a sequence of elements of A that converges to x. By Theorem 15.2, $(x_n)_{n \in \mathbf{N}}$ is a Cauchy sequence. Since A is complete, $(x_n)_{n \in \mathbf{N}}$ converges to some $x' \in A$. By Theorem 14.3, $x = x'$; hence $x \in A$. Since $x \in \bar{A}$ was arbitrary, we deduce that $A = \bar{A}$.

Example 5.—(i) If $A \subset \mathbf{R}^n$ is a bounded n-dimensional closed interval (see 7.14), then A is complete. (ii) If A is a closed ball in a metric complete space, then A is complete.

It is enough to notice that a closed bounded n-dimensional interval is closed and that a closed ball is closed.

Example 6.—On the set $\mathbf{R}$, consider the metrics d and δ defined by

$$d(x, y) = |x - y| \quad \text{for} \quad (x, y) \in \mathbf{R} \times \mathbf{R}$$

and

$$\delta(x, y) = \left| \frac{x}{1 + |x|} - \frac{y}{1 + |y|} \right| \quad \text{for} \quad (x, y) \in \mathbf{R} \times \mathbf{R}$$

(see Examples 1 and 5, Chapter 13). In this manner, we obtain two metric spaces. We shall show that:

(i) $\mathscr{T}_\delta = \mathscr{T}_d$;
(ii) $(\mathbf{R}, \delta)$ *is not complete.*

Since we have indicated in Example 4 that $(\mathbf{R}, d)$ is complete, this shows that the fact that a metric space is or is not complete depends essentially on its metric and not on its topology.

Denote by f the mapping $x \mapsto x/(1 + |x|)$ of $\mathbf{R}$ into $\mathbf{R}$; note that f is a homeomorphism of $\mathbf{R}$ onto $f(\mathbf{R})$. Hence the mapping $x \mapsto f^{-1}(x)$ of $f(\mathbf{R})$ into $\mathbf{R}$ is continuous on $f(\mathbf{R})$.

Now let $x_0 \in \mathbf{R}$. Then (since f is continuous on $\mathbf{R}$), given $\beta > 0$, there exists $\alpha > 0$ such that

$$|x - x_0| < \alpha \Rightarrow |f(x) - f(x_0)| < \beta;$$

that is,

15.5 $$d(x, x_0) < \alpha \Rightarrow \delta(x, x_0) < \beta.$$

Furthermore, (since $x \mapsto f^{-1}(x)$ is continuous on $f(\boldsymbol{R})$), given $\gamma > 0$, there exists $\eta > 0$ such that

$$|f(x) - f(x_0)| < \eta \Rightarrow |x - x_0| < \gamma;$$

that is,

15.6 $$\delta(x, x_0) < \eta \Rightarrow d(x, x_0) < \gamma.$$

Now let $V \in \mathscr{N}_{\mathscr{T}_\delta}(x_0)$. Then there is $\beta > 0$ such that

$$\{x \mid \delta(x, x_0) < \beta\} \subset V.$$

By 15.5, there is $\alpha > 0$ such that

$$\{x \mid d(x, x_0) < \alpha\} \subset \{x \mid \delta(x, x_0) < \beta\},$$

and hence

$$\{x \mid d(x, x_0) < \alpha\} \subset V.$$

Hence $V \in \mathscr{N}_{\mathscr{T}_d}(x_0)$. We deduce that

15.7 $$\mathscr{N}_{\mathscr{T}_\delta}(x_0) \subset \mathscr{N}_{\mathscr{T}_d}(x_0).$$

In the same way (use 15.6), we deduce that

15.8 $$\mathscr{N}_{\mathscr{T}_d}(x_0) \subset \mathscr{N}_{\mathscr{T}_\delta}(x_0).$$

Combining 15.7 and 15.8, we deduce that $\mathcal{N}_{\mathcal{T}_d}(x_0) = \mathcal{N}_{\mathcal{T}_\delta}(x_0)$. Since $x_0 \in \mathbf{R}$ was arbitrary, we obtain $\mathcal{T}_d = \mathcal{T}_\delta$ (see Theorem 2.5), and hence (i) is proved.

Now consider the sequence $(n)_{n\in\mathbf{N}}$. Then

$$d(n, m) = |n - m| \geq 1 \quad \text{if} \quad n \neq m.$$

Hence $(n)_{n\in\mathbf{N}}$ is not a Cauchy sequence in $(\mathbf{R}, d)$, and hence $(n)_{n\in\mathbf{N}}$ is not convergent for the topology $\mathcal{T}_d$.

Since $(1/p)_{n\in\mathbf{N}}$ converges to zero (see Example 2, Chapter 14), and since the mapping $x \mapsto 1/(1 + |x|)$ is continuous on $\mathbf{R}$, we deduce that

$$\lim_{n\in\mathbf{N}} f(n) = \lim_{n\in\mathbf{N}} \frac{n}{1+n} = \lim_{n\in\mathbf{N}} \frac{1}{1 + 1/n} = 1.$$

Since

$$\delta(n, m) = \left| \frac{n}{1+n} - \frac{m}{1+m} \right| = \left| \frac{1}{1+1/n} - \frac{1}{1+1/m} \right|$$

for n and m in $\mathbf{N}$, we conclude that $(n)_{n\in\mathbf{N}}$ is a Cauchy sequence in $(\mathbf{R}, \delta)$. If $(\mathbf{R}, \delta)$ were complete, then $(n)_{n\in\mathbf{N}}$ would converge to some $x_0 \in \mathbf{R}$ for the topology $\mathcal{T}_\delta$. Since $\mathcal{T}_\delta = \mathcal{T}_d$, it would converge for $\mathcal{T}_d$. But we have seen that this is not true. Hence, $(\mathbf{R}, \delta)$ is not complete, and hence (ii) is proved.

In Chapter 17, we shall show that *any metric space can be embedded in a complete metric space.* So that we do not repeat certain definitions, we postpone the exact statement of this result until Chapter 17.

Exercises for Chapter 15

1. Consider the sets $[0, 1) \subset \mathbf{R}$, $\mathbf{N} \subset \mathbf{R}$, and $\mathbf{Z} \subset \mathbf{R}$. Are these sets complete? (Here, $\mathbf{R}$ is endowed with the usual metric d.)

2. Let $(x_n)_{n\in\mathbf{N}}$ be a Cauchy sequence of elements of a metric space (X, d). Show that $(x_n)_{n\in\mathbf{N}}$ is convergent if and only if there exists a convergent subsequence $(x_{\varphi(n)})_{n\in\mathbf{N}}$ of $(x_n)_{n\in\mathbf{N}}$.

3. Any subsequence of a Cauchy sequence is a Cauchy sequence.

4. Let (X, d) be a metric space and suppose that $(x_n)_{n\in\mathbf{N}}$ is a sequence of elements of X that converges to $a \in X$. Show that $\{x_n \mid n \in \mathbf{N}\} \cup \{a\}$ is complete.

5. Let (X_n, d_n) be a metric space for each n in $\mathbf{N}$. On $X = \prod_{n\in\mathbf{N}} X_n$, consider the metric d defined by

$$d(x, y) = \sum_{n=1}^{\infty} \frac{1}{2^n} \inf (1, d_n(x_n, y_n))$$

(see Example 8, Chapter 13). Show that (X, d) is complete if (X_n, d_n) is complete for every n in $\mathbf{N}$.

6. A metric space is complete if and only if every decreasing sequence $(F_n)_{n\in\mathbf{N}}$ of closed subsets of X such that $(\delta(F_n))_{n\in\mathbf{N}}$ tends to zero has a non-void intersection.

* 7. Let (X, d) be a complete metric space, $k \in (0, 1)$ and f a mapping of X into X such that $d(f(x), f(y)) \leq k\, d(x, y)$ whenever x, y belong to X. Then there is x in X such that $f(x) = x$.

Chapter 16

Compact Metrizable Spaces

Compact sets were defined and discussed in Chapter 9. In this chapter, we shall give some results concerning subsets of a metrizable space. The main result of this section is Theorem 16.4. For its statement and proof, we need a definition and several preliminary results.

Let $(X, \mathscr{T})$ be a separated topological space.

16.1 Definition (*The Bolzano-Weierstrass property*).—*A set* $A \subset X$ *has the Bolzano-Weierstrass property if for every sequence* $(a_n)_{n \in \mathbf{N}}$ *of elements of* A *there exists a subsequence that converges to some* $a \in A$.

Instead of "the Bolzano-Weierstrass property," we shall usually say "the B-W property."

Recall that by Example 6, Chapter 14, if a sequence $(x_n)_{n \in \mathbf{N}}$ converges to x, then every subsequence of $(x_n)_{n \in \mathbf{N}}$ converges to x.

We denote now by $(X, \mathscr{T})$ a metrizable topological space and by d a metric on X such that $\mathscr{T} = \mathscr{T}_d$.

16.2 Theorem.—*Let* $A \subset X$ *be a set having the B-W property and let* $(U_i)_{i \in I}$ *be an open covering of* A. *There is then an* $\varepsilon > 0$ *such that every ball* $V_\varepsilon(a)$, *with* $a \in A$, *is contained in at least one* U_i, *with* $i \in I$.

Proof.—Suppose that the conclusion of the theorem does not hold. Then for every $\varepsilon > 0$ there is an $a_\varepsilon \in A$ such that $V_\varepsilon(a_\varepsilon)$ is *not* contained in any U_i with $i \in I$. Hence, for every $n \in \mathbf{N}$ there is $x_n \in A$ such that $V_{1/n}(x_n)$ is *not* contained in any U_i, with $i \in I$. By the B-W property, there exists a subsequence $(x_{\varphi(n)})_{n \in \mathbf{N}}$ that converges to some $x_0 \in A$. Since $(U_i)_{i \in I}$ is a covering of A, there is $i_0 \in I$ such that $x_0 \in U_{i_0}$. Since U_{i_0} is *open*, there exists $\varepsilon > 0$ such that $V_\varepsilon(x_0) \subset U_{i_0}$. Now let $p \in \mathbf{N}$ be such that

$$1/p < \varepsilon/2 \quad \text{and} \quad d(x_{\varphi(p)}, x_0) < \varepsilon/2.$$

Then $x \in V_{1/\varphi(p)}(x_{\varphi(p)})$ implies*

$$\begin{aligned} d(x, x_0) &\leq d(x, x_{\varphi(p)}) + d(x_{\varphi(p)}, x_0) \\ &< 1/\varphi(p) + \varepsilon/2 \leq 1/p + \varepsilon/2 \\ &< \varepsilon/2 + \varepsilon/2 = 2(\varepsilon/2) = \varepsilon. \end{aligned}$$

Hence, $x \in V_{1/\varphi(p)}(x_{\varphi(p)})$ implies that $x \in V_\varepsilon(x_0)$; that is,

$$V_{1/\varphi(p)}(x_{\varphi(p)}) \subset V_\varepsilon(x_0) \subset U_{i_0}.$$

But this is a contradiction. Hence, Theorem 16.2 is proved.

16.3 Theorem.—*Let $A \subset X$ be a set having the B-W property. Then for every $\varepsilon > 0$ there exists a finite family $(a_j)_{1 \leq j \leq n}$ of elements of A such that $(V_\varepsilon(a_j))_{1 \leq j \leq n}$ is a covering of A.*

Proof.—Let $\varepsilon > 0$ and suppose that the conclusion of Theorem 16.3 does not hold for this ε. Let $a_1 \in A$; then $A \not\subset V_\varepsilon(a_1)$. Let

$$a_2 \in A - V_\varepsilon(a_1);$$

then $(V_\varepsilon(a_j))_{1 \leq j \leq 2}$ is not a covering of A. Let

$$a_3 \in A - \bigcup_{i=1}^{2} V_\varepsilon(a_i);$$

then $(V_\varepsilon(a_j))_{1 \leq j \leq 3}$ is not a covering of A. Proceeding this way, we construct a sequence $(a_j)_{j \in \mathbf{N}}$ of elements of A such that

$$a_{n+1} \in A - \bigcup_{j=1}^{n} V_\varepsilon(a_j) \quad \text{for all} \quad n \in \mathbf{N}.$$

Clearly, this implies that $a_m \notin V_\varepsilon(a_n)$; that is, $d(a_m, a_n) \geq \varepsilon$ if $m > n$; hence

$$d(a_n, a_m) \geq \varepsilon \quad \text{if} \quad n \neq m.$$

By the B-W property, there is a subsequence $(a_{\varphi(j)})_{j \in \mathbf{N}}$ of $(a_n)_{n \in \mathbf{N}}$ that is convergent. Then $(a_{\varphi(j)})_{j \in \mathbf{N}}$ is a Cauchy sequence (see Theorem 15.2). However, this is a contradiction, since

$$d(a_{\varphi(i)}, a_{\varphi(j)}) \geq \varepsilon \quad \text{whenever} \quad i \neq j.$$

Hence, Theorem 16.3 is proved.

Let $(X, \mathscr{T})$ be a metrizable topological space and let d be a metric on X such that $\mathscr{T} = \mathscr{T}_d$.

---

* Note that $\varphi(p) \geq p$.

16.4 Theorem.—*Let $A \subset X$. Then the following two assertions are equivalent:*

(a) *A is compact;*
(b) *A has the B-W property.*

Proof of (a) $\Rightarrow$ (b).—Let $(x_n)_{n \in \mathbf{N}}$ be a sequence of elements of A. For each $p \in \mathbf{N}$, let

$$A_p = \{x_n \mid n \geq p\};$$

since A is closed (see Theorem 9.7), $\bar{A}_p \subset A$. Clearly,

$$\bar{A}_1 \supset \bar{A}_2 \supset \ldots \supset \bar{A}_p \supset \ldots .$$

Since A is compact, there exists (see Theorem 9.15)

$$x_0 \in \bigcap_{p \in N} \bar{A}_p.$$

Notice that $V \cap A_p \neq \varnothing$ for all $V \in \mathcal{N}(x_0)$ and $p \in \mathbf{N}$.

Let $\varphi(1) \in \mathbf{N}$ be such that

$$x_{\varphi(1)} \in V_1(x_0) \cap A_1.$$

Let $\varphi(2) \in \mathbf{N}$ be such that

$$x_{\varphi(2)} \in V_{1/2}(x_0) \cap A_{\varphi(1)+1}$$

and $\varphi(2) \geq \varphi(1) + 1$; that is, $\varphi(2) > \varphi(1)$. Suppose that we have already defined $\varphi(1), \varphi(2), \ldots, \varphi(p)$. Let $\varphi(p+1) \in \mathbf{N}$ be such that

$$x_{\varphi(p+1)} \in V_{1/(p+1)}(x_0) \cap A_{\varphi(p)+1}$$

and $\varphi(p+1) \geq \varphi(p) + 1$; that is, $\varphi(p+1) > \varphi(p)$.

This way we define a mapping $\varphi: \mathbf{N} \to \mathbf{N}$ such that $\varphi(n) < \varphi(m)$ if $n < m$. Moreover, $x_{\varphi(n)} \in V_{1/n}(x_0)$; that is,

$$d(x_{\varphi(n)}, x_0) < 1/n \quad \text{for all} \quad n \in \mathbf{N}.$$

By 14.6, we deduce that $(x_{\varphi(n)})_{n \in \mathbf{N}}$ converges to $x_0 \in A$. Since $(x_n)_{n \in \mathbf{N}}$ was arbitrary, we deduce that A has the B-W property.

Proof of (b) $\Rightarrow$ (a).—Let $(U_i)_{i \in I}$ be an open covering of A. By Theorem 16.2, there is an $\varepsilon > 0$ such that every $V_\varepsilon(a)$, with $a \in A$, is contained in a set U_i, with $i \in I$. By Theorem 16.3, there exists a finite family $(a_j)_{1\leq j\leq n}$ of elements of A such that $(V_\varepsilon(a_j))_{1\leq j\leq n}$ is a covering of A.

For each $1 \leq j \leq n$, let $i_j \in I$ be such that $V_\varepsilon(a_j) \subset U_{i_j}$. Let $J = \{i_j \mid 1 \leq j \leq n\}$; then

$$V_\varepsilon(a_j) \subset \bigcup_{j\in J} U_j \quad \text{for all} \quad 1 \leq j \leq n,$$

and hence

$$A \subset \bigcup_{j\in J} U_j\,.$$

Since J is finite, and since the covering $(U_i)_{i\in I}$ was arbitrary, we deduce that A is compact.

▼ We note here that if $(X, \mathscr{T})$ is not metrizable, then the implications (a) ⇒ (b) and (b) ⇒ (a) do not necessarily hold. For instance, if $(X, \mathscr{T})$ is the space considered in Appendix II, Chapter 13, then the set A of all $f \in X$ such that $|f(x)| \leq 1$ for all $x \in R$ is compact, but does not have the B-W property. If B is the set of all $f \in A$ such that $\{x \mid f(x) \neq 0\}$ is finite, then B has the B-W property but is not compact. ▲

Let (X, d) be a metric space. The following result is often useful.

16.5 Theorem.—*Let $(x_n)_{n\in\mathbf{N}}$ be a sequence of elements of X. Suppose that:*

(a) *$(x_n)_{n\in\mathbf{N}}$ is a Cauchy sequence;*

(b) *there exists a subsequence $(x_{\varphi(n)})_{n\in\mathbf{N}}$ of $(x_n)_{n\in\mathbf{N}}$ that converges to some $x_0 \in X$.*

Then $(x_n)_{n\in\mathbf{N}}$ converges to x_0.

Proof.—Let $\varepsilon > 0$. Since $(x_n)_{n\in\mathbf{N}}$ is a Cauchy sequence, there is $A_1 \in \mathbf{N}$ such that

$$d(x_n, x_m) < \varepsilon/2 \quad \text{whenever} \quad n \geq A_1,\ m \geq A_1.$$

Since $(x_{\varphi(x)})_{n\in\mathbf{N}}$ converges to x_0, there is $A_2 \in \mathbf{N}$ such that

$$d(x_{\varphi(n)}, x_0) < \varepsilon/2 \quad \text{whenever} \quad n \geq A_2.$$

Let $A = \sup\{A_1, A_2\}$. Then, if $n \geq A$,

$$\begin{aligned} d(x_n, x_0) &\leq d(x_n, x_{\varphi(A)}) + d(x_{\varphi(A)}, x_0) \\ &< 2(\varepsilon/2) = \varepsilon. \end{aligned}$$

Since $\varepsilon > 0$ was arbitrary, it follows that $(x_n)_{n\in\mathbf{N}}$ converges to x_0.

16.6 Theorem.—*If $A \subset X$ is compact, then A is complete.*

Proof.—Let $(x_n)_{n \in \mathbf{N}}$ be a Cauchy sequence of elements of A. Since A is compact, A has the B-W property; whence there is a subsequence $(x_{\varphi(n)})_{n \in \mathbf{N}}$ of $(x_n)_{n \in \mathbf{N}}$ that converges to some $x_0 \in A$. By 16.5, $(x_n)_{n \in \mathbf{N}}$ converges to x_0. Hence A is complete.

Example 1.—Let (X, d) be a metric space such that every bounded closed part is compact. Then (X, d) is complete.

It follows from Theorem 16.6 that every bounded closed part is complete. The completeness of (X, d) follows then from the result in Example 3, Chapter 15.

Recall that a set $A \subset \mathbf{R}^n$ $(n \geq 1)$ is bounded if and only if A is contained in an n-dimensional interval (open or closed).

Example 2.—The space $(\mathbf{R}^n, d_n)$ $(n \geq 1)$ is complete.

By the result in Example 7, Chapter 9, a bounded n-dimensional closed interval is compact. The completeness of $(\mathbf{R}^n, d_n)$ follows from the above remark and from the result in Example 1.

Let (X, d) be a metric space and let $A \subset X$. We shall show before proceeding further that *the following assertions are equivalent:*

16.7 *for every $\varepsilon > 0$, there exists a finite family $(a_j)_{1 \leq j \leq n}$ of elements of A such that $(V_\varepsilon(a_j))_{1 \leq j \leq n}$ is a covering of A;*

16.8 *for every $\varepsilon > 0$, there exists a finite family $(x_j)_{1 \leq j \leq m}$ of elements of X such that $(V_\varepsilon(x_j))_{1 \leq j \leq m}$ is a covering of A;*

16.9 *for every $\varepsilon > 0$, there exists a finite covering of A consisting of sets having diameter $< \varepsilon$.*

Proof of 16.7 $\Rightarrow$ 16.8.—Obvious.

Proof of 16.8 $\Rightarrow$ 16.9.—Obvious.

Proof of 16.9 $\Rightarrow$ 16.7.—Let $\varepsilon > 0$ and let $(U_j)_{1 \leq j \leq p}$ be a finite covering of A such that $\delta(U_j) < \varepsilon$ for every $1 \leq j \leq p$. We may and shall suppose that

$$U_j \cap A \neq \varnothing$$

for every $1 \leq j \leq p$. Let $a_j \in U_j \cap A$. Then $(V_\varepsilon(a_j))_{1 \leq j \leq p}$ is a finite covering of A. In fact, if $x \in A$, then $x \in U_j$ for some $1 \leq j \leq p$;

hence

$$d(x, a_j) \leq \delta(U_j) < \varepsilon,$$

so that $x \in V_\varepsilon(a_j)$.

16.10 Definition.—*Let (X, d) be a metric space and let $A \subset X$. We say that A is totally bounded if A satisfies Condition* 16.7, 16.8, *or* 16.9.

We leave it to the reader to show that *a set $A \subset X$ is totally bounded if and only if $\bar{A}$ is totally bounded.*

We also notice that *if A is totally bounded, then there is a countable set $D \subset A$ dense in A.*

16.11 Theorem.—*Let (X, d) be a metric space and let $A \subset X$. Then the following assertions are equivalent:*

(a) *A is compact;*
(b) *A is complete and totally bounded.*

Proof of (a) $\Rightarrow$ (b).—This follows from 16.6, 16.4, and 16.3.

Proof of (b) $\Rightarrow$ (a).—Let $(U_i)_{i \in I}$ be an open covering of A. Suppose that there is no finite set $F \subset I$ such that $(U_i)_{i \in F}$ is a covering of A. By induction, we show* then that there is a sequence $(C_n)_{n \in \mathbf{N}}$ of non-void parts of A, none of them contained in the union of a finite family of the form $(U_i)_{i \in H}$ with $H \subset I$, and such that

$$C_1 \supset C_2 \supset \cdots \supset C_n \supset \cdots$$

and

$$\lim_{n \in \mathbf{N}} \delta(C_n) = 0.$$

If for each $n \in \mathbf{N}$ we chose $x_n \in C_n$, then $(x_n)_{n \in \mathbf{N}}$ is a Cauchy sequence consisting of elements of A. Since A is complete, $(x_n)_{n \in \mathbf{N}}$ converges to some $a \in A$. Since $(U_i)_{i \in I}$ is an *open* covering of A, there is $\alpha \in I$ and $r > 0$ such that $V_r(a) \subset U_\alpha$. Since $\lim_{n \in \mathbf{N}} \delta(C_n) = 0$, we deduce that $C_n \subset V_r(a) \subset U_\alpha$ for n large enough. But this is a contradiction. Hence, there is a finite set $F \subset I$ such that $(U_i)_{i \in F}$ is a covering of A. Since $(U_i)_{i \in I}$ was arbitrary, we deduce that A is compact.

Remarks.—Let (X, d) be a metric space. Then:

(i) if B is totally bounded and $A \subset B$, A is totally bounded;
(ii) if $A_1, \ldots, A_n$ are totally bounded, $A_1 \cup \ldots \cup A_n$ is totally bounded;
(iii) if A is relatively compact, A is totally bounded;
(iv) if A is totally bounded, A is bounded;
(v) if (X, d) is complete, then A is relatively compact if and only if A is totally bounded.

* The details are left to the reader.

Exercises for Chapter 16

1. Let $((x_n, y_n))_{n\in\mathbf{N}}$ be a sequence of elements of $\mathbf{R}^2$ such that $|x_n| \leq L$, $|y_n| \leq L$ for some L and all $n \in \mathbf{N}$. Show that there exists a subsequence $(x_{\varphi(n)}, y_{\varphi(n)})_{n\in\mathbf{N}}$ that converges to some $(x_0, y_0) \in \mathbf{R}^2$.

2. Give several examples showing that a bounded subset of a metric space need not be totally bounded.

* 3. Let (X, d) be a compact metric space. Suppose that for every pair of points x, y in X and every $\varepsilon > 0$ there is a finite sequence $z_0, \ldots, z_n$ such that $z_0 = x$, $z_n = y$, and $d(z_{i-1}, z_i) < \varepsilon$ whenever $1 \leq i \leq n$. Then $(X, \mathscr{T}_d)$ is connected.

* 4. Let $(X, \mathscr{T})$ be a metrizable topological space. Then $(X, \mathscr{T})$ is compact if and only if for every metric d on X such that $\mathscr{T}_d = \mathscr{T}$, (X, d) is complete.

Chapter 17

Uniformly Continuous Functions

In this chapter, we shall discuss uniformly continuous functions having metric spaces for domain and range. We shall also prove that any metric space can be embedded in a complete metric space.

We denote by (X_1, d_1) and (X_2, d_2) two metric spaces.

17.1 Definition.—*We say that* $f: X_1 \to X_2$ *is uniformly continuous on* X_1 *if for every* $\varepsilon > 0$ *there exists some* $\delta > 0$ *such that*

$$d_2(f(x_1), f(x_2)) < \varepsilon$$

whenever x_1 *and* x_2 *belong to* X_1 *and* $d_1(x_1, x_2) < \delta$.

We obtain, obviously, an *equivalent* definition if "$d_2(f(x_1), f(x_2)) < \varepsilon$ and $d_1(x_1, x_2) < \delta$" are replaced by

$$\text{“}d_2(f(x_1), f(x_2)) \leq \varepsilon \quad \text{and} \quad d_1(x_1, x_2) \leq \delta.\text{”}$$

Since δ depends on ε, we often use some notation to indicate this. For instance, we may write δ_ε or $\delta(\varepsilon)$ instead of δ.

If $A \subset X_1$, we say that $f: A \to X_2$ is *uniformly continuous on* A if considered as a mapping "of the metric space* $(A, (d_1)_A)$ into (X_2, d_2)," f is uniformly continuous in the sense of Definition 17.1.

Hence, the function $f: A \to X_2$ is uniformly continuous on A if and only if, for every $\varepsilon > 0$ there exists some $\delta > 0$ such that

$$d_2(f(x_1), f(x_2)) < \varepsilon$$

whenever x_1 and x_2 belong to A and $d(x_1, x_2) < \delta$.

* Recall that $(d_1)_A = d \mid A \times A$.

17.2 Theorem.—*Let $A \subset X_1$ and let $f: A \to X_2$ be uniformly continuous. Then f is continuous on A.*

Proof.—Let $\varepsilon > 0$. Since f is uniformly continuous, there is $\delta_\varepsilon > 0$ such that

$$d_2(f(x_1), f(x_2)) < \varepsilon$$

whenever x_1 and x_2 belong to A and $d(x_1, x_2) < \delta_\varepsilon$. Let $a \in A$; we deduce that

$$d_2(f(a), f(x)) < \varepsilon$$

whenever $x \in A$ and $d(a, x) < \delta_\varepsilon$. Since $\varepsilon > 0$ was arbitrary, we deduce that f is continuous at a (see 13.18). Since $a \in A$ was arbitrary, f is continuous on A.

We have shown, therefore, that a uniformly continuous function is continuous. We shall see in Example 2 that the converse is not true; that is, that there exist continuous functions that are not uniformly continuous.

If we compare 17.1 with the definition of continuity, we see that in the definition of continuity of f at a (see, for instance, 13.18), given $\varepsilon > 0$, the corresponding neighborhood of a depends not only on ε but also on $a \in A$ (although we usually do not indicate this dependence by any supplementary symbol). Expressed with inequalities, this means that "δ_ε" depends not only on ε, but also on a. In the definition of uniform continuity, δ_ε *does not depend on* a.

Example 1.—Let (X_1, d_1) and (X_2, d_2) be two metric spaces, $A \subset X_1$ totally bounded and $f: A \to X_2$ uniformly continuous. Then the set $f(A)$ is *bounded.*

Since f is uniformly continuous, there is $\delta > 0$ such that

$$d_2(f(x), f(y)) < 1$$

whenever x and y belong to A and $d_1(x, y) < \delta$. Since A is totally bounded, there is a finite family $(a_j)_{1 \le j \le n}$ of elements of A such that $(V_\delta(a_j))_{1 \le j \le n}$ is a covering of A. Let

$$L = \sup \{d(f(a_i), f(a_j)) \mid 1 \le i \le n,\ 1 \le j \le n\}.$$

If x and y belong to A, there are $1 \le i \le n$ and $1 \le j \le n$ such that $x \in V_\delta(a_i)$ and $y \in V_\delta(a_j)$. Hence

$$\begin{aligned} d(f(x), f(y)) &\le d(f(x), f(a_i)) + d(f(a_i), f(a_j)) \\ &\quad + d(f(a_j), f(y)) \le 2 + L. \end{aligned}$$

Since x and y were arbitrary elements of A, we deduce that $\delta(f(A)) \leq 2 + L$. Hence, $f(A)$ is bounded.

If (X_2, d_2) is the metric space $\mathbf{R}$, then $f(A)$ is bounded if and only if there is $K \in \mathbf{R}_+$ such that $|f(x)| \leq K$ for all $x \in A$.

Example 2.—The mapping $x \mapsto 1/x$ of $(0, 1)$ into $\mathbf{R}$ is *not* uniformly continuous.

Let $L \geq 0$. By Archimedes' property, there is $p \in \mathbf{N}$ satisfying $p > L$. Hence $x_0 \in (0, 1)$ and $|1/x_0| = p > L$ if $x_0 = 1/p$. Hence there is no L satisfying $|1/x| \leq L$ for all $x \in (0, 1)$, and hence (by the result in Example 1), the mapping $x \mapsto 1/x$ of $(0, 1)$ into $\mathbf{R}$ is not uniformly continuous.

Example 3.—The mapping $x \mapsto 1/x$ of $\mathbf{R}^*$ into $\mathbf{R}$ is not uniformly continuous.

We leave the details to the reader.

17.3 Theorem.—*Let (X_1, d_1) and (X_2, d_2) be two metric spaces, $A \subset X_1$ compact and $f: A \to X_2$ continuous. Then f is uniformly continuous.*

Proof.—Let $\varepsilon > 0$ and let $t \in A$. Since $f: A \to X_2$ is continuous, there is $\delta(t) > 0$ such that $x \in V_{\delta(t)}(t)$ implies that

$$d_2(f(x), f(t)) < \varepsilon/2.$$

Clearly, $(V_{\delta(t)/2}(t))_{t \in A}$ is an *open* covering of A. Since A is compact, there exists a finite family $(t_j)_{1 \leq j \leq n}$ of elements of A such that $(V_{\delta(t_j)/2}(t_j))_{1 \leq j \leq n}$ is a covering of A. Let $\delta = \inf\{\delta(t_1)/2, \ldots, \delta(t_n)/2\}$. Now let x and y in A be such that $d(x, y) < \delta$. Since $(V_{\delta(t_j)/2}(t_j))_{1 \leq j \leq n}$ is a covering of A, there exists $1 \leq j_0 \leq n$ such that $x \in V_{\delta(t_{j_0})/2}(t_{j_0})$. Hence,

$$d_1(t_{j_0}, x) < \delta(t_{j_0})/2 < \delta(t_{j_0})$$

and

$$\begin{aligned} d_1(t_{j_0}, y) &< d_1(t_{j_0}, x) + d_1(x, y) < \delta(t_{j_0})/2 + \delta \\ &< \delta(t_{j_0})/2 + \delta(t_{j_0})/2 = \delta(t_{j_0}). \end{aligned}$$

Hence, $d_1(t_{j_0}, x) < \delta(t_{j_0})$ and $d_1(t_{j_0}, y) < \delta(t_{j_0})$. We obtain, then,

$$\begin{aligned} d_2(f(x), f(y)) &< d_2(f(x), f(t_{j_0})) + d_2(f(t_{j_0}), f(y)) \\ &< \varepsilon/2 + \varepsilon/2 = \varepsilon. \end{aligned}$$

Since $\varepsilon > 0$ was arbitrary, we conclude that f is uniformly continuous.

Exercise.—Let X be a metric space, $A \subset X$, and let $C^u_{\mathbf{R}}(A)$ be the set of all mappings $f: A \to \mathbf{R}$ uniformly continuous on A. (a) Show that, endowed with the usual addition and scalar multiplication, $C^u_{\mathbf{R}}(A)$ is a vector space. (b) Show that if f and g belong to $C^u_{\mathbf{R}}(A)$, it is not necessarily true that fg belongs to A. (Hint: Let $A = X = \mathbf{R}$ and let j be the mapping $x \mapsto x$ of $\mathbf{R}$ into $\mathbf{R}$; then $j \in C^u_{\mathbf{R}}(A)$, whereas $j^2 \notin C^u_{\mathbf{R}}(A)$.)

Again, let (X_1, d_1) and (X_2, d_2) be two metric spaces.

17.4 Definition.—*Let $A \subset X_1$ and $f: A \to X_2$. We say that f is a Lipschitz function* if there is $L \geq 0$ such that*

$$d_2(f(x), f(y)) \leq L d_1(x, y)$$

for all x and y in A.

It is easy to see that a Lipschitz function is uniformly continuous. In fact, let $\varepsilon > 0$ and let $\delta_\varepsilon = \varepsilon/(L+1)$. Then x and y in A and $d_1(x, y) \leq \delta_\varepsilon$ imply that

$$d_2(f(x), f(y)) \leq L\varepsilon/(L+1) \leq \varepsilon.$$

Since $\varepsilon > 0$ was arbitrary, it follows that f is uniformly continuous on A.

Example 4.—For each non-void set $A \subset X_1$, the mapping $f: x \mapsto d_1(x, A)$ of X_1 into $\mathbf{R}$ is a Lipschitz function.

In fact, (see 13.6),

$$|f(x) - f(y)| \leq d_1(x, y)$$

for all x and y in X_1.

Recall that we have already proved that f is continuous on X_1 (see 13.19).

Before proceeding farther, we shall make the following remarks: Let (X, d) be a metric space, $(x_n)_{n \in \mathbf{N}}$ a sequence of elements of X that converges to a, and $(y_n)_{n \in \mathbf{N}}$ a sequence of elements of X that converges to b. Then:

17.5(i) $\lim_{n \in \mathbf{N}} d(x_n, y_n) = d(a, b)$;

(ii) $a = b \Leftrightarrow \lim_{n \in \mathbf{N}} d(x_n, y_n) = 0$.

---

* Lipschitz functions having for domain a part of **R** have been introduced in Example 5, Chapter 5.

We leave to the reader the proof of 17.5 (hint: Use Theorem 13.21).

Let (X_1, d_1) and (X_2, d_2) be two metric spaces.

17.6 Definition.—Suppose $A \subset X_1$. *A mapping $\varphi : A \to X_2$ is an isometry if*

$$d_2(\varphi(x), \varphi(y)) = d_1(x, y)$$

for all x and y belonging to A.

The metric spaces (X_1, d_1) and (X_2, d_2) are said to be *isometric* if there is an isometry of X_1 onto X_2 or an isometry of X_2 onto X_1.

We notice that:

17.7 *If $\varphi : A \to X_2$ is an isometry, then φ is an injection.*

In fact, if x and y belong to A, and $x \neq y$, then

$$d_2(\varphi(x), \varphi(y)) = d_1(x, y) \neq 0,$$

so that $\varphi(x) \neq \varphi(y)$. Since x and y were arbitrary, φ is injective.

17.8 *If $\varphi : A \to X_2$ is an isometry, then φ is a Lipschitz function.*

17.9 Let $\varphi : A \to X_2$ be an isometry. Consider φ as a bijection of A onto $\varphi(A)$. If φ^{-1} is the inverse of this mapping, then $\psi : x \mapsto \varphi^{-1}(x)$, *on $\varphi(A)$ to X_1, is an isometry.*

17.10 Let $\varphi : A \to X_2$ be an isometry and $(x_n)_{n \in \mathbf{N}}$ a sequence of elements of A. Then $(x_n)_{n \in \mathbf{N}}$ *is a Cauchy sequence if and only if* $(\varphi(x_n))_{n \in \mathbf{N}}$ *is a Cauchy sequence.*

This follows immediately from the equation,

$$d_2(\varphi(x_n), \varphi(x_m)) = d_1(x_n, x_m)$$

for $n \in \mathbf{N}$, $m \in \mathbf{N}$.

17.11 Let $\varphi : A \to X_2$ be an isometry, $(x_n)_{n \in \mathbf{N}}$ and $(y_n)_{n \in \mathbf{N}}$ two sequences of elements of A that converge to an element* $a \in X_1$. Assume that (X_2, d_2) is *complete*. Then $(\varphi(x_n))_{n \in \mathbf{N}}$ and $(\varphi(y_n))_{n \in \mathbf{N}}$ *are convergent and*

(*) $\lim_{n \in \mathbf{N}} \varphi(x_n) = \lim_{n \in \mathbf{N}} \varphi(y_n)$.

* Notice that we do not assume that $a \in A$.

Since $(x_n)_{n \in \mathbf{N}}$ is convergent, it is Cauchy. By 17.10, $(\varphi(x_n))_{n \in \mathbf{N}}$ is a Cauchy sequence. Since (X_2, d_2) is complete, $(\varphi(x_n))_{n \in \mathbf{N}}$ is convergent. In the same way, we show that $(\varphi(y_n))_{n \in \mathbf{N}}$ is convergent. Since

$$\lim_{n \in \mathbf{N}} x_n = \lim_{n \in \mathbf{N}} y_n = a,$$

we deduce from 17.5(ii) that

$$\lim_{n \in \mathbf{N}} d_1(x_n, y_n) = 0.$$

Since φ is an isometry, we have

$$\lim_{n \in \mathbf{N}} d_2(\varphi(x_n), \varphi(y_n)) = 0.$$

Using 17.5(ii), again, we conclude that

$$\lim_{n \in \mathbf{N}} \varphi(x_n) = \lim_{n \in \mathbf{N}} \varphi(y_n).$$

17.12 Theorem.—*Let (X_1, d_1) and (X_2, d_2) be two complete metric spaces, A_1 a dense part of X_1, A_2 a dense part of X_2, and $\varphi: A_1 \to X_2$ an isometry such that $\varphi(A_1) = A_2$. Then there exists an isometry $\bar{\varphi}$ of X_1 onto X_2 such that $\bar{\varphi} \mid A_1 = \varphi$.*

Proof.—Let $x \in X_1$ and let $(x_n)_{n \in \mathbf{N}}$ be a sequence of elements of A_1 that converges to x. Then $(x_n)_{n \in \mathbf{N}}$ is a Cauchy sequence. By 17.10, $(\varphi(x_n))_{n \in \mathbf{N}}$ is a Cauchy sequence; since (X_2, d_2) is complete, $(\varphi(x_n))_{n \in \mathbf{N}}$ is convergent. Let

$$\bar{\varphi}(x) = \lim_{n \in \mathbf{N}} \varphi(x_n).$$

By 17.11, $\bar{\varphi}(x)$ does not depend on the particular sequence that converges to x. Since $x \in X_1$ was arbitrary, we defined this way a mapping $\bar{\varphi}: X_1 \to X_2$. Clearly, $\bar{\varphi} \mid A_1 = \varphi$.

Let x and y be two elements belonging to X_1. Let $(x_n)_{n \in \mathbf{N}}$ and $(y_n)_{n \in \mathbf{N}}$ be sequences of elements of A_1 that converge to x and y, respectively. Using 17.5(i), we obtain

$$\begin{aligned} d_2(\bar{\varphi}(x), \bar{\varphi}(y)) &= \lim_{n \in \mathbf{N}} d_2(\bar{\varphi}(x_n), \bar{\varphi}(y_n)) \\ &= \lim_{n \in \mathbf{N}} d_1(x_n, y_n) = d_1(x, y). \end{aligned}$$

Since x and y were arbitrary, it follows that $\bar{\varphi}$ is an isometry.

Now let $y \in X_2$ and let $(y_n)_{n \in \mathbf{N}}$ be a sequence of elements of A_2 that converges to y. Since $\varphi(A_1) = A_2$, we deduce that for each $n \in \mathbf{N}$ there is $x_n \in A_1$ such that $\varphi(x_n) = y_n$. By 17.10, $(x_n)_{n \in \mathbf{N}}$ is a Cauchy sequence. Since (x_1, d_1) is complete, $(x_n)_{n \in \mathbf{N}}$ converges to some element $x \in X_1$. We deduce that

$$y = \lim_{n\in\mathbf{N}} \varphi(x_n) = \bar{\varphi}(x).$$

Since $y \in X_2$ was arbitrary, we conclude that $\bar{\varphi}(x_1) = x_2$.

The uniqueness of $\bar{\varphi}$ follows from the fact that $\bar{A}_1 = X_1$ and from Corollary 7.21.

Hence, Theorem 17.2 is completely proved.

Exercise.—Let (X_1, d_1) and (X_2, d_2) be two metric spaces, A_1 a dense part of X_1 and $f: A_1 \to X_2$ a uniformly continuous function. Suppose that (X_2, d_2) is *complete*. Then there exists a unique uniformly continuous mapping $\bar{f}: X_1 \to X_2$ such that $\bar{f} \mid A_1 = f$.

We shall prove the result stated at the end of Chapter 15; namely, that any metric space can be embedded in a complete metric space. We shall first establish certain results that will be used in the proof.

Let (X, d) be a metric space and let $\mathscr{C}$ be the set of all *Cauchy sequences* of elements of X. For

$$\mathbf{x} = (x_n)_{n\in\mathbf{N}} \in \mathscr{C} \quad \text{and} \quad \mathbf{y} = (y_n)_{n\in\mathbf{N}} \in \mathscr{C},$$

we write

$$\mathbf{x} \equiv \mathbf{y}(S) \Leftrightarrow \lim_{n\in\mathbf{N}} d(x_n, y_n) = 0;$$

we define thus an *equivalence relation* in $\mathscr{C}$.

We shall now show that:

17.13 *If* $\mathbf{x} = (x_n)_{n\in\mathbf{N}}$ *and* $\mathbf{y} = (y_n)_{n\in\mathbf{N}}$ *are elements of* $\mathscr{C}$, *then* $(d(x_n, y_n))_{n\in\mathbf{N}}$ *is convergent.*

17.14 *If* $\mathbf{x} = (x_n)_{n\in\mathbf{N}}$, $\mathbf{x}' = (x'_n)_{n\in\mathbf{N}}$, $\mathbf{y} = (y_n)_{n\in\mathbf{N}}$, *and* $\mathbf{y}' = (y'_n)_{n\in\mathbf{N}}$ *are elements of* $\mathscr{C}$, *and if* $\mathbf{x} \equiv \mathbf{x}'$ *and* $\mathbf{y} \equiv \mathbf{y}'$, *then*

$$\lim_{n\in\mathbf{N}} d(x_n, y_n) = \lim_{n\in\mathbf{N}} d(x'_n, y'_n).$$

Proof of 17.13.—For all $n \in \mathbf{N}$ and $m \in \mathbf{N}$, we have

$$d(x_n, y_n) \leq d(x_n, x_m) + d(x_m, y_m) + d(y_m, y_n)$$

and

$$d(x_m, y_m) \leq d(x_m, x_n) + d(x_n, y_n) + d(y_n, y_m);$$

hence

$$|d(x_n, y_n) - d(x_m, y_m)| \leq d(x_n, x_m) + d(y_n, y_m).$$

We deduce that $(d(x_n, y_n))_{n\in\mathbf{N}}$ is a Cauchy sequence, and hence convergent.

Proof of 17.14.—For all $n \in \mathbf{N}$ and $m \in \mathbf{N}$, we have

$$d(x_n, y_n) \leq d(x_n, x'_n) + d(x'_n, y'_n) + d(y'_n, y_n)$$

and

$$d(x'_n, y'_n) \leq d(x'_n, x_n) + d(x_n, y_n) + d(y_n, y'_n);$$

hence

$$|d(x_n, y_n) - d(x'_n, y'_n)| \leq d(x_n, x'_n) + d(y_n, y'_n).$$

We deduce that

$$\lim_{n\in\mathbf{N}} |d(x_n, y_n) - d(x'_n, y'_n)| = 0.$$

By 17.13, $(d(x_n, y_n))_{n\in\mathbf{N}}$ and $(d(x'_n, y'_n))_{n\in\mathbf{N}}$ are convergent, and then, by 17.5(ii),

$$\lim_{n\in\mathbf{N}} d(x_n, y_n) = \lim_{n\in\mathbf{N}} d(x'_n, y'_n).$$

17.15 Theorem.—*Let (X, d) be a metric space. Then there exists a complete metric space $(\hat{X}, \hat{d})$ and an isometry φ of X into $\hat{X}$ such that $\overline{\varphi(X)} = \hat{X}$.*

Proof.—Let $\hat{X} = \mathscr{C}/S$ and let $\mathbf{x} \mapsto \hat{\mathbf{x}}$ be the canonical mapping of $\mathscr{C}$ onto $\hat{X}$.

If $\hat{\mathbf{x}}$ and $\hat{\mathbf{y}}$ belong to $\hat{X}$ (so that $\mathbf{x} = (x_n)_{n\in\mathbf{N}} \in \mathscr{C}$ and $\mathbf{y} = (y_n)_{n\in\mathbf{N}} \in \mathscr{C}$), we define

$$\hat{d}(\hat{\mathbf{x}}, \hat{\mathbf{y}}) = \lim_{n\in\mathbf{N}} d(x_n, y_n).$$

By 17.13 and 17.14, $\hat{d}$ is well-defined as a mapping of $\hat{X} \times \hat{X}$ into $\mathbf{R}$.

It is easy to see that $\hat{d}$ is a *metric* on $\hat{X}$. *Thus the metric space $(\hat{X}, \hat{d})$ is defined.*

For each $x \in X$, let $\mathbf{x}^*$ be the Cauchy sequence $(x_n)_{n\in\mathbf{N}}$ where $x_n = x$ for all $n \in \mathbf{N}$. Let $\varphi(x) = \hat{\mathbf{x}}^*$ for $x \in X$. If x and y belong to X, we deduce, from the definition of $\hat{d}$, that

$$\hat{d}(\varphi(x), \varphi(y)) = \hat{d}(\hat{\mathbf{x}}^*, \hat{\mathbf{y}}^*) = d(x, y).$$

Since x and y were arbitrary, $\varphi: X \to \hat{X}$ is an isometry.

Now let $\hat{\mathbf{x}} \in \hat{X}$, where $\mathbf{x} = (x_n)_{n\in\mathbf{N}}$, and let $\varepsilon > 0$. Since $(x_n)_{n\in\mathbf{N}}$ is a Cauchy sequence, there is $p \in \mathbf{N}$ such that $m \geq p$ and $n \geq p$ imply that

$$d(x_m, x_n) \leq \varepsilon.$$

Then

$$\hat{d}(\hat{\mathbf{x}}, \hat{\mathbf{x}}_p^*) = \lim_{n\in\mathbf{N}} d(x_n, x_p) \leq \varepsilon.$$

Since $\hat{\mathbf{x}} \in \hat{X}$ and $\varepsilon > 0$, we deduce that $\overline{\varphi(X)} = \hat{X}$.

Thus the isometry $\varphi: X \to \hat{X}$ is defined and we have shown that $\overline{\varphi(X)} = \hat{X}$.

Now let $(\hat{\mathbf{x}}_n)_{n\in\mathbf{N}}$ be a Cauchy sequence of elements of $\hat{X}$. For each $n \in \mathbf{N}$, let $y_n \in X$ be such that

$$\hat{d}(\hat{\mathbf{x}}_n, \varphi(y_n)) \leq 1/n.$$

Then $(\varphi(y_n))_{n\in\mathbf{N}}$ is a Cauchy sequence of elements of $\hat{X}$. Since φ is an isometry, we deduce (see 17.10) that $\mathbf{y} = (y_n)_{n\in\mathbf{N}}$ is a Cauchy sequence of elements of X.

Let $\varepsilon > 0$ and let $n_\varepsilon \in \mathbf{N}$ be such that $1/n_\varepsilon \leq \varepsilon/2$ and $d(y_n, y_m) \leq \varepsilon/2$ if $n \geq n_\varepsilon$, $m \geq n_\varepsilon$. Then, if $n \geq n_\varepsilon$

$$\begin{aligned}\hat{d}(\hat{\mathbf{x}}_n, \hat{\mathbf{y}}) &\leq \hat{d}(\hat{\mathbf{x}}_n, \varphi(y_n)) + \hat{d}(\varphi(y_n), \hat{\mathbf{y}}) \\ &\leq \frac{1}{n} + \hat{d}(\varphi(y_n), \hat{\mathbf{y}}) \\ &= \frac{1}{n} + \lim_{m\in\mathbf{N}} d(y_n, y_m) \\ &\leq 2(\varepsilon/2) = \varepsilon.\end{aligned}$$

Hence, $\hat{d}(\hat{\mathbf{x}}_n, \hat{\mathbf{y}}) \leq \varepsilon$ if $n \geq n_\varepsilon$; since $\varepsilon > 0$ was arbitrary, $(\hat{\mathbf{x}}_n)_{n\in\mathbf{N}}$ converges to $\hat{\mathbf{y}}$.

Thus we have shown that $(\hat{X}, \hat{d})$ is complete, and hence Theorem 17.15 is proved.

Remarks.—The metric space $(\hat{X}, \hat{d})$ is called the *completion* of (X, d). If we *identify* X with $\varphi(X)$, then:

(i) $X \subset \hat{X}$;

(ii) $\hat{d} \mid X \times X = d$;

(iii) $\overline{X} = \hat{X}$.

We shall show below that, in a certain sense, $(\hat{X}, \hat{d})$ is "unique."

17.16 Theorem.—*Let (X, d) be a metric space, (X', d') a complete metric space, φ' an isometry of X into X' such that $\overline{\varphi'(X)} = X'$, (X'', d'') a complete metric space, and φ'' an isometry of X into X'' such that $\overline{\varphi''(X)} = X''$. Then (X', d') and (X'', d'') are isometric.*

Proof.—Let $A' = \varphi'(X)$, $A'' = \varphi''(X)$, and let $\psi: A' \to A''$ be defined by

$$\psi(\varphi'(x)) = \varphi''(x)$$

for $x \in X$. Then ψ is an isometry of A' into X'' such that $\psi(A') = A''$. By Theorem 17.12, (X', d') and (X'', d'') are isometric.

Exercises for Chapter 17

1. Let (X_1, d_1) and (X_2, d_2) be metric spaces and let $f: X_1 \to X_2$ be a Lipschitz function. If A is a totally bounded subset of X_1, then $f(A)$ is totally bounded in X_2.

2. Let d be the usual metric on $\mathbf{R}$ and δ the metric on $\mathbf{R}$ defined in Example 6, Chapter 15. Show that the mapping $f: \mathbf{R} \to \mathbf{R}$ defined by $f(x) = x$ for $x \in \mathbf{R}$ is uniformly continuous as a mapping of $(\mathbf{R}, d)$ into $(\mathbf{R}, d)$, but is *not* uniformly continuous as a mapping of $(\mathbf{R}, \delta)$ into $(\mathbf{R}, d)$.

3. Let f be the mapping $x \mapsto 1/x$ of $(0, \infty)$ into $\mathbf{R}$. Show that for every $n > 0$, $f \mid [\eta, +\infty)$ is uniformly continuous (although f is not uniformly continuous).

4. Generalize the results stated in Exercises 5, 6, and 9 at the end of Chapter 5 to Lipschitz functions having for domain a subset of a metric space and for range $\mathbf{R}$ (see Example 4).

Chapter 18

Normed Spaces

Let X be a (real) vector space.*

18.1 Definition.—*A norm on X is a mapping $p: X \to \mathbf{R}$ having the properties:*

(N_1) $p(x) = 0$ *if and only if* $x = 0$;

(N_2) $p(\lambda x) = |\lambda|\, p(x)$ *for* $\lambda \in \mathbf{R}$ *and* $x \in X$;

(N_3) $p(x + y) \leq p(x) + p(y)$ *for* $x \in X$ *and* $y \in X$.

Exercise.—If p is a norm on X, then

$$p(x_1 + \ldots + x_n) \leq p(x_1) + \ldots + p(x_n)$$

for every $x_1, \ldots, x_n$ belonging to X.

If p is a *norm on* X, then $p(x) \geq 0$ for every $x \in X$; in fact, we have

$$0 = p(0) = p(x - x) \leq p(x) + p(x) = 2p(x).$$

Let us also notice that if $x \in X$ and $y \in X$, then

18.2 $$|p(x) - p(y)| \leq p(x - y).$$

In fact, we have

$$p(x) = p(y + (x - y)) \leq p(y) + p(x - y);$$

* We assume that the elementary algebraic properties of vector spaces and algebras are known to the reader.

that is,

(i) $$p(x) - p(y) \leq p(x - y).$$

In the same way, we see that

(ii) $$-(p(x) - p(y)) \leq p(x-y).$$

Comparing (i) and (ii), we deduce 18.2.

If p is a norm on X, then the mapping $d: X \times X \to \mathbf{R}$ defined by

18.3 $$d(x, y) = p(x - y)$$

for $(x, y) \in X \times X$ is a distance on X (we leave to the reader the verification of the properties given in 13.1 to 13.3).

A norm on a set X is often denoted $x \mapsto \|x\|$; for each $x \in X$, the real number $\|x\|$ is read "norm of x."

18.4 Definition.—*A couple (X, p), where X is a vector space and p is a norm on X, is called a normed space.*

If (X, p) is a normed space, we always suppose that X is endowed with the distance d (defined by 18.3). We shall often say that d is the metric of the normed space (X, p) and $\mathscr{T}_d$ the topology of (X, p). The topology $\mathscr{T}_d$ is often denoted $\mathscr{T}_p$ if d is defined by 18.3.

When there is no ambiguity, we shall often say the "normed space X" instead of "the normed space (X, p)."

Exercise 1.—Let p be a norm on X and $(x_n)_{n \in \mathbf{N}}$ a sequence of elements of X that converges to $x \in X$. Then $(p(x_n))_{n \in \mathbf{N}}$ converges to $p(x)$ (hint: Use 18.2).

18.5 Theorem.—*If X is a normed space, then:*

(a) *$\varphi: (\lambda, x) \mapsto \lambda x$ on $\mathbf{R} \times X$ to X is continuous on $\mathbf{R} \times X$;*

(b) *$\psi: (x, y) \mapsto x + y$ on $X \times X$ to X is continuous on $X \times X$.*

Proof of (a).—Let $(\lambda_0, x_0) \in \mathbf{R} \times X$. Then for every $(\lambda, x) \in \mathbf{R} \times X$,

$$\begin{aligned} \|\varphi(\lambda, x) - \varphi(\lambda_0, x_0)\| &= \|\lambda x - \lambda_0 x_0\| \\ &= \|\lambda(x - x_0) + (\lambda - \lambda_0)x_0\| \leq |\lambda|\ \|x - x_0\| + |\lambda - \lambda_0|\ \|x_0\|. \end{aligned}$$

Now let $\varepsilon > 0$ and let $r = \inf\{1, \varepsilon/2\ \|x_0\|\}$ and $\delta = \varepsilon/2(1 + |\lambda_0|)$. Let $U = W_r(\lambda_0)$ and $V = W_\delta(x_0)$. Then $U \times V \in \mathscr{N}((\lambda_0, x_0))$ and $(\lambda, x) \in U \times V$ imply $|\lambda - \lambda_0| \leq r$ and $\|x - x_0\| \leq \delta$; hence

$$\begin{aligned} \|\varphi(\lambda, x) - \varphi(\lambda_0, x_0)\| &\leq (1 + |\lambda_0|)\|x - x_0\| + |\lambda - \lambda_0|\|x_0\| \\ &\leq \frac{\varepsilon}{2(1 + |\lambda_0|)}(1 + |\lambda_0|) + \frac{\varepsilon}{2\ \|x_0\|}\|x_0\| = 2\frac{\varepsilon}{2} = \varepsilon. \end{aligned}$$

Since $\varepsilon > 0$ was arbitrary, φ is continuous at (λ_0, x_0). Since $(\lambda_0, x_0) \in \mathbf{R} \times X$ was arbitrary, φ is continuous on $\mathbf{R} \times X$.

Proof of (b).—Let $(x_0, y_0) \in X \times X$. Then for every $(x, y) \in X \times X$,

$$\begin{aligned}\|\psi(x,y) - \psi(x_0,y_0)\| &= \|(x+y) - (x_0+y_0)\| \\ &= \|(x-x_0) + (y-y_0)\| \leq \|x - x_0\| + \|y - y_0\|.\end{aligned}$$

Now let $\varepsilon > 0$ and let $U = W_{\varepsilon/2}(x_0)$ and $V = W_{\varepsilon/2}(y_0)$. Then $U \times V \in \mathcal{N}((x_0, y_0))$ and $(x, y) \in U \times V$ imply that $\|x - x_0\| \leq \varepsilon/2$ and $\|y - y_0\| \leq \varepsilon/2$; hence

$$\|\psi(x,y) - \psi(x_0,y_0)\| \leq 2(\varepsilon/2) = \varepsilon.$$

Since $\varepsilon > 0$ was arbitrary, ψ is continuous at (x_0, y_0). Since $(x_0, y_0) \in X \times X$ was arbitrary, ψ is continuous on $X \times X$.

Exercise 2.—(i) The mapping $(\lambda, \mu, x, y) \mapsto \lambda x + \mu y$, on $\mathbf{R} \times \mathbf{R} \times X \times X$ to X, is continuous on $\mathbf{R} \times \mathbf{R} \times X \times X$. (ii) For every $\lambda \in \mathbf{R}$, $\lambda \neq 0$ and $y \in X$, the mapping $x \mapsto \lambda x + y$ on X to X is a homeomorphism of X onto X.

A set $F \subset X$ is a *subspace* of X if $(\lambda, x) \in \mathbf{R} \times F$ implies that $\lambda x \in F$ and $(x, y) \in F \times F$ implies that $x + y \in F$. Hence (with the notations of Theorem 18.5), F is a subspace if and only if

$$\varphi(\mathbf{R} \times F) \subset F \quad \text{and} \quad \psi(F \times F) \subset F.$$

18.6 Theorem.—*If X is a normed space, and if $F \subset X$ is a subspace, then $\overline{F}$ is a subspace.*

Proof.—We have (see 7.10′ and 5.11(d))

$$\varphi(\mathbf{R} \times \overline{F}) = \varphi(\overline{\mathbf{R} \times F}) \subset \overline{\varphi(\mathbf{R} \times F)} \subset \overline{F}$$

and

$$\psi(\overline{F} \times \overline{F}) = \psi(\overline{F \times F}) \subset \overline{\psi(F \times F)} \subset \overline{F}.$$

Hence, $\overline{F}$ is a subspace.

A *normed space* (X, p) is said to be *complete* if the corresponding metric space (X, d), where d is the metric defined by 18.3, is complete.

18.7 Definition.—*A couple (X, p) is a Banach space if it is a complete normed space.*

Example 1.—By Theorem 13.29, the mapping

$$p : x \mapsto \|x\| = (x_1^2 + \ldots + x_n^2)^{1/2}$$

($x = (x_1, \ldots, x_n)$) of $\mathbf{R}^n$ ($n \geq 1$) into $\mathbf{R}$ is a norm on $\mathbf{R}^n$. When endowed with this norm (called the Euclidean norm), $\mathbf{R}^n$ is complete (see Example 4, Chapter 15). Hence, $(\mathbf{R}^n, p)$ *is a Banach space.*

Let (X, p) be a *normed space* and let

18.8 $$(x_n)_{n \in \mathbf{N}}$$

be a sequence of elements of X. For $n \in \mathbf{N}$, we write

$$s_n = x_1 + \ldots + x_n;$$

we obtain this way a new sequence, $(s_n)_{n \in \mathbf{N}}$. The *pair* of sequences

18.9 $$((x_n)_{n \in \mathbf{N}}, (s_n)_{n \in \mathbf{N}})$$

is said to be *the series generated by* $(x_n)_{n \in \mathbf{N}}$.

We say that the series given in 18.9 *converges* (or is *convergent*) if the sequence $(s_n)_{n \in \mathbf{N}}$ converges to some $S \in X$. In this case, we say that S is the *sum* of the series, and we write

$$S = \sum_{n \in \mathbf{N}} x_n, \qquad S = \sum_{n=1}^{+\infty} x_n \quad \text{or} \quad S = x_1 + x_2 + \ldots + x_n + \ldots.$$

The notations*

$$\sum_{n \in \mathbf{N}} x_n, \qquad \sum_{n=1}^{+\infty} x_n \quad \text{or} \quad x_1 + x_2 + \ldots + x_n + \ldots$$

are often used (instead of 18.9) to designate the series generated by the sequence $(x_n)_{n \in \mathbf{N}}$.

In the same way, we define series generated by sequences of the form $(x_n)_{n \in I}$, where $I = \{n \mid n \in \mathbf{Z}, n \geq p\}$ for some $p \in \mathbf{Z}$.

The series $\sum_{n \in \mathbf{N}} x_n$ generated by the sequence $(x_n)_{n \in \mathbf{N}}$ of elements of X is *absolutely convergent* if $\sum_{n \in \mathbf{N}} p(x_n)$ is convergent.

18.10 Theorem.—*If* (X, p) *is a Banach space and* $\sum_{n \in \mathbf{N}} x_n$ *is absolutely convergent, then* $\sum_{n \in \mathbf{N}} x_n$ *is convergent and*

---

* Observe that these notations are used to denote either the series given in 18.9 or its sum (if it is convergent).

18.11 $$p(\sum_{n\in\mathbf{N}} x_n) \le \sum_{n\in\mathbf{N}} p(x_n).$$

Proof.—For each $n \in \mathbf{N}$, let

$$s_n = \sum_{j=1}^{n} x_j \quad \text{and} \quad t_n = \sum_{j=1}^{n} p(x_j).$$

Notice that for every $n \in \mathbf{N}$, $m \in \mathbf{N}$,

$$p(s_n - s_m) \le |t_n - t_m|.$$

Since $\sum_{n\in\mathbf{N}} p(x_n)$ is convergent, the sequence $(t_n)_{n\in\mathbf{N}}$ is convergent and hence Cauchy. We deduce that $(s_n)_{n\in\mathbf{N}}$ is a Cauchy sequence. Since (X, p) is a Banach space, $(s_n)_{n\in\mathbf{N}}$ converges to some $S \in X$. Hence $\sum_{n\in\mathbf{N}} x_n$ *is convergent.* Moreover,

$$\begin{aligned} p(\textstyle\sum_{n\in\mathbf{N}} x_n) &= p(S) = \lim_{n\in\mathbf{N}} p(s_n) \\ &\le \lim_{n\in\mathbf{N}} \textstyle\sum_{j=1}^{n} p(x_j) \le \lim_{n\in\mathbf{N}} t_n = \lim_{n\in\mathbf{N}} \textstyle\sum_{j=1}^{n} p(x_j); \end{aligned}$$

hence 18.11 is proved.

Exercise 3.—Let $0 \le \beta < 1$ and $(x_n)_{n\in\mathbf{Z}_+}$ ($\mathbf{Z}_+ = \{n \mid n \in \mathbf{Z}, n \ge 0\}$) be a sequence of elements of X such that $p(x_n) \le \beta^n$ for all $n \in \mathbf{Z}_+$. Then $\sum_{n\in\mathbf{Z}_+} x_n$ is absolutely convergent and

$$p(\sum_{n\in\mathbf{Z}_+} x_n) \le \sum_{n\in\mathbf{Z}_+} p(x_n) \le \frac{1}{1-\beta}.$$

18.12 Definition.—*A Banach algebra is a couple (X, p) where X is an algebra and p a norm on* X such that (X, p) is complete and*

$$p(xy) \le p(x)\, p(y)$$

for all x and y belonging to X.

If X has a unit element† e, we always assume that $p(e) = 1$. When there is no ambiguity, we shall say "the Banach algebra X" instead of "the Banach algebra (X, p)."

Exercise 4.—Let (X, p) be a Banach algebra. If $x_1, \ldots, x_n$ are elements belonging to X, then

$$p(x_1, \ldots, x_n) \le p(x_1) \ldots p(x_n).$$

---

* Observe that an algebra is, in particular, a vector space. A Banach algebra is a Banach space.

† We say that $e \in X$ is a unit element of X if $ex = xe = x$ for every $x \in X$. There is at most one unit element of X.

If $x \in X$ and $n \in \mathbf{N}$, then

$$p(x^n) \leq p(x)^n.$$

18.13 Theorem.—*If X is a Banach algebra, then the mapping $\gamma:(x,y) \mapsto xy$ on $X \times X$ to X is continuous on X.*

Proof.—Let $(x_0,y_0) \in X \times X$. Then for every $(x,y) \in X \times X$,

$$\begin{aligned}\|\gamma(x,y) - \gamma(x_0,y_0)\| &= \|xy - x_0y_0\| \\ &= \|xy - xy_0 + xy_0 - x_0y_0\| \leq \|x\|\,\|y - y_0\| + \|x - x_0\|\,\|y_0\|.\end{aligned}$$

Now let $\varepsilon > 0$ and let $\gamma = \{1, \varepsilon/2\,\|x_0\|\}$ and $\delta = \varepsilon/2(1 + \|x_0\|)$. Let $U = W_\gamma(x_0)$ and $V = W_\delta(y_0)$. Then $U \times V \in \mathscr{N}((x_0,y_0))$ and $(x,y) \in U \times V$ imply that $\|x - x_0\| \leq \gamma$ and $\|y - y_0\| \leq \delta$; hence

$$\begin{aligned}\|\gamma(x,y) - \gamma(x_0,y_0)\| &\leq (1 + \|x_0\|)\|y - y_0\| + \gamma\|y_0\| \\ &\leq \frac{\varepsilon}{2} + \frac{\varepsilon}{2} = \varepsilon_0.\end{aligned}$$

Since $\varepsilon > 0$ was arbitrary, γ is continuous at (x_0,y_0). Since $(x_0,y_0) \in X \times X$ was arbitrary, γ is continuous on $X \times X$.

Let X be an algebra. A set $F \subset X$ is a *subalgebra* of X if F is a subspace and if $(x,y) \in F \times F$ implies that $xy \in F$. Hence (with the notations of Theorem 18.13), F is a subalgebra if and only if F is a subspace and

$$\gamma(F \times F) \subset F.$$

18.14 Theorem.—*If X is a Banach algebra, and if $F \subset X$ is a subalgebra, then $\overline{F}$ is a subalgebra.*

Proof.—By Theorem 18.6, $\overline{F}$ is a subspace. Furthermore (see 7.10′ and 5.11(d))

$$\gamma(\overline{F} \times \overline{F}) = \gamma(\overline{F \times F}) \subset \overline{\gamma(F \times F)} \subset \overline{F}.$$

Hence, $\overline{F}$ is a subalgebra.

The Banach Algebra $C^b_{\mathbf{R}}(X)$

Let (X, T) be a topological space and let $C^b_{\mathbf{R}}(X)$ be the set of all *bounded** continuous functions on X to $\mathbf{R}$. It is easy to see that $C^b_{\mathbf{R}}(X)$ is an *algebra* having the constant function 1 for unit element.

---

* A function $f: X \to \mathbf{R}$ is bounded if there is $L \in \mathbf{R}$ such that for every $x \in X$,

$$|f(x)| \leq L.$$

For $f \in C^b_{\mathbf{R}}(X)$, define

$$\|f\| = \sup_{x \in X} |f(x)|.$$

Then $p : f \mapsto \|f\|$ is a mapping of $C^b_{\mathbf{R}}(X)$ into $\mathbf{R}$. Clearly, $\|f\| = 0$ if and only if $|f(x)| = 0$ for all $x \in X$. Hence, $\|f\| = 0$ if and only if $f = 0$, and hence *p satisfies* (N_1). If $\lambda \in \mathbf{R}$ and $f \in C^b_{\mathbf{R}}(X)$, then

$$\begin{aligned}\|\lambda f\| &= \sup_{x \in X} |\lambda f(x)| = \sup_{x \in X} |\lambda|\, |f(x)| \\ &= |\lambda| \sup_{x \in X} |f(x)| = |\lambda|\, \|f\|;\end{aligned}$$

hence *p satisfies* (N_2). If f and g belong to $C^b_{\mathbf{R}}(X)$, then

$$\begin{aligned}\|f + g\| &= \sup_{x \in X} |(f + g)(x)| \leq \sup_{x \in X} (|f(x)| + |g(x)|) \\ &\leq \sup_{x \in X} |f(x)| + \sup_{x \in X} |g(x)| = \|f\| + \|g\|;\end{aligned}$$

hence *p satisfies* (N_3). Hence *p is a norm on* $C^b_{\mathbf{R}}(X)$. In the same way, we show that

$$\|fg\| \leq \|f\|\, \|g\|$$

for $f \in C^b_{\mathbf{R}}(X)$ and $g \in C^b_{\mathbf{R}}(X)$.

18.15 Theorem.—*The couple* $(C^b_{\mathbf{R}}(X), p)$ *is a Banach algebra.*

Proof.—We have shown above that $p : f \mapsto \|f\|$ is a norm on $C^b_{\mathbf{R}}(X)$ satisfying $\|fg\| \leq \|f\|\, \|g\|$ for $f \in C^b_{\mathbf{R}}(X)$ and $g \in C^b_{\mathbf{R}}(X)$. Since, obviously, $\|1\| = 1$, in order to show that $(C^b_{\mathbf{R}}(X), p)$ is a Banach algebra it only remains to be shown that $(C^b_{\mathbf{R}}(X), p)$ is complete.

Let $(f_n)_{n \in \mathbf{N}}$ be a Cauchy sequence of elements of $C^b_{\mathbf{R}}(X)$. For $x \in X$ and $n \in \mathbf{N}$, $m \in \mathbf{N}$, we have

$$|f_n(x) - f_m(x)| \leq \|f_n - f_m\|.$$

Hence, for each $x \in X$, $(f_n(x))_{n \in \mathbf{N}}$ is a Cauchy sequence of real numbers. Since $\mathbf{R}$ is complete, we may define $f : X \to \mathbf{R}$ by

$$f(x) = \lim_{n \in \mathbf{N}} f_n(x) \quad \text{for} \quad x \in X.$$

We shall show now that $f \in C^b_{\mathbf{R}}(X)$ and that $(f_n)_{n \in \mathbf{N}}$ converges to f.

For every $\varepsilon > 0$, there is $p(\varepsilon) \in \mathbf{N}$ such that

$$\|f_n - f_m\| \leq \varepsilon \quad \text{if} \quad n \geq p(\varepsilon), \qquad m \geq p(\varepsilon);$$

hence

$$|f_n(x) - f_m(x)| \leq \varepsilon$$

for all $x \in X$, if $n \geq p(\varepsilon)$, $m \geq p(\varepsilon)$. We deduce that

$$(*) \qquad |f(x) - f_m(x)| = \lim_{n\in\mathbf{N}} |f_n(x) - f_m(x)| \leq \varepsilon$$

for all $x \in X$ if $m \geq p(\varepsilon)$.

Let $x_0 \in X$. Then, by (*),

$$|f(x) - f(x_0)| \leq |f(x) - f_{p(\varepsilon)}(x)| + |f_{p(\varepsilon)}(x) - f_{p(\varepsilon)}(x_0)| + |f_{p(\varepsilon)}(x_0) - f(x_0)| \leq 2\varepsilon + |f_{p(\varepsilon)}(x) - f_{p(\varepsilon)}(x_0)|$$

for all $x \in X$. Since $f_{p(\varepsilon)}$ is continuous, there is $V \in \mathcal{N}(x_0)$ such that $|f_{p(\varepsilon)}(x) - f_{p(\varepsilon)}(x_0)| \leq \varepsilon$ if $x \in V$; hence

$$|f(x) - f(x_0)| \leq 3\varepsilon \quad \text{if} \quad x \in V.$$

Since $\varepsilon > 0$ was arbitrary, f is continuous at x_0; since $x_0 \in X$ was arbitrary, *f is continuous on X.*

From (*), we also obtain

$$|f(x)| \leq |f_{p(1)}(x)| + 1 \leq \|f_{p(1)}\| + 1$$

for all $x \in X$. Hence f is bounded, and hence $f \in C^b_{\mathbf{R}}(X)$.

Using (*) again, we deduce that

$$\|f - f_m\| = \sup_{x\in X} |f(x) - f_m(x)| \leq \varepsilon$$

for $m \geq p(\varepsilon)$. Hence *$(f_m)_{m\in\mathbf{N}}$ converges to f.* Since $(f_n)_{n\in\mathbf{N}}$ was an arbitrary Cauchy sequence, we deduce that *$(C^b_{\mathbf{R}}(X), p)$ is complete.*

18.16 Theorem.—*The mapping $(f, x) \mapsto f(x)$ on $C^b_{\mathbf{R}}(X) \times X$ to* **R** *is continuous on $C^b_{\mathbf{R}}(X) \times X$.*

Proof.—Let $(h, a) \in C^b_{\mathbf{R}}(X) \times X$ and let $\varepsilon > 0$. Let $V \in \mathcal{N}(a)$ be such that $|h(x) - h(a)| \leq \varepsilon/2$ for $x \in V$ and let $W = W_{\varepsilon/2}(h) \times V$. Then $W \in \mathcal{N}((h, a))$ and, if $(f, x) \in W$, we have

$$|f(x) - h(x)| \leq |f(x) - h(x)| + |h(x) - h(a)| \leq \|f - h\| + |h(x) - h(a)| \leq 2(\varepsilon/2) = \varepsilon.$$

Since $\varepsilon > 0$ was arbitrary, we deduce that $(f, x) \mapsto f(x)$ is continuous at (h, a). Since (h, a) was arbitrary, $(f, x) \mapsto f(x)$ is continuous on $C^b_{\mathbf{R}}(X) \times X$.

18.17 Corollary.—*For every $a \in X$, the mapping $u_a : f \mapsto f(a)$ on $C^b_{\mathbf{R}}(X)$ to R is continuous on $C^b_{\mathbf{R}}(X)$.*

Proof.—See 7.24.

For each $a \in X$, let $C^b_{\mathbf{R},a}(X)$ be the set of all $f \in C^b_{\mathbf{R}}(X)$ such that $f(a) = 0$. Then:

18.18 Corollary.—*For each* $a \in X$, $C^b_{\mathbf{R},a}(X)$ *is a closed subalgebra* $C^b_{\mathbf{R}}(X)$.

Proof.—Clearly,

$$C^b_{\mathbf{R},a}(X) = \{f \mid f \in C^b_{\mathbf{R}}(X), u_a(f) = 0\}.$$

Since u_a is continuous, we deduce that $C^b_{\mathbf{R},a}(X)$ is closed. We leave it to the reader to verify that $C^b_{\mathbf{R},a}(X)$ is a subalgebra.*

Notations.—If X is compact, we have $C^b_{\mathbf{R}}(X) = C_{\mathbf{R}}(X)$ (see 9.11). In this case, for $a \in X$, we write $C_{\mathbf{R},a}(X)$ instead of $C^b_{\mathbf{R},a}(X)$.

The Banach Algebra $C_{\mathbf{R},\infty}(X)$

Let $(X, \mathscr{T})$ be a *locally compact space.* We say that $f: X \to \mathbf{R}$ *vanishes at infinity* if for every $\varepsilon > 0$ there is a compact $K_\varepsilon \subset X$ such that

$$|f(x)| \leq \varepsilon \quad \text{whenever} \quad x \in X - K_\varepsilon.$$

Let X' be the one-point compactification of X and let ω be the point at infinity. For each mapping $f: X \to \mathbf{R}$, let $f': X' \to \mathbf{R}$ be defined by

$$f'(x) = \begin{cases} f(x) & \text{if} \quad x \in X \\ 0 & \text{if} \quad x = \omega. \end{cases}$$

Notice that $f \mapsto f'$ is an injection of† $\mathscr{F}(X, \mathbf{R})$ into $\mathscr{F}(X', \mathbf{R})$, and that

18.19 $\quad (\lambda f + \mu g)' = \lambda f' + \mu g' \quad \text{and} \quad (fg)' = f'g'$

for $\lambda \in \mathbf{R}$, $\mu \in \mathbf{R}$, $f \in \mathscr{F}(X, \mathbf{R})$, and $g \in \mathscr{F}(X, \mathbf{R})$.

We denote by $C_{\mathbf{R},\infty}(X)$ the set of all functions $f \in C_{\mathbf{R}}(X)$ that vanish at infinity. Then

18.20 Theorem.—*We have* $f \in C_{\mathbf{R},\infty}(X)$ *if and only if* $f' \in C_{\mathbf{R},\omega}(X')$.

---

* Notice, for instance, that $u_a(\lambda f + \mu g) = \lambda u_a(f) + \mu u_a(g)$ and that $u_a(fg) = u_a(f)u_a(g)$ for every $\lambda \in \mathbf{R}$, $\mu \in \mathbf{R}$, $f \in C^b_{\mathbf{R}}(X)$, and $g \in C^b_{\mathbf{R}}(X)$.

† For a set E, we denote by $\mathscr{F}(X, \mathbf{R})$ the *algebra* of all mappings of X into $\mathbf{R}$.

Proof.—*Assume* $f \in C_{\mathbf{R},\infty}(X)$. Let $\varepsilon > 0$ and let $K_\varepsilon \subset X$ be a compact set such that $|f(x)| \leq \varepsilon$ if $x \in X - K_\varepsilon$. Then $|f'(x)| \leq \varepsilon$ if $x \in X' - K_\varepsilon$. Since $X' - K_\varepsilon \in \mathcal{N}_{X'}(\omega)$, and since $\varepsilon > 0$ was arbitrary, we deduce that f' is continuous at ω. Since f' is obviously continuous at every $x \in X$ (X is open in X'), we conclude that $f' \in C_{\mathbf{R},\omega}(X')$.

Assume $f' \in C_{\mathbf{R},\omega}(X')$. Let $\varepsilon > 0$ and let $U \in \mathcal{N}_{X'}(\omega)$, *open*, be such that $|f'(x)| \leq \varepsilon$ if $x \in U$. Let $K_\varepsilon = X' - U$. Then K_ε is a compact part of X and $|f(x)| \leq \varepsilon$ if $x \notin K_\varepsilon$. Since $\varepsilon > 0$ was arbitrary, f vanishes at infinity. Since f is obviously continuous at every $x \in X$, $f \in C_{\mathbf{R},\infty}(X)$.

18.21 Corollary.—*The set* $C_{\mathbf{R},\infty}(X)$ *is a subalgebra of* $C^b_{\mathbf{R}}(X)$.

Proof.—Let $f \in C_{\mathbf{R},\infty}(X)$. By Theorem 18.20, $f' \in C_{\mathbf{R}}(X')$, and hence f' is bounded. We deduce that f is bounded, so that $f \in C^b_{\mathbf{R}}(X)$. Since f was arbitrary, we conclude that $C_{\mathbf{R},\infty}(X) \subset C^b_{\mathbf{R}}(X)$.

The fact that $C_{\mathbf{R},\infty}(X)$ is a subalgebra of $C^b_{\mathbf{R}}(X)$ follows from 18.18, 18.19, and the relation

$$C_{\mathbf{R},\infty}(X) = \{f \mid f' \in C_{\mathbf{R},\omega}(X')\}.$$

The norm on $C_{\mathbf{R},\infty}(X)$ is the restriction to $C_{\mathbf{R},\infty}(X)$ of the norm of $C^b_{\mathbf{R}}(X)$; hence, for $f \in C_{\mathbf{R},\infty}(X)$,

18.22 $$\|f\| = \sup_{x \in X} |f(x)|.$$

Notice that

18.23 $$\|f\| = \|f'\|$$

for every $f \in C_{\mathbf{R},\infty}(X)$.

18.24 Theorem.—*The algebra* $C_{\mathbf{R},\infty}(X)$, *endowed with the norm* $f \mapsto \|f\|$ (*defined by* 18.22) *is a Banach algebra.*

Proof.—We know already that $C_{\mathbf{R},\infty}(X)$ is an algebra, that $f \mapsto \|f\|$ is a norm on $C_{\mathbf{R}\infty,}(X)$, and that $\|fg\| \leq \|f\|\,\|g\|$ for $f \in C_{\mathbf{R},\infty}(X)$ and $g \in C_{\mathbf{R},\infty}(X)$. It remains to show that $C_{\mathbf{R},\infty}(X)$ is complete. Let $(f_n)_{n \in \mathbf{N}}$ be a Cauchy sequence of elements of $C_{\mathbf{R},\infty}(X)$. Then $(f'_n)_{n \in \mathbf{N}}$ is a Cauchy sequence of elements of $C_{\mathbf{R}}(X')$ (see 18.23). Hence, $(f'_n)_{n \in \mathbf{N}}$ converges to a function $g \in C_{\mathbf{R}}(X')$. Since $f'_n \in C_{\mathbf{R},\omega}(X')$ for all $n \in \mathbf{N}$, and since $C_{\mathbf{R},\omega}(X')$ is closed, we deduce that $g \in C_{\mathbf{R},\omega}(X')$. Let $f \in C_{\mathbf{R},\infty}(X)$ be such that $f' = g$. Then

$$\lim_{n \in \mathbf{N}} \|f_n - f\| = \lim_{n \in \mathbf{N}} \|f'_n - f'\| = 0.$$

Hence $(f_n)_{n\in\mathbf{N}}$ is convergent in $C_{\mathbf{R},\infty}(X)$. Since $(f_n)_{n\in\mathbf{N}}$ was arbitrary, we conclude that $C_{\mathbf{R},\infty}(X)$ is complete.

Compact Parts of $C^b_{\mathbf{R}}(X)$

We denote by $(X, \mathcal{T})$ a *topological space*.

A set $\mathcal{H} \subset C_{\mathbf{R}}(X)$ is *equicontinuous at* $a \in X$ if for every $\varepsilon > 0$ there is $V \in \mathcal{N}_X(a)$ such that for *all* $f \in \mathcal{H}$,

$$|f(x) - f(a)| \leq \varepsilon$$

whenever $x \in V$ (the neighborhood V does not depend on $f \in \mathcal{H}$). The set $\mathcal{H} \subset C_{\mathbf{R}}(X)$ is *equicontinuous on* X if $\mathcal{H}$ is equicontinuous at *every* $a \in X$.

Notice that a set $\mathcal{H} \subset C^b_{\mathbf{R}}(X)$ is *bounded* if and only if there is $L \in \mathbf{R}$ such that

$$|f(x)| \leq L$$

for all $f \in \mathcal{H}$ and $x \in X$.

Let $\mathcal{H} \subset C^b_{\mathbf{R}}(X)$ and consider the following two assertions:

(A) *$\mathcal{H}$ is relatively compact;*

(B) *$\mathcal{H}$ is bounded and equicontinuous.*

18.25 Theorem.—*The assertion (A) implies (B).*

Proof.—Since $\mathcal{H}$ is relatively compact, $\mathcal{H}$ is totally bounded, whence bounded (see Remarks (iii) and (iv) at the end of Chapter 16).

Since $\mathcal{H}$ is totally bounded, given $\varepsilon > 0$, there are $f_1, \ldots, f_p$ in $\mathcal{H}$ such that

$$\mathcal{H} \subset V_{\varepsilon/3}(f_1) \cup \ldots \cup V_{\varepsilon/3}(f_p).$$

Now let $a \in X$ and let $V \in \mathcal{N}(a)$ be such that

$$|f_j(x) - f_j(a)| \leq \varepsilon/3$$

for all $j = 1, \ldots, p$ and $x \in V$. Let $f \in \mathcal{H}$. Then $f \in V_{\varepsilon/3}(f_j)$ for some $j = 1, \ldots, p$. Hence

$$\begin{aligned} |f(x) - f(a)| &\leq |f(x) - f_j(x)| + |f_j(x) - f_j(a)| + |f_j(a) - f_j(x)| \\ &\leq 2\,\|f - f_j\| + |f_j(a) - f_j(x)| \\ &\leq 2(\varepsilon/3) + \varepsilon/3 = \varepsilon. \end{aligned}$$

Since $\varepsilon > 0$ was arbitrary, $\mathcal{H}$ is equicontinuous at a. Since a was arbitrary, $\mathcal{H}$ is equicontinuous on X.

18.26 Theorem.—*If $(X, \mathscr{T})$ is compact, then (B) implies (A).*

Proof.—Let $\varepsilon > 0$. For each $z \in X$, there is $V_z \in \mathscr{N}(z)$ such that

$$|f(x) - f(z)| \leq \varepsilon/3$$

for all $f \in \mathscr{H}$ if $x \in V_z$. Since X is compact, there is a finite set $\{z_1, \ldots, z_n\} \subset X$ such that

$$V_{z_1} \cup \ldots \cup V_{z_n} = X.$$

Since $\mathscr{H}$ is bounded, the set $S = \{f(x) \mid (f, x) \in \mathscr{H} \times X\}$ is bounded (in $\mathbf{R}$), and hence it is totally bounded. Hence there is a finite set $\{r_1, \ldots, r_p\} \subset S$ such that

$$S \subset V_{\varepsilon/6}(r_1) \cup \ldots \cup V_{\varepsilon/6}(r_p).$$

Now let $\mathscr{F}$ be the set of all mappings of $\{1, \ldots, n\}$ *into* $\{1, \ldots, p\}$. Notice that for each $f \in \mathscr{H}$, there is $\alpha \in \mathscr{F}$ satisfying $f(z_i) \in V_{\varepsilon/6}(r_{\alpha(i)})$ for all $i = 1, \ldots, n$. Let $\mathscr{G}$ be the set of all $\alpha \in \mathscr{F}$ for which there is $f \in \mathscr{H}$ satisfying

18.27 $\quad f(z_i) \in V_{\varepsilon/6}(r_{\alpha(i)}) \quad$ for all $\quad i = 1, \ldots, n.$

For each $\alpha \in \mathscr{G}$, let $f_\alpha \in \mathscr{H}$ be such that the relations in 18.27 are satisfied. Then

18.28 $$\mathscr{H} \subset \bigcup_{\alpha \in \mathscr{G}} V_\varepsilon(f_\alpha).$$

In fact, let $f \in \mathscr{H}$ and let $\alpha \in \mathscr{G}$ be such that $f(z_i) \in V_{\varepsilon/6}(r_{\alpha(i)})$ for all $i = 1, \ldots, n$. Let $x \in X$ and let $i = 1, \ldots, n$ be such that $x \in V_{z_i}$; then

$$\begin{aligned} |f(x) - f_\alpha(x)| &= |f(x) - f(z_i)| + |f(z_i) - f_\alpha(z_i)| + |f_\alpha(z_i) - f_\alpha(x)| \\ &\leq \varepsilon/3 + |f(z_i) - f_\alpha(z_i)| + \varepsilon/3 \\ &\leq \varepsilon/3 + 2(\varepsilon/6) + \varepsilon/3 = \varepsilon. \end{aligned}$$

Hence, $f \in V_\varepsilon(f_\alpha)$. Since $f \in \mathscr{H}$ was arbitrary, 18.28 is proved. Since $\varepsilon > 0$ was arbitrary, we deduce that $\mathscr{H}$ is totally bounded (notice that $\mathscr{G}$ is a finite set). Since $C_{\mathbf{R}}(X)$ is complete, we conclude that $\mathscr{H}$ is relatively compact (see Remark (v) at the end of Chapter 16).

Exercises for Chapter 18

1. Let (X, p) and (Y, q) be normed spaces and f a linear mapping of X into Y. Then the following assertions are equivalent:

(a) f is continuous;
(b) f is continuous at 0;
(c) there is a constant $k > 0$ such that $q(f(x) \leq kp(x)$ for all x in X.

* 2. Let (X, p) and (Y, q) be normed spaces and denote by $L(X, Y)$ the space of continuous linear mappings of X into Y. Then:

A. $L(X, Y)$ is a vector space under the operations defined by:
$(f + g)(x) = f(x) + g(x)$ for all x in X, f and g in $L(X, Y)$;
$(\lambda f)(x) = \lambda(f(x))$ for all λ in R, x in X, and f in $L(X, Y)$.

B. The mapping $h:f \mapsto \sup \{q(f(x)) \mid p(x) \leq 1\}$ is a norm on $L(X, Y)$.

C. $(L(X, Y), h)$ is a Banach space if Y is.

D. $L(X, X)$ is an algebra under the operations given in A and the multiplication defined by $f \circ g(x) = f(g(x))$ for all x in X, f and g in $L(Y, Y)$.

E. $(L(X, X), h)$ is a Banach algebra (with the multiplication introduced in D) if X is a Banach space.

3. Let (X, p) and (Y, q) be normed spaces and f a continuous linear mapping of X into Y. Then (see 2B)

$$h(f) = \inf \{k \mid g(f(x)) \leq kp(x) \text{ for all } x \text{ in } X\}$$

(see 1(c)).

* 4. Let (X, p), (Y, q), and (Z, r) be normed spaces and consider a bilinear map $f: X \times Y \to Z$; that is, a function such that $x \mapsto f(x, y)$ is linear for each y in Y and $y \mapsto f(x, y)$ is linear for each x in X. Then f is continuous if and only if there is $k > 0$ such that $r(f(x, y)) \leq kp(x)q(y)$ for every x in X and y in Y.

5. A subset A of a vector space X is convex if $tx + (1 - t)y \in A$ whenever $t \in [0, 1]$, $x \in A$, and $y \in A$. Let (X, p) be a normed space. Show that every convex subset of X is arcwise connected.

6. Let (X, p) be a normed space. Then for each a in X there is a fundamental system of convex neighborhoods.

7. Every normed space is locally connected.

8. Let $(X, \mathscr{T})$ be a topological space and $\mathscr{F}$ an equicontinuous family of mappings of X into $\mathbf{R}$. Suppose that for each t in X there is a constant M_t such that $f(t) \leq M_t$ for all f in $\mathscr{F}$. Show that the mapping

$$t \mapsto \sup \{f(t) \mid f \in \mathscr{F}\}$$

is continuous.

9. Let X be a compact space and $(f_n)_{n \in \mathbf{N}}$ a sequence of elements of $C_{\mathbf{R}}(X)$ such that $f_n \geq f_{n+1}$ for all $n \in \mathbf{N}$ and $\lim_{n \in \mathbf{N}} f_n(x) = 0$ for all $x \in X$. Then $\lim_{n \in \mathbf{N}} \|f_n\| = 0$.

Chapter 19

The Stone-Weierstrass Theorem*

The main results of this chapter are Theorem 19.5 and Corollary 19.8. For the statements and proofs of these results, we need several definitions and auxiliary results.

We notice that for every $t \in [0, 1]$,

$$(1) \qquad 1 - \tfrac{1}{2}\sqrt{t} \leq \frac{2}{1 + \sqrt{t}},$$

and that for $n \in \mathbf{N}$ and $t \in [0, 1]$,

$$(2) \qquad \frac{2\sqrt{t}}{2 + n\sqrt{t}}\left(1 - \frac{\sqrt{t}}{2}\right) \leq \frac{2\sqrt{t}}{2 + (n + 1)\sqrt{t}}.$$

The relation in (1) is immediate. To prove (2), we observe that, for $n \in \mathbf{N}$ and $t \in [0, 1]$, we have

$$1 - \frac{\sqrt{t}}{2} \leq 1 - \frac{\sqrt{t}}{2 + (n + 1)\sqrt{t}} = \frac{2 + (n + 1)\sqrt{t} - \sqrt{t}}{2 + (n + 1)\sqrt{t}}$$

$$= \frac{2 + n\sqrt{t} + \sqrt{t} - \sqrt{t}}{2 + (n + 1)\sqrt{t}} = \frac{2 + n\sqrt{t}}{2 + (n + 1)\sqrt{t}}.$$

19.1 Theorem.—*Let β be the mapping $t \mapsto \sqrt{t}$ of $[0, 1]$ into $\mathbf{R}$. Then there is a sequence of polynomial functions† belonging to $C_{\mathbf{R}}([0, 1])$ that converges to β in $C_{\mathbf{R}}([0, 1])$.*

---

* This chapter can be omitted in a first reading.

† By a (real) polynomial function on $E \subset \mathbf{R}$ to $\mathbf{R}$, we mean a function $p: E \to \mathbf{R}$ defined by

$$p(x) = a_n x^n + \ldots + a_1 x + a_0 \quad \text{for} \quad x \in E,$$

where $a_n, \ldots, a_1, a_0$ are $n + 1$ real numbers.

Proof.—We define the sequence $(p_n)_{n \in \mathbf{N}}$ of polynomial functions belonging to $C_{\mathbf{R}}([0, 1])$ as follows: Let $p_1(t) = (\frac{1}{2})t$ for $t \in [0, 1]$. Assume now that $n \in \mathbf{N}$ and that $p_n \in C_{\mathbf{R}}([0, 1])$ has been defined. We write, then,

$$p_{n+1}(t) = p_n(t) + \tfrac{1}{2}(t - p_n(t)^2) \quad \text{for} \quad t \in [0, 1].$$

Notice that for all $n \in \mathbf{N}$ and $t \in [0, 1]$,

$$\begin{aligned}(3)\quad (\sqrt{t} - p_n(t))(1 - \tfrac{1}{2}(\sqrt{t} + p_n(t))) &= \sqrt{t} - p_n(t) - \tfrac{1}{2}(t - p_n(t)^2) \\ &= \sqrt{t} - (p_n(t) + \tfrac{1}{2}(t - p_n(t)^2)) = \sqrt{t} - p_{n+1}(t).\end{aligned}$$

We shall now show that

$$(4) \qquad 0 \leq \sqrt{t} - p_n(t) \leq \frac{2\sqrt{t}}{2 + n\sqrt{t}}$$

for all $n \in \mathbf{N}$ and $t \in [0, 1]$. In fact, let $\mathbf{A}$ be the set of *all* $n \in \mathbf{N}$ such that (4) is satisfied for all $t \in [0, 1]$. By (1), $1 \in \mathbf{A}$. Assume now that $n \in \mathbf{A}$. By (3) and (2),

$$0 \leq \sqrt{t} - p_{n+1}(t) \leq (\sqrt{t} - p_n(t))(1 - \tfrac{1}{2}\sqrt{t})$$

$$\leq \frac{2\sqrt{t}}{2 + n\sqrt{t}}(1 - \tfrac{1}{2}\sqrt{t}) \leq \frac{2\sqrt{t}}{2 + (n + 1)\sqrt{t}}$$

for $t \in [0, 1]$. By induction, $\mathbf{A} = \mathbf{N}$. Hence

$$|\sqrt{t} - p_n(t)| \leq \frac{2\sqrt{t}}{2 + n\sqrt{t}} \leq \frac{2}{n}$$

for all $t \in [0, 1]$ and $n \in \mathbf{N}$, and hence $(p_n)_{n \in \mathbf{N}}$ converges to β in $C_{\mathbf{R}}([0, 1])$.

19.2 Theorem.—*Let* $\mathscr{A}$ *be a closed subalgebra of* $C^b_{\mathbf{R}}(X)$. *Then* $|f| \in \mathscr{A}$ *for every* $f \in \mathscr{A}$.

Proof.—Let $f \in \mathscr{A}$ and let $a > 0$ be such that $|f(x)| \leq a$ for $x \in X$. For each $n \in \mathbf{N}$, define the polynomial function q_n on $[-a, +a]$ to $\mathbf{R}$ by

$$q_n(t) = |a| \, p_n((1/a^2)t^2)$$

for $t \in [-a, +a]$.

Let $\varepsilon > 0$ and let $n_\varepsilon \in \mathbf{N}$ be such that $n \geq n_\varepsilon$ implies

$$|\sqrt{t} - p_n(t)| \leq \varepsilon/a$$

for all $t \in [0, 1]$. Then

$$\big||t| - q_n(t)\big| = |a|\,|\sqrt{t^2/a^2} - p_n(t^2/a^2)| \leq \varepsilon$$

if $n \geq n_\varepsilon$ and $t \in [-a, +a]$. Hence

$$\big||f|(x) - q_n \circ f(x)\big| \leq \varepsilon$$

if $n \geq n_\varepsilon$ and $x \in X$. Hence $(q_n \circ f)_{n\in\mathbf{N}}$ converges to $|f|$ in $C_{\mathbf{R}}([0, 1])$. Since $q_n \circ f \in \mathscr{A}$ for every $n \in \mathbf{N}$, and since $\mathscr{A}$ is closed, we deduce that $|f| \in \mathscr{A}$.

Remarks.—It follows that, if $\mathscr{A}$ satisfies the hypotheses of Theorem 19.2, and f and g belong to $\mathscr{A}$, then

$$\inf\{f, g\} = \tfrac{1}{2}(f + g - |f - g|)$$

and

$$\sup\{f, g\} = \tfrac{1}{2}(f + g + |f - g|)$$

belong to $\mathscr{A}$. By induction, we deduce that if $n > 2$, and $f_1, \ldots, f_n$ belong to $\mathscr{A}$, then

$$\inf\{f_1, \ldots, f_n\} \quad \text{and} \quad \sup\{f_1, \ldots, f_n\}$$

belong to $\mathscr{A}$.

19.3 Definition.—*Let X be a set and $\mathscr{F}$ a set of functions on X to* $\mathbf{R}$. *We say that $\mathscr{F}$ separates the points of X if for every $x \in X$, $y \in X$, $x \neq y$ there is f* in *$\mathscr{F}$ satisfying $f(x) \neq f(y)$.*

Example 1.—(i) It will follow from results in Chapter 20 that if X is compact, then $C_{\mathbf{R}}(X)$ separates the points of X.

(ii) If X is a set, $\mathscr{F}$ a set of functions on X to R that separates the points of X, and $\mathscr{F}'$ a set of functions on X to R that contains $\mathscr{F}$, then $\mathscr{F}'$ separates the points of X.

(iii) If $E \subset \mathbf{R}^n$ $(n \geq 1)$, and if $\mathscr{P}_0$ is the set of all polynomial functions* $p: E \to \mathbf{R}$ *without constant term*, then $\mathscr{P}_0$ is a subalgebra of $C_{\mathbf{R}}(E)$ that separates the points of E.

---

* By a (real) polynomial function on $E \subset \mathbf{R}^n$ to $\mathbf{R}$, we mean a function $p: E \to \mathbf{R}$ defined by

$$p(x_1, \ldots, x_n) = \sum_{(p_1, \ldots, p_n) \in I} a_{p_1, \ldots, p_n} x_1^{p_1} \cdots x_n^{p_n}$$

for $(x_1, \ldots, x_n) \in E$, where I is a finite set of n-tuples of positive integers and $a_{p_1, \ldots, p_n} \in \mathbf{R}$ for every $(p_1, \ldots, p_n) \in I$.

(iv) If $E \subset \mathbf{R}^n$ $(n \geq 1)$, and if $\mathscr{P}$ is the set of all polynomial functions $p: E \to \mathbf{R}$, then $\mathscr{P}$ is a subalgebra of $C_{\mathbf{R}}(E)$ that separates the points of E and contains 1.

19.4 Theorem.—*Let X be a set and $\mathscr{A}$ an algebra of functions on X to $\mathbf{R}$. Assume that $\mathscr{A}$ separates the points of X and that for every $x \in X$ there is $f \in \mathscr{A}$ such that $f(x) \neq 0$. Then for every $x \in X$, $y \in X$, $x \neq y$, and $\alpha \in \mathbf{R}$, $\beta \in \mathbf{R}$, there is $f \in \mathscr{A}$ such that $f(x) = \alpha$ and $f(y) = \beta$.*

Proof.—Notice that for every $x \in X$, $y \in X$, $x \neq y$, there is $u \in \mathscr{A}$ such that

$$u(x) = 1 \quad \text{and} \quad u(y) = 0. \tag{5}$$

In fact, let $f \in \mathscr{A}$ such that $f(x) \neq f(y)$ and $g \in \mathscr{A}$ such that $g(x) = 1$. Then

$$u = \frac{f - f(y)}{f(x) - f(y)} g$$

belongs to $\mathscr{A}$ and satisfies (5). Now let $x \in X$, $y \in X$, $x \neq y$, and $\alpha \in \mathbf{R}$, $\beta \in \mathbf{R}$. Let $u \in \mathscr{A}$ be such that $u(x) = 1$ $u(y) = 0$, and let $v \in \mathscr{A}$ be such that $v(x) = 0$ and $v(y) = 1$. Then

$$f = \alpha u + \beta v.$$

belongs to $\mathscr{A}$, $f(x) = \alpha$ and $f(y) = \beta$.

19.5 Theorem (*Stone-Weierstrass*).—*Let X be compact and $\mathscr{A}$ a subalgebra of $C_{\mathbf{R}}(X)$. Suppose that $\mathscr{A}$ separates the points of X and that for every $x \in X$ there is $f \in \mathscr{A}$ such that $f(x) \neq 0$. Then*

$$\bar{\mathscr{A}} = C_{\mathbf{R}}(X).$$

Proof.—Let $\mathscr{B} = \bar{\mathscr{A}}$. We shall show that for every $f \in C_{\mathbf{R}}(X)$ and $\varepsilon > 0$ there is $g \in \mathscr{B}$ such that $\|f - g\| < \varepsilon$. This will show that $\mathscr{B}$ is dense in $C_{\mathbf{R}}(X)$; that is (since $\mathscr{B}$ is closed),

$$\bar{\mathscr{A}} = \mathscr{B} = C_R(X).$$

By 18.14, $\mathscr{B}$ is a subalgebra of $C_{\mathbf{R}}(X)$. By the remarks following Theorem 19.2, for any functions $f_1, \ldots, f_n$ belonging to $\mathscr{B}$,

$$\inf\{f_1, \ldots, f_n\} \in B \quad \text{and} \quad \sup\{f_1, \ldots, f_n\} \in \mathscr{B}.$$

Clearly, $\mathscr{B}$ separates the points of X, and for every $x \in X$ there is $f \in \mathscr{B}$ such that $f(x) \neq 0$. By Theorem 19.4, for every $x \in X$, $y \in X$ $x \neq y$, and $\alpha \in \mathbf{R}$, $\beta \in \mathbf{R}$, there is $f \in C_{\mathbf{R}}(X)$ such that $f(x) = \alpha$ and $f(y) = \beta$.

Now let $f \in C_{\mathbf{R}}(X)$ and $\varepsilon > 0$. For every $x \in X$ and $y \in X$, let $u_{x,y} \in \mathscr{B}$ be such that $u_{x,y}(x) = f(x)$ and $u_{x,y}(y) = f(y)$.

Let $t \in X$. For each $s \in X$, define

$$G_s = \{x \mid u_{s,t}(x) > f(x) - \varepsilon\};$$

then G_s is open and $G_s \ni s$. Hence, there is a finite set $\{s_1, \ldots, s_n\} \subset X$ such that $G_{s_1} \cup \ldots \cup G_{s_n} = X$. Let

$$u_t = \sup \{u_{s_1,t}, \ldots, u_{s_n,t}\}.$$

Then $u_t \in \mathscr{B}$ and, if $x \in G_{s_j}$ $(1 \leq j \leq n)$,

$$u_t(x) \geq u_{s_j,t}(x) > f(x) - \varepsilon;$$

that is,

$$u_t(x) > f(x) - \varepsilon.$$

Since $G_{s_1} \cup \ldots \cup G_{s_n} = X$, we obtain

$$(6) \qquad u_t(x) > f(x) - \varepsilon \quad \text{for} \quad x \in X.$$

Moreover, since $u_{s,t}(t) = f(t)$ for all $s \in X$, we have $u_t(t) = f(t)$.

For each $t \in X$, define

$$H_t = \{x \mid u_t(x) < f(x) + \varepsilon\};$$

then H_t is open and $H_t \ni t$. Hence there is a finite set $\{t_1, \ldots, t_m\} \subset X$ such that $G_{t_1} \cup \ldots \cup G_{t_m} = X$. Let

$$g = \inf \{u_{t_1}, \ldots, u_{t_m}\}.$$

Then $g \in \mathscr{B}$ and, if $x \in H_{t_j}$ $(1 \leq j \leq m)$,

$$g(x) \leq u_{t_j}(x) < f(x) + \varepsilon;$$

that is,

$$g(x) < f(x) + \varepsilon.$$

Since $H_{t_1} \cup \ldots \cup H_{t_m} = X$, we obtain

$$(7) \qquad g(x) < f(x) + \varepsilon \quad \text{for} \quad x \in X.$$

From (6), we deduce

$$(8) \qquad g(x) > f(x) - \varepsilon \quad \text{for} \quad x \in X.$$

Comparing (7) and (8), we conclude $\|f - g\| \leq \varepsilon$.
Hence, Theorem 19.5 is proved.

Remark.—If $\mathscr{A}$ separates the points of X, and $1 \in \mathscr{A}$, then $\mathscr{A}$ satisfies the conditions of Theorem 19.5.

Exercise.—If $\mathscr{A} \subset C_{\mathbf{R}}(X)$ and $\overline{\mathscr{A}} = C_{\mathbf{R}}(X)$, then $\mathscr{A}$ separates the points of X, and for every $x \in X$ there is $f \in \mathscr{A}$ such that $f(x) \neq 0$.

Example 2.—Let E be a compact part of $\mathbf{R}^n$ $(n \geq 1)$. Then (see Example 1(iv)) $\overline{\mathscr{P}} = C_{\mathbf{R}}(E)$.

Example 3.—Let $(X_i)_{i \in I}$ be a family of compact spaces and X its product. For each family $(\varphi_i)_{i \in I}$, where $\varphi_i \in C_{\mathbf{R}}(X_i)$ for all $i \in I$ and $\{i \mid \varphi_i \neq 1\}$ is finite, let $\bigotimes_{i \in I} \varphi_i$ be defined by

$$\bigotimes_{i \in I} \varphi_i((x_i)_{i \in I}) = \prod_{i \in I} \varphi_i(x_i)$$

for all $(x_i)_{i \in I} \in X$. Let $\mathscr{H}$ be the set of all functions of the form $\bigotimes_{i \in I} \varphi_i$. Notice that if $f \in \mathscr{H}$ and $g \in \mathscr{H}$, then $fg \in \mathscr{H}$, and that the constant functions belong to $\mathscr{H}$. Let $\mathscr{A}$ be the set of all functions on X to $\mathbf{R}$ of the form $\sum_{i \in I} \lambda_i f_i$, where I is finite, $\lambda_i \in \mathbf{R}$ for all $i \in I$, and $f_i \in \mathscr{H}$ for all $i \in I$. Then $\mathscr{A}$ is a subalgebra of $C_{\mathbf{R}}(X)$ and, clearly, $\mathscr{A}$ separates the points of X. By Theorem 19.4, $\overline{\mathscr{A}} = C_{\mathbf{R}}(X)$.

19.6 Corollary.—*Let X be a compact space and let $\mathscr{H} \subset C_{\mathbf{R}}(X)$ be a set separating the points of X. Let $\mathscr{A}$ be the set of all functions of the form $p(f_1, \ldots, f_n)$, where $n \in \mathbf{N}, f_1, \ldots, f_n$ are elements of $\mathscr{H}$, and $p: \mathbf{R}^n \to \mathbf{R}$ is a polynomial function. Then*

$$\overline{\mathscr{A}} = C_{\mathbf{R}}(X).$$

Proof.—It is easy to see that $\mathscr{A}$ is a subalgebra and that $1 \in \mathscr{A}$. The conclusions follow from Theorem 19.5 (see also the remark at the end of Theorem 19.5).

For every topological space X and $a \in X$, we denote by $C_{\mathbf{R},a}(X)$ the *subalgebra* of $C_{\mathbf{R}}(X)$ consisting of all functions belonging to $C_{\mathbf{R}}(X)$ and vanishing at a. If X is compact, it is obvious that $C_{\mathbf{R},a}(X)$ is closed.

19.7 Corollary.—*Let X be a compact space, $a \in X$, and $\mathscr{H} \subset C_{\mathbf{R},a}(X)$ a subalgebra separating the points of X. Then*

$$\overline{\mathscr{H}} = C_{\mathbf{R},a}(X).$$

Proof.—Let $\mathscr{A}$ be the set of all functions of the form $p(f_1, \ldots, f_n)$, where $n \in \mathbf{N}$, $f_1, \ldots, f_n$ are elements of $\mathscr{H}$ and $p\colon \mathbf{R}^n \to \mathbf{R}$ is a polynomial function.* It is easy to see that a function g belongs to $\mathscr{A}$ if and only if $g = u + \lambda$, where $u \in \mathscr{H}$ and $\lambda \in \mathbf{R}$.

Now let $f \in C_{\mathbf{R},a}(X)$. By 19.6 there is a sequence $(g_n)_{n\in\mathbf{N}}$ of functions belonging to $\mathscr{A}$ such that

$$\lim_{n\in\mathbf{N}} \|f - g_n\| = 0.$$

For each $n \in \mathbf{N}$, let $u_n \in \mathscr{H}$ and $\lambda_n \in \mathbf{R}$ be such that $g_n = u_n + \lambda_n$ Then

$$\lim_{n\in\mathbf{N}} |\lambda_n| = \lim_{n\in\mathbf{N}} |f(a) - (u_n + \lambda_n)(a)| = 0.$$

Since

$$\|f - u_n\| = \|f - (g_n - \lambda_n)\| \le \|f - g_n\| + |\lambda_n|$$

for all $n \in \mathbf{N}$, we deduce

$$\lim_{n\in\mathbf{N}} \|f - u_n\| = 0.$$

Since $f \in C_{\mathbf{R},a}(X)$ was arbitrary, Corollary 19.7 is proved.

19.8 Corollary.—*Let X be locally compact and $\mathscr{H}$ a subalgebra of $C_{\mathbf{R},\infty}(X)$. Suppose that $\mathscr{H}$ separates the points of X and that for every $x \in X$ there is $f \in \mathscr{H}$ such that $f(x) \neq 0$. Then*

$$\overline{\mathscr{H}} = C_{\mathbf{R},\infty}(X).$$

Proof.—Let X' be the one-point compactification of X and let ω be the point at infinity. For each $f \in C_{\mathbf{R},\infty}(X)$, let $f'\colon X' \to \mathbf{R}$ be defined by

$$f'(x) = \begin{cases} f(x) & \text{if } x \in X \\ 0 & \text{if } x = \omega. \end{cases}$$

Then $f \mapsto f'$ is an injection of $C_{\mathbf{R},\infty}(X)$ onto $C_{\mathbf{R},\omega}(X')$ (see Chapter 18).

Let $\mathscr{A} = \{f' \mid f \in \mathscr{H}\}$. Clearly, $\mathscr{A} \subset C_{\mathbf{R},\omega}(X)$ and $\mathscr{A}$ separates the points of X'. By Corollary 19.7, $\overline{\mathscr{A}} = C_{\mathbf{R},\omega}(X')$. Hence, for $f \in C_{\mathbf{R},\infty}(X)$ and $\varepsilon > 0$, there is $g \in \mathscr{A}$ such that

$$\|f - g\| = \|f' - g'\| \le \varepsilon.$$

Hence $\overline{\mathscr{H}} = C_{\mathbf{R},\infty}(X)$.

---

* If p is a polynomial function without constant term, then $p(f_1, \ldots, f_n) \in \mathscr{H}$.

Complex Valued Functions

We have defined Banach spaces having the real numbers for scalars, in particular the spaces $C_{\mathbf{R}}(X)$ (X compact) and $C_{\mathbf{R},\infty}(X)$ (X locally compact). In the same way, we may define Banach spaces having the complex numbers for scalars. In particular, for *X compact*, we may define the space $C_{\mathbf{C}}(X)$ of all continuous functions on X to $\mathbf{C}$ ($\mathbf{C}$ = the field of complex numbers), and for *X locally compact*, we may define the space $C_{\mathbf{C},\infty}(X)$ of all continuous functions on X to $\mathbf{C}$ that vanish at the infinity. We leave to the reader the complete formulation and discussion of these definitions.

For every complex number $z = x + iy$ ($x \in \mathbf{R}, y \in \mathbf{R}$), we write $\mathscr{R}z = x$, $\mathscr{I}z = y$, and $\bar{z} = x - iy$. If $f: X \to \mathbf{C}$, we denote by $\bar{f}$ the mapping $x \mapsto \overline{f(x)}$ of X into $\mathbf{C}$. If $f \in C_{\mathbf{C}}(X)$, then $\bar{f} \in C_{\mathbf{C}}(X)$ and $\|f\| = \|\bar{f}\|$. This result holds if were place $C_{\mathbf{C}}(X)$ by $C_{\mathbf{C},\infty}(X)$. For every $f: X \to \mathbf{C}$, we denote by $\mathscr{R}f$ the mapping $x \mapsto \mathscr{R}f(x)$ of X into $\mathbf{C}$, and by $\mathscr{I}f$ the mapping $x \mapsto \mathscr{I}f(x)$ of X into $\mathbf{C}$. If $f \in C_{\mathbf{C}}(X)$, then $\mathscr{R}f$ and $\mathscr{I}f$ belong to $C_{\mathbf{C}}(X)$; if $f \in C_{\mathbf{C},\infty}(X)$, then $\mathscr{R}f$ and $\mathscr{I}f$ belong to $C_{\mathbf{C},\infty}(X)$.

The Stone-Weierstrass theorem (19.5) was stated and proved in the setting of real-valued functions. We observe here that without some modifications, this theorem does not hold in the setting of complex-valued functions. This is shown in the following example.

Example 4.—Let $X = \{z \mid z \in \mathbf{C}, |z| \leq 1\}$ and let $\mathscr{A}$ be the set of all $f \in C_{\mathbf{C}}(X)$ such that $f \mid \mathring{X}$ is holomorphic on $\mathring{X}$. Since the mapping $z \mapsto z$ of X into $\mathbf{C}$ belongs to $\mathscr{A}$, $\mathscr{A}$ separates the points of X. Since $1 \in \mathscr{A}$, $\mathscr{A}$ satisfies the hypotheses of Theorem 19.5. However,

$$\mathscr{A} \neq C_{\mathbf{C}}(X).$$

In fact, if $\overline{\mathscr{A}} = C_{\mathbf{C}}(X)$, we would deduce that for every function $f \in C_{\mathbf{C}}(X), f \mid \mathring{X}$ is holomorphic on $\mathring{X}$, which is a contradiction.*

* If f is the mapping $z \mapsto \bar{z}$ of X into $\mathbf{C}$, then $f \mid \mathring{X}$ is not holomorphic on $\mathring{X}$.

Let X be compact. A subalgebra $\mathscr{A}$ of $C_{\mathbf{C}}(X)$ is said to be *involutive* if $f \in \mathscr{A}$ implies that $\bar{f} \in \mathscr{A}$. Notice that if $\mathscr{A}$ is involutive, and $f \in \mathscr{A}$, then $\mathscr{R}f \in \mathscr{A}$ and $\mathscr{I}f \in \mathscr{A}$. In the same way, we define involutive subalgebras of $C_{\mathbf{C},\infty}(X)$.

We may now state and prove the following theorem, which corresponds to 19.5.

19.9 Theorem (*Stone-Weierstrass*).—*Let X be compact and $\mathscr{A}$ an involutive subalgebra of $C_{\mathbf{C}}(X)$. Suppose that $\mathscr{A}$ separates the points of X and that for every $x \in X$ there is $f \in \mathscr{A}$ such that $f(x) \neq 0$. Then*

$$\bar{\mathscr{A}} = C_{\mathbf{C}}(X).$$

Proof.—Let $\mathscr{A}_{\mathbf{R}} = \{\mathscr{R}f \mid f \in \mathscr{A}\}$. Since $\mathscr{A}$ is involutive, $\mathscr{A}_{\mathbf{R}} \subset \mathscr{A}$ and $\mathscr{A}_{\mathbf{R}}$ is a subalgebra. We leave it to the reader to establish that $\mathscr{A}_{\mathbf{R}}$ separates the points of X and that for every $x \in X$ there is $f \in \mathscr{A}_{\mathbf{R}}$ such that $f(x) \neq 0$. By 19.5, $\bar{\mathscr{A}}_{\mathbf{R}} = C_{\mathbf{R}}(X)$. Now let $f \in C_{\mathbf{C}}(X)$ and $\varepsilon > 0$. Then there are u and v in $\mathscr{A}_{\mathbf{R}}$ such that $\|\mathscr{R}f - u\| \leq \varepsilon/2$ and $\|\mathscr{I}f - v\| \leq \varepsilon/2$. Hence $(f = \mathscr{R}f + i\mathscr{I}f)$,

$$\|f - (u + iv)\| \leq \varepsilon.$$

Since $u \in \mathscr{A}_{\mathbf{R}}$, $v \in \mathscr{A}_{\mathbf{R}}$ and since $\varepsilon > 0$ was arbitrary, Theorem 19.9 is proved.

The following result corresponds to 19.8. Its proof is left to the reader.

19.10 Corollary.—*Let X be locally compact and let $\mathscr{H}$ be an involutive subalgebra of $C_{\mathbf{C},\infty}(X)$. Suppose that $\mathscr{H}$ separates the points of X and that for every $x \in X$ there is $f \in \mathscr{H}$ such that $f(x) \neq 0$. Then*

$$\bar{\mathscr{H}} = C_{\mathbf{C},\infty}(X).$$

Exercises for Chapter 19

1. Let I be a non-void set, and for each i in I let $X_i = [0, 1]$. Let $X = \prod_{i \in I} X_i$ with the product topology. Show that for each g in $C_{\mathbf{R}}(X)$ and $\varepsilon > 0$ there are n indices $i_1, \ldots, i_n$ and a polynomial p on $\mathbf{R}^n$ such that

$$\sup_{\mathbf{x} \in X} |g(\mathbf{x}) - p \circ (pr_{i_1}, \ldots, pr_{i_n})(\mathbf{x})| < \varepsilon.$$

2. Show that the set of all functions of the form

$$x \mapsto \sum_{n=1}^{k} \alpha_n \exp(-\beta_n x),$$

with $k \in \mathbf{N}$, $\alpha_n \in \mathbf{R}$, $\beta_n \in \mathbf{R}_+$, is dense in $C_{\mathbf{R},\infty}([0, \infty))$.

3. Consider the locally compact space $(\mathbf{N}, \mathscr{P}(\mathbf{N}))$. Show that the set of all functions of the form $f: n \mapsto \sum_{j=1}^{k} \alpha_j(1/n^j)$ $(n \in \mathbf{N})$, where $k \in \mathbf{N}$ and $\alpha_j \in \mathbf{R}$ for each j, $1 \leq j \leq k$, is dense in $C_{\mathbf{R},\infty}(\mathbf{N})$.

4. Consider the compact subset of the complex plane $\{e^{i\theta} \mid \theta \in \mathbf{R}\} = \Gamma$. Show that the set of all functions of the form

$$x \mapsto \sum_{n \in J} \alpha_n \exp(in \mathscr{R}x), \ (x \in \Gamma),$$

where J is a finite subset of $\mathbf{Z}$ and $\alpha_n \in \mathbf{C}$, for each n in J, is dense in $C_{\mathbf{C}}(\Gamma)$.

Chapter 20

Normal Spaces

Let $(X, \mathscr{T})$ be a topological space and $f: X \to R$. We call "*support* of f" and we denote by Supp (f) the *adherence* of the set

$$\{x \mid f(x) \neq 0\}.$$

20.1 Definition.—*A topological space* $(X, \mathscr{T})$ *is normal if it is separated and if:*

(N_1) *For every closed set* F *and open set* U *containing* F *there is* $f \in C_{\mathbf{R}}(X)$ *such that:*

(ε.1) $0 \leq f(x) \leq 1$ *for* $x \in X$;
(ε.2) $f(x) = 1$ *for* $x \in F$;
(ε.3) Supp $(f) \subset U$.

Various examples of normal spaces will be given later.

20.2 Theorem.—*Let* $(X, \mathscr{T})$ *be a topological space. Then* (N_1) *is equivalent with each of the following conditions:*

(N_2) *for any closed disjoint parts of* X, A *and* B, *there is* $f \in C_{\mathbf{R}}(X)$ *such that* $0 \leq f(x) \leq 1$ *for all* $x \in X$, $f(x) = 1$ *for* $x \in A$, *and* $f(x) = 0$ *for* $x \in B$;

(N_3) *for any closed disjoint parts of* X, A, B, *and real numbers* a *and* b *there is* $f \in C_{\mathbf{R}}(X)$ *such that* $a \leq f(x) \leq b$ *for all* $x \in X$, $f(x) = a$ *for* $x \in A$, *and* $f(x) = b$ *for* $x \in B$.

Proof of $(N_1) \Rightarrow (N_2)$.—Let A and B be closed disjoint parts of X. Let $F = A$ and $U = \mathbf{C}B$. Then F is closed, U is open, and $F \subset U$. By (N_1), there is $f \in C_{\mathbf{R}}(X)$ such that $0 \leq f(x) \leq 1$ for $x \in X$, $f(x) = 1$ for $x \in A$, and Supp $(f) \subset U$. Hence $f(x) = 0$ for $x \in B$, and hence (N_2) is satisfied.

Proof of $(N_2) \Rightarrow (N_1)$.—Let F be a closed set and U an open set containing F. Let $A = F$ and $B = \mathbf{C}U$. By (N_2), there is $g \in C_{\mathbf{R}}(X)$ such that $0 \leq g(x) \leq 1$ for $x \in X$, $g(x) = 1$ for $x \in F$, and $g(x) = 0$ for $X \in B$. Let

$$C = \{x \mid g(x) \leq \tfrac{1}{2}\} \quad \text{and} \quad D = \{x \mid g(x) \geq \tfrac{1}{2}\}.$$

Then D is closed and contained in U. Also, C is closed and $A \cap C = \varnothing$. By (N_2), there is $f \in C_{\mathbf{R}}(X)$ such that $0 \leq f(x) \leq 1$ for $x \in X, f(x) = 1$ for $x \in F$, and $f(x) = 0$ for $x \in C$. Hence, if $f(x) \neq 0$, then $x \notin C$, so that $x \in D$. Hence

$$\{x \mid f(x) \neq 0\} \subset D.$$

Since D is closed, Supp $(f) \subset D$. Since $D \subset U$, we deduce that (N_1) is satisfied.

Proof of $(N_2) \Rightarrow (N_3)$.—Let A and B be closed disjoint parts of X and let $g \in C_{\mathbf{R}}(X)$ be such that $0 \leq g(x) \leq 1$ for all $x \in X$, $g(x) = 1$ for $x \in A$, and $g(x) = 0$ for $x \in B$. The function $f = a + (b - a)g \in C_{\mathbf{R}}(X)$ (see Example 10, Chapter 5), and clearly $a \leq f(x) \leq b$ for $x \in X$, $f(x) = a$ for $x \in A$, and $f(x) = b$ for $x \in B$. Hence (N_3) is satisfied.

Proof of $(N_3) \Rightarrow (N_2)$.—Obvious.

Further conditions that are equivalent with (N_1) are given in Theorem 20.5. First, we will prove Theorems 20.3 and 20.4, which are used in the proof of Theorem 20.5.

20.3 Theorem.—*Let* $(X, \mathscr{T})$ *be a topological space and* $D \subset [0, 1]$ *dense in* $[0, 1]$. *Let* A *and* B *be disjoint closed parts of* X *and* $U: D \to \mathscr{T}$ *a function having the following properties:*

(a) $A \subset U(s) \subset \mathbf{C}B$ *for all* $s \in D$;
(b) $\overline{U(s)} \subset U(t)$ *for* s *and* t *in* D *such that* $s < t$.
Then there is $f \in C_{\mathbf{R}}(X)$ *such that* $0 \leq f(x) \leq 1$ *for* $x \in X$, $f(x) = 1$ *for* $x \in A$, *and* $f(x) = 0$ *for* $x \in B$.

Proof.—Define $V: \mathbf{R} \to \mathscr{T}$ by

$$V(r) = \begin{cases} \bigcup_{s \in D, s < r} U(s) & \text{if } r \leq 1 \\ X & \text{if } r > 1. \end{cases}$$

Clearly, $V(r) = \varnothing$ if $r < 0$, $V(r) = X$ if $r > 1$, and $\overline{V(r')} \subset V(r'')$ if

$r' < r''$. Define now $f: X \to \mathbf{R}$ by

$$f(x) = \inf \{r \mid V(r) \ni x\}.$$

Since $V(r) = \varnothing$ if $r < 0$, and $V(r) = X$ if $r > 1$, we have $0 \leq f(x) \leq 1$ for all $x \in X$. Since $V(r) = \varnothing$ if $r < 0$, and $V(r) \supset A$ if $r > 0$, we deduce that $f(x) = 0$ for $x \in A$. Since $U(s) \subset \mathbf{C}B$ for all $s \in D$, we deduce that $V(r) \subset \mathbf{C}B$ for $r \leq 1$; hence $f(x) = 1$ for $x \in B$. We still need to prove that f is *continuous*. Let $t \in X$, $c = f(t)$, $\varepsilon > 0$, and

$$V = V(c + \varepsilon) \cap \mathbf{C}\overline{V(c - \varepsilon)};$$

clearly $V \in \mathscr{N}_X(t)$. If $x \in V(c + \varepsilon)$, then $f(x) \leq c + \varepsilon$. If $x \in \mathbf{C}\overline{V(c - \varepsilon)}$, then $x \notin \overline{V(c - \varepsilon)}$, whence $x \notin V(c - \varepsilon)$. Hence $x \notin V(r)$ if $r \leq c - \varepsilon$, so that $f(x) \geq c - \varepsilon$. We deduce that

$$|f(x) - f(t)| \leq \varepsilon$$

if $x \in V$. Since $\varepsilon > 0$ and $t \in X$ were arbitrary, we deduce that $f \in C_{\mathbf{R}}(X)$.

Let

$$D_n = \{k/2^n \mid 0 \leq k \leq 2^n\}$$

for $n \in N$ and *$D = \bigcup_{n \in N} D_n$. Clearly, $D \subset [0, 1]$. We leave it to the reader to show that *D is dense in* $[0, 1]$.

Let $n \in \mathbf{N}$. Then $D_n \subset D_{n+1}$. In fact, if $x \in D_n$, then $x = k/2^n$ with $0 \leq k \leq 2^n$, so that $x = 2k/2^{n+1}$; hence $x \in D_{n+1}$. Now let $x \in D_{n+1}$; then $x = k'/2^{n+1}$ with $0 \leq k' \leq 2^{n+1}$. If $k' = 2m$, with m positive numbers, then $x = m/2^n$ with $0 \leq m \leq 2^n$, so that $x \in D_n$. If $k' = 2m + 1$, with m positive numbers, then $0 \leq m < 2^n$ and

$$\frac{m}{2^n} = \frac{2m}{2^{n+1}} < \frac{2m+1}{2^{n+1}} < \frac{2(m+1)}{2^{n+1}} = \frac{m+1}{2^n},$$

so that $x \notin D_n$. It follows that $x \in D_{n+1} - D_n$ if and only if

$$x = \frac{2m+1}{2^{n+1}} \text{ with } 0 \leq m < 2^n.$$

Moreover, for each $0 \leq k < 2^n$, the number $x = (2k + 1)/2^{n+1}$ is the *only* number in D_{n+1} such that

$$\frac{k}{2^n} < \frac{2k+1}{2^{n+1}} < \frac{k+1}{2^n}.$$

---

* The numbers of the form $k/2^n$, with $k \in \mathbf{Z}$ and $n \in \mathbf{N}$, are called *dyadic numbers*. The set of all dyadic numbers is dense in $\mathbf{R}$.

20.4 Theorem.—*Let $(X, \mathscr{T})$ be a topological space satisfying the condition:*

(N_4) for any closed set F and open set U containing F, there is an open set V such that $F \subset V \subset \overline{V} \subset U$.

Then, for any closed disjoint parts of X, A, and B, there is a function $U: D \to \mathscr{T}$ having the following properties:

(a) *$A \subset U(s) \subset \mathbf{C}B$ for all $s \in D$;*

(b) *$\overline{U(s)} \subset U(t)$ for s and t in D such that $s < t$.*

Proof.—To prove the theorem, it is enough to show that for each $n \in \mathbf{N}$ there is $U_n: D_n \to \mathscr{T}$ such that:

(j) $A \subset U_n(s) \subset \mathbf{C}B$ for all $n \in \mathbf{N}$ and $s \in D$.

(jj) $\overline{U_n(s)} \subset U_n(t)$ for s and t in D_n such that $s < t$.

(jjj) $U_{n+1} \mid D_n = U_n$ for all $n \in \mathbf{N}$.

Then the mapping $U: D \to \mathscr{T}$ defined by $U(s) = U_n(s)$ if $s \in D_n$ clearly satisfies 20.4(a) and (b).

To establish the existence of the sequence $(U_n)_{n \in N}$, we proceed as follows: We define $U_1(1) = \mathbf{C}B$ and we choose $U_1(0) \in \mathscr{T}$ such that

$$A \subset U_1(0) \subset \overline{U_1(0)} \subset U_1(1).$$

Assume now that $p \in \mathbf{N}$ and that we have defined $U_p: D_p \to \mathscr{T}$ satisfying (j) and (jj). We define then U_{p+1} as follows: Let $s \in D_{p+1}$. If $s \in D_p$, we write $U_{p+1}(s) = U_p(s)$. If $s \in D_{p+1} - D_p$, then $s = (2k+1)/2^{n+1}$ for some $0 \leq k < 2^n$. In this case, we choose for $U_{p+1}(s)$ an open set such that*

$$\overline{U_p\left(\frac{k}{2^n}\right)} \subset U_{p+1}\left(\frac{2k+1}{2^{n+1}}\right) \subset \overline{U_{p+1}\left(\frac{2k+1}{2^{n+1}}\right)} \subset U_p\left(\frac{k+1}{2^n}\right).$$

Then $U_{p+1}: D_{p+1} \to \mathscr{T}$ is well-defined and satisfies (j) and (jj), and $U_{p+1} \mid D_p = U_p$.

The main result of this chapter is the following theorem.

20.5 Theorem.—*Let $(X, \mathscr{T})$ be a topological space. Then (N_1) is equivalent with each of the following conditions:*

(N_4) for any closed set F and open set U containing F, there is an open set V such that $F \subset V \subset \overline{V} \subset U$;

---

* Recall that if $0 \leq k < 2^p$, then $(2k+1)/2^{p+1}$ is the only element in D_{p+1} satisfying

$$\frac{k}{2^p} < \frac{2k+1}{2^{p+1}} < \frac{k+1}{2^p}.$$

(N_5) *for any closed disjoint parts of* X, A, *and* B, *there are open sets* U *and* V *such that* $A \subset U$, $B \subset V$, *and* $U \cap V = \varnothing$;

(N_6) *for any closed set* F *and* $f \in C_{\mathbf{R}}^b(F)$, *there is* $\bar{f} \in C_{\mathbf{R}}^b(X)$ *such that* $\bar{f}|F = f$ *and* $\|\bar{f}\| = \|f\|$;

(N_7) *for any closed set* F *and* $f \in C_{\mathbf{R}}(F)$, *there is* $\bar{f} \in C_{\mathbf{R}}(X)$ *such that* $\bar{f}|F = f$.

Remarks.—The equivalence between (N_1) and (N_7) is known as the Tietze-Urysohn theorem.

We have already shown that (N_1), (N_2), and (N_3) are equivalent.

To prove Theorem 20.5, we show first that

$$(N_1) \Rightarrow (N_4) \Rightarrow (N_1) \quad \text{and} \quad (N_4) \Rightarrow (N_5) \Rightarrow (N_4).$$

This will show that (N_1), (N_2), (N_3), (N_4), and (N_5) are equivalent. Afterwards, we will show that

$$(N_1) \Rightarrow (N_6) \Rightarrow (N_1) \quad \text{ana} \quad (N_6) \Rightarrow (N_7) \Rightarrow (N_6).$$

Proof of $(N_1) \Rightarrow (N_4)$.—Let F be a closed set and U an open set containing F. By (N_1), there is $f \in C_{\mathbf{R}}(X)$ such that $0 \leq f(x) \leq 1$, $f(x) = 1$ for $x \in F$ and $f(x) = 0$ for $x \notin U$. Let

$$V = \{x \mid f(x) > \tfrac{1}{2}\} \quad \text{and} \quad V_1 = \{x \mid f(x) \geq \tfrac{1}{2}\}.$$

Then V is open and V_1 is closed, so that $\bar{V} \subset V_1$. Moreover,

$$F \subset V \subset \bar{V} \subset V_1 \subset U.$$

Hence, $(N_1) \Rightarrow (N_4)$ is proved.

Proof of $(N_4) \Rightarrow (N_1)$.—From Theorems 20.4 and 20.3, it follows immediately that $(N_4) \Rightarrow (N_1)$.

Proof of $(N_4) \Rightarrow (N_5)$.—Let A and B be closed disjoint parts of X. By (N_4), there is V open such that $A \subset V \subset \bar{V} \subset B$. Then, if $W = \mathbf{C}\bar{V}$, the set V and W are open, $A \subset V$, $B \subset V$, and $V \cap W = \varnothing$. Hence, $(N_4) \Rightarrow (N_5)$.

Proof of $(N_5) \Rightarrow (N_4)$.—Let F be a closed set and U an open set containing F. By (N_5), there are open sets V_1 and V_2 such that $F \subset V_1$, $\mathbf{C}U \subset V_2$, and $V_1 \cap V_2 = \varnothing$. Again by (N_5), there are open sets W_1

and W_2 such that $F \subset W_1$, $\mathbf{C}V_1 \subset W_2$, and $W_1 \cap W_2 = \varnothing$. Let $V = W_1$. Then $F \subset V$, and

$$V = W_1 \subset \mathbf{C}W_2 \subset V_1 \subset \mathbf{C}V_2 \subset U.$$

Since $\mathbf{C}W_2$ is closed, $\bar{V} \subset \mathbf{C}W_2$. Hence

$$F \subset V \subset \bar{V} \subset U,$$

and hence $(N_5) \Rightarrow (N_4)$.

Proof of $(N_1) \Rightarrow (N_6)$.—Recall that $(N_1) \Leftrightarrow (N_3)$. Let $\alpha \in (0, \frac{1}{3})$ and $\beta = 1 - \alpha$. To each $g \in C^b_{\mathbf{R}}(F)$, we shall associate a function $g' \in C^b_{\mathbf{R}}(X)$ in the following way. If $g = 0$, we take $g' = 0$. If $g \neq 0$, we write

$$A = \{x \mid g(x) \leq -\alpha \, \|g\|\} \quad \text{and} \quad B = \{x \mid g(x) \geq \alpha \, \|g\|\}.$$

Then A and B are disjoint closed parts of X, and hence there is $g' \in C^b_{\mathbf{R}}(X)$ such that $-\alpha \, \|g\| \leq g'(x) \leq \alpha \, \|g\|$ for $x \in X$, $g'(x) = -\alpha \, \|g\|$ for $x \in A$, and $g'(x) = \alpha \, \|g\|$ for $x \in B$.

Define now $T: C^b_{\mathbf{R}}(F) \to C^b_{\mathbf{R}}(X)$ and $S: C^b_{\mathbf{R}}(F) \to C^b_{\mathbf{R}}(F)$ by

$$T(g) = g' \quad \text{and} \quad S(g) = g - (Tg) \mid F$$

for $g \in C^b_{\mathbf{R}}(F)$. Notice that

(i) $\|T(g)\| \leq \alpha \, \|g\|$ and $\|S(g)\| \leq \beta \, \|g\|$ for $g \in C^b_{\mathbf{R}}(F)$.
Let $f \in {}^b_{\mathbf{R}}C(F)$. By induction, it is easily established that*

(ii) $S^n(f) = f - (\sum_{0 \leq i \leq n-1} T \circ S^i(g)) \mid F$
for $n \in \mathbf{N}$. Since

$$\|T \circ S^i(f)\| \leq \alpha\beta^i \, \|f\|$$

(use (i)) for all $i \in \mathbf{N}$, it follows that the series $\sum_{i=0}^{+\infty} T \circ S^i(f)$ is absolutely convergent in $C^b_{\mathbf{R}}(X)$. Let $\bar{f}$ be the sum of this series. Since $\|S^n(f)\| \leq \beta^n \, \|f\|$ (use (i)) for all $n \in \mathbf{N}$, the sequence $(S^n(f))_{n\in\mathbf{N}}$ converges to zero. We deduce from (ii) that

$$f = \bar{f} \mid F.$$

Moreover,

$$\|\bar{f}\| \leq \sum_{i=0}^{+\infty} \|T \circ S^i(f)\| \leq \sum_{i=0}^{+\infty} \alpha \, \beta^i \, \|f\| = \frac{\alpha}{1-\beta} \, \|f\| = \|f\|;$$

* We write $S^\circ(f) = f$.

since, obviously, $\|f\| \leq \|\bar{f}\|$, we deduce that $\|f\| = \|\bar{f}\|$. Hence $(N_1) \Rightarrow (N_6)$.

Proof of $(N_6) \Rightarrow (N_1)$.—Recall that $(N_1) \Leftrightarrow (N_5)$. Let A and B be closed disjoint parts of X. Let $f: A \cup B \to \mathbf{R}$ be defined by $f(x) = 1$ for $x \in A$ and $f(x) = 0$ for $x \in B$. Then $f \in C_{\mathbf{R}}^b(A \cup B)$. By (N_6), there is $\bar{f} \in C_{\mathbf{R}}(X)$ such that $\bar{f} \mid A \cup B = f$. Let

$$U = \bar{f}^{-1}((\tfrac{1}{2}, +\infty)) \quad \text{and} \quad V = \bar{f}^{-1}((-\infty, \tfrac{1}{2})).$$

Then U and V are open, $A \subset U$, $B \subset V$, and $U \cap V = \varnothing$. Hence, $(N_6) \Rightarrow (N_5)$, and hence $(N_6) \Rightarrow (N_1)$.

Proof of $(N_6) \Rightarrow (N_7)$.—Recall that $(N_6) \Leftrightarrow (N_2)$. The mapping $h: x \mapsto x/(1 + |x|)$ is a homeomorphism of $\mathbf{R}$ onto the subspace $(-1, +1)$ *of* $\mathbf{R}$. Now let F be a closed subset of X and $f \in C_{\mathbf{R}}(F)$. Then $h \circ f \in C_{\mathbf{R}}^b(F)$. By (N_6), there is $g \in C_{\mathbf{R}}(X)$ such that $g \mid F = h \circ f$. Since $g(F) \subset (-1, +1)$, the closed sets

$$F \quad \text{and} \quad B = \{x \mid |g(x)| \geq 1\}$$

are disjoint. By (N_2), there is $\varphi \in C_{\mathbf{R}}^b(X)$ such that $0 \leq \varphi(x) \leq 1$ for $x \in X$, $\varphi \mid F = 1$ and $\varphi \mid B = 0$. Then $\varphi g(X) \subset (-1, +1)$ and $\varphi g \mid F = g \mid F = h \circ f$. We deduce that $\bar{f} = h^{-1} \circ (\varphi g) \in C_{\mathbf{R}}(X)$ and $\bar{f} \mid F = h^{-1}(h \circ f) = f$. Hence $(N_6) \Rightarrow (N_7)$.

Proof of $(N_7) \Rightarrow (N_6)$.—We show that $(N_7) \Rightarrow (N_5)$ in the same way as we proved that $(N_6) \Rightarrow (N_5)$. Since $(N_5) \Rightarrow (N_1) \Rightarrow (N_6)$, it follows that $(N_7) \Rightarrow (N_6)$.

20.6 Corollary.—*A normal space is regular.*

Proof.—The assertion follows from the fact that in a separated space, a set of the form $\{x\}$, $x \in X$, is closed, and from the fact that $(N_1) \Leftrightarrow (N_4)$.

Theorem 9.8 and the equivalence between (N_1) and (N_5) show that *a compact space is normal.* This is, however, a particular case of the following theorem.

20.7 Theorem.—*A paracompact space* $(X, \mathscr{T})$ *is normal.*

Proof.—Since a paracompact space is separated, it is enough to show that (N_5) is satisfied.

Let, then, A and B be two closed disjoint parts of X. Since $(X, \mathscr{T})$ is regular (see 20.12), and $\mathbf{C}B$ is an open set containing A, we deduce that for every $x \in A$ there is a closed $W_x \in \mathscr{N}_X(x)$ contained in $\mathbf{C}B$. If we define

$$U_x = \mathring{W}_x \quad \text{and} \quad V_x = \mathbf{C}W_x,$$

then $U_x \ni x$ and $V_x \supset B$ are open sets and $U_x \cap V_x = \varnothing$. By 20.11, there are open sets $U_A \supset A$ and $V_B \supset B$ such that $U_A \cap V_B = \varnothing$.

Since A and B were arbitrary, we deduce that (N_5) is satisfied. Hence $(X, \mathscr{T})$ is normal.

20.8 Theorem.—*A metrizable topological space $(X, \mathscr{T})$ is normal.*

Proof.—Since a metrizable space is separated, it is enough to show that (N_5) is satisfied.

Let d be a metric on X such that $\mathscr{T} = \mathscr{T}_d$. Let A and B be two closed disjoint parts of X. The mappings $x \mapsto d(x, A)$ and $x \mapsto d(x, B)$ on X to $\mathbf{R}$ are continuous (see 13.19). If

$$U_A = \{x \mid d(x, A) < d(x, B)\}$$

and

$$V_B = \{x \mid d(x, B) < d(x, A)\},$$

then U_A and V_B are open,* $U_A \supset A$, $U_B \supset B$, and $U_A \cap U_B = \varnothing$. Since A and B were arbitrary, we deduce that (N_5) is satisfied. Hence $(X, \mathscr{T})$ is normal.

Remark.—It can be shown that a metrizable topological space is paracompact.† The fact that a metrizable space is normal follows then from Theorem 20.7.

20.9 Theorem.—*Let $(X, \mathscr{T})$ be a locally compact space, F a compact part, and U an open set containing F. Then there is $f \in C_{\mathbf{R}}(X)$ such that:*

(ε.1) $0 \leq f(x) \leq 1$ *for* $x \in X$;

(ε.2) $f(x) = 1$ *for* $x \in F$;

(ε.3) $\operatorname{Supp}(f) \subset U$.

Proof.—We may and shall suppose that U is relatively compact (see 10.15). Let $(X', \mathscr{T}')$ be a one-point compactification of $(X, \mathscr{T})$ (we assume that $X \subset X'$). Then $(X', \mathscr{T}')$ is normal, and hence, by

---

* Recall that if $F \subset X$ is a non-void set, then $d(x, F) = 0$ if and only if $x \in \bar{F}$.

† See, for instance, [2], Chapter 9 (second edition), pp. 92–95.

(N_1), there is $f' \in C_{\mathbf{R}}(X')$ such that $0 \leq f'(x) \leq 1$ for $x \in X'$, $f'(x) = 1$ for $x \in F$, and Supp $(f') \subset U$. Clearly, then, $f = f' \mid X$ belongs to $C_{\mathbf{R}}(X)$ and satisfies 20.9.

Remark.—Notice that we proved that the function f in Theorem 20.9 can be chosen so that Supp (f) is compact.

20.10 Theorem.—*Let $(X, \mathscr{T})$ be a normal space and $(U_i)_{i \in I}$ a locally finite open covering of X. Then there is an open covering $(V_i)_{i \in I}$ of X such that $\overline{V}_i \subset U_i$ for all $i \in I$.*

Proof.—Let $(U_i)_{i \in I}$ be a locally finite open covering of X. For each $i \in I$, let $\mathscr{U}(i)$ be the set of all open parts V of X such that $\overline{V} \subset U_i$.

Let $\mathcal{O}$ be the set of all *open coverings* $(A_i)_{i \in I}$ of X, where for each $i \in I$ we have either $A_i \in \mathscr{U}(i)$ or $A_i = U_i$. We shall now *order* $\mathcal{O}$ as follows: For any two families $\mathscr{A}' = (A'_i)_{i \in I}$ and $\mathscr{A}'' = (A''_i)_{i \in I}$, we write

$$\mathscr{A}' \leq \mathscr{A}''$$

if and only if $A'_i = A''_i$ *for all $i \in I$ such that $A'_i \in \mathscr{U}(i)$.*

We notice that if $\mathscr{A} = (A_i)_{i \in I} \in \mathcal{O}$, and if $\{i \mid A_i \notin \mathscr{U}(i)\} \neq \varnothing$, then there is $\mathscr{B} \in \mathcal{O}$ such that $\mathscr{A} \leq \mathscr{B}$. In fact, let $\alpha \in I$ be such that $A_\alpha = U_\alpha$ and let

$$F = X - \bigcup_{i \in I - \{\alpha\}} A_i.$$

Then F is closed and, obviously, $F \subset A_\alpha$. By (N_4), there is an open set V such that $F \subset V \subset \overline{V} \subset A_\alpha$. If $B_i = A_i$ for $i \neq \alpha$, and $B_\alpha = V$, then $\mathscr{B} = (B_i)_{i \in I} \in \mathcal{O}$ and $\mathscr{A} < \mathscr{B}$.

We shall now show that $\mathcal{O}$ is *inductive.* Let $(\mathscr{A}(t))_{t \in T}$ be a *totally ordered* family of elements of $\mathcal{O}$ (we assume that T is totally ordered and that $t' \leq t''$ if and only if $\mathscr{A}(t') \leq \mathscr{A}(t'')$). For each $t \in T$, let $\mathscr{A}(t) = (A_i(t))_{i \in I}$. Define $\mathscr{A} = (A_i)_{i \in I}$ by

$$A_i = \bigcap_{t \in T} A_i(t) \quad \text{for all} \quad i \in I.$$

Let $i \in I$. If $A_i(t) = U_i$ for all $t \in T$, then $A_i = U_i$. If for some $t_0 \in T$, $A_i(t_0) \in \mathscr{U}(i)$, then $A_i(t) = A_i(t_0)$ for $t \geq t_0$ (since $\mathscr{A}(t) \geq \mathscr{A}(t_0)$), so that $A_i = A_i(t_0) \in \mathscr{U}(i)$. Now $\mathscr{A}$ is a *covering* of X (see below), so that $\mathscr{A} \in \mathcal{O}$. Moreover, clearly $\mathscr{A}(t) \leq \mathscr{A}$ for all $t \in T$ (in fact, $\mathscr{A} = \sup_{t \in T} \mathscr{A}(t)$). Since $(\mathscr{A}(t))_{t \in T}$ was arbitrary, $\mathcal{O}$ is inductive. By Zorn's lemma, $\mathcal{O}$ contains a maximal element $(V_i)_{i \in I}$. But we have noticed that if $\{i \mid V_i \notin \mathscr{U}(i)\} \neq \varnothing$, then $(V_i)_{i \in I}$ cannot be maximal. Hence, $V_i \in \mathscr{U}(i)$ for all $i \in I$, so that $\overline{V}_i \subset U_i$ for all $i \in I$. Hence, Theorem 20.10 is proved.

We shall show here that $\mathscr{A}$ *is a covering of* X. Let $a \in X$. If there is $(i, t) \in I \times T$ such that $A_i(t) \in \mathscr{U}(i)$ and $a \in A_i(t)$, then $a \in A_i = A_i(t)$, so that $a \in \bigcup_{i \in I} A_i$. Suppose there is no such (i, t). Let $I(a) = \{i \mid a \in U_i\}$; since $(U_i)_{i \in I}$ is locally finite, $I(a)$ is finite. For each $t \in T$, we have $A_i(t) = U_i$ for at least *one* $i \in I(a)$. For each $i \in I(a)$, let $T(i)$ be the set of all $t \in T$ such that $A_i(t) = U_i$. Then at least one of the $T(i)$, $i \in I(a)$, is cofinal, and then for that i, $x \in A_i = U_i$. Hence, $a \in \bigcup_{i \in I} A_i$, so that $\mathscr{A}$ is a covering of X.

Theorem 20.10 will be used in the study of partitions of unity.

In this appendix, we shall prove two results that were used in the proof of Theorem 20.7.

20.11 Theorem.—*Let $(X, \mathscr{T})$ be a paracompact space and let F and T be two closed parts of X. Suppose that for each $x \in F$ there is an open set U_x containing x and an open set V_x containing T such that $U_x \cap V_x = \varnothing$. Then there is an open set U_F containing F and an open set V_T containing T such that*

$$U_F \cap V_T = \varnothing.$$

▼ *Proof.*—Since $\mathbf{C}F \cup \bigcup_{x \in F} U_x = X$, and since $(X, \mathscr{T})$ is paracompact, there is a locally finite open covering of X, $(A_i)_{i \in I}$ such that $i \in I$ and $A_i \cap F \neq \varnothing$ implies that $A_i \subset U_{x(i)}$ for some $x(i) \in F$. We define this way a mapping $\varphi : i \mapsto x(i)$ of the set

$$H = \{i \mid A_i \cap F \neq \varnothing\} \quad \text{into} \quad F.$$

Let $U_F = \bigcup_{i \in H} A_i$; then U_F is an open set containing F.

For each $x \in T$, let S_x be an open neighborhood of x such that

$$J(x) = \{i \in H \mid A_i \cap S_x \neq \varnothing\}$$

is finite. Let

$$W_x = S_x \cap \bigcap_{y \in \varphi(J(x))} V_y.$$

Since $i \in J(x)$ implies $A_i \subset U_{\varphi(i)}$, and since

$$U_{\varphi(i)} \cap \bigcap_{y \in \varphi(J(x))} V_y = \varnothing,$$

we deduce that $A_i \cap W_x = \varnothing$ for $i \in J(x)$ and hence for all $i \in H$. If we define

$$V_T = \bigcup_{x \in T} W_x,$$

then V_T is an open set containing T and $U_F \cap V_T = \varnothing$. ▲

20.12 Corollary.—*A paracompact space $(X, \mathscr{T})$ is regular.*

Proof.—Let $t \in X$ and $V \in \mathscr{N}_X(t)$. Let $T = \{t\}$ and $F = \mathbf{C}\mathring{V}$. Then F and T are closed. Since $(X, \mathscr{T})$ is separated, the hypotheses of Theorem 20.11 are satisfied. Hence, there is an open set $U \ni t$ and an open set $W \supset F$ such that $U \cap W = \varnothing$. Then

$$t \in U \subset \mathbf{C}W \subset V.$$

Since $t \in X$ and $V \in \mathscr{N}_X(t)$ were arbitrary, and since $\mathbf{C}W$ is closed, we deduce that $(X, \mathscr{T})$ is regular.

Exercises for Chapter 20

1. A closed subspace of a normal space is normal.

2. Give an example of a normal space with a subspace that is not normal.

* 3. Let $(X, \mathscr{T})$ be a regular space and suppose there is a countable base for $\mathscr{T}$. Then every subspace of $(X, \mathscr{T})$ is normal.

4. A topological space $(X, \mathscr{T})$ is said to be a T_4 space if $(X, \mathscr{T})$ is both separated and normal. If $(X, \mathscr{T})$ and $(Y, \mathscr{S})$ are topological spaces such that $(X, \mathscr{T})$ is a T_4 space and $f: X \to Y$ is a continuous closed surjection, then $(Y, \mathscr{S})$ is a T_4 space.

5. Let $(X, \mathscr{T})$ be a compact space. The following assertions are equivalent:

(i) $(X, \mathscr{T})$ is metrizable;
(ii) there is a countable base for $\mathscr{T}$;
(iii) $C_{\mathbf{R}}(X)$ with the sup-norm topology has a countable dense set.

6. Let $(X, \mathscr{T})$ be a separated space. Suppose that for every locally finite open covering $(U_i)_{i \in I}$ of X there is an open covering $(V_i)_{i \in I}$ of X such that $\overline{V_i} \subset U_i$ for each i in I. Then $(X, \mathscr{T})$ is normal.

Chapter 21

Partitions of Unity

Let X be a topological space and $\mathscr{F}$ a set of functions on X to $\mathbf{R}$.

21.1 Definition.—*An $\mathscr{F}$-partition of unity of X is a family of positive functions $(f_i)_{i\in I}$ belonging to $\mathscr{F}$ such that* (Supp $(f_i))_{i\in I}$ *is locally finite and** $\sum_{i\in I} f_i = 1$.

If $(A_i)_{i\in I}$ is a family of parts of X and $(f_i)_{i\in I}$ is a family of functions on X to $\mathbf{R}$, we say that $(f_i)_{i\in I}$ is *subordinated* to $(A_i)_{i\in I}$ if

$$\text{Supp } (f_i) \subset A_i$$

for all $i \in I$.

We shall denote by $(X, \mathscr{T})$ a topological space and by $\mathscr{A}$ an *algebra* of functions on X to $\mathbf{R}$ having the following properties:

(i) if $f \in \mathscr{A}$ and $f(x) \neq 0$ for all $x \in X$, then $1/f \in \mathscr{A}$;

(ii) if $(f_i)_{i\in I}$ is a family of functions belonging to $\mathscr{A}$ such that (Supp $(f_i))_{i\in I}$ is locally finite, then $\sum_{i\in I} f_i \in \mathscr{A}$;

(iii) for every closed set $F \subset X$ and open set U containing F, there is $f \in \mathscr{A}$ such that:

(a) $0 \leq f \leq 1$;

(b) $f(x) = 1$ for $x \in F$;

(c) Supp $(f) \subset U$.

Example 1.—If $(X, \mathscr{T})$ is *normal*, and $\mathscr{A} = C_{\mathbf{R}}(X)$, then $\mathscr{A}$ is an algebra (see 8.15) and Conditions (i), (ii), and (iii) are satisfied. In fact, (i) follows from 8.13. To prove (ii), we notice that if $t \in X$, then there is $V \in \mathscr{N}(t)$ such that

$$J = \{i \mid V \cap \text{Supp } (f_i) \neq \varnothing\}$$

* Since (supp $(f_i))_{i\in I}$ is locally finite, $\sum_{i\in I} f_i$ is well-defined as a mapping of X into $\mathbf{R}$.

is finite, whence

$$(\sum_{i \in I} f_i) \mid V = (\sum_{i \in J} f_i) \mid V.$$

From 5.3 and 8.15, we deduce that $\sum_{i \in I} f_i$ is continuous at t. Since $t \in X$ was arbitrary, $\sum_{i \in I} f_i$ is continuous on X. Finally, (iii) follows from (N_1).

Instead of saying "*$C_{\mathbf{R}}(X)$-partition of unity*," we often say "*continuous partition of unity*."

Other examples will be discussed later.

21.2 Theorem.—*Let $(X, \mathscr{T})$ be normal and $(U_i)_{i \in I}$ a locally finite open covering of X. Then there exists an $\mathscr{A}$-partition of unity of X subordinated to $(U_i)_{i \in I}$.*

Proof.—Let $(F_i)_{i \in I}$ be a covering of X consisting of *closed sets* such that $F_i \subset U_i$ for all $i \in I$. For each $i \in I$, let $g_i \in \mathscr{A}$ be such that $0 \leq g_i \leq 1$, $g_i(x) = 1$ for $x \in F_i$, and Supp $(g_i) \subset U_i$. Since Supp $(g_i) \subset U_i$ for all $i \in I$, it follows that (Supp $(g_i))_{i \in I}$ is locally finite. Hence, $g = \sum_{i \in I} g_i \in \mathscr{A}$. Moreover, $g(x) > 0$ for all $x \in X$. Using the fact that $\mathscr{A}$ is an algebra and using (i), we deduce that $f_i = g_i/g \in \mathscr{A}$ for all $i \in I$. Clearly, now,

$$\text{Supp } (f_i) \subset \text{Supp } (g_i) \subset U_i$$

for all $i \in I$ and $\sum_{i \in I} f_i = 1$. Hence, $(f_i)_{i \in I}$ is an $\mathscr{A}$-partition of unity of X subordinated to $(U_i)_{i \in I}$.

21.3 Corollary.—*Let $(X, \mathscr{T})$ be a normal space, $F \subset X$ a closed set, U an open set containing F, and $(U_i)_{i \in I}$ a locally finite open covering of F. Then there exists a family $(f_i)_{i \in I}$ of positive functions belonging to $\mathscr{A}$ subordinated to $(U_i)_{i \in I}$ and such that*

$$\sum_{i \in I} f_i \leq 1, \qquad \text{Supp } (\sum_{i \in I} f_i) \subset U$$

and

$$(\sum_{i \in I} f_i)(x) = 1 \quad \textit{for all} \quad x \in F.$$

Proof.—Let $\alpha \notin I$, $I' = I \cup \{\alpha\}$, and

$$U_i' = \begin{cases} U_i \cap U & \text{if } \; i \in I \\ \mathbf{C}F & \text{if } \; i = \alpha. \end{cases}$$

Then $(U_i')_{i \in I'}$ is a locally finite covering of X, and hence there is an

$\mathscr{A}$-partition $(f_i)_{i\in I'}$ subordinated to $(U'_i)_{i\in I'}$. Then $(f_i)_{i\in I}$ is obviously subordinated to $(U_i)_{i\in I}$,

$$\textstyle\sum_{i\in I} f_i \leq 1, \quad \text{and} \quad (\sum_{i\in I} f_i)(x) = 1 \quad \text{for all} \quad x \in F.$$

Since $(\operatorname{Supp}(f_i))_{i\in I}$ is locally finite, it follows that $S = \bigcup_{i\in I} \operatorname{Supp}(f_i)$ is closed (see 2.9). Hence

$$\operatorname{Supp}(\textstyle\sum_{i\in I} f_i) \subset S \subset \bigcup_{i\in I} U'_i \subset U,$$

and hence Corollary 21.3 is proved.

21.4 Corollary.—*Let $(X, \mathscr{T})$ be paracompact* and $(V_j)_{j\in J}$ an open covering of X. Then there exists an $\mathscr{A}$-partition of unity of X subordinated to $(V_j)_{j\in J}$.*

Proof.—Let $(U_i)_{i\in I}$ be a locally finite covering of X *finer* than $(V_j)_{j\in J}$. Then for every $i \in I$ there is $\varphi(i) \in J$ such that $U_i \subset V_{\varphi(i)}$; we defined this way a mapping $\varphi: I \to J$. Now let $(f_i)_{i\in I}$ be an $\mathscr{A}$-partition of X subordinated to $(U_i)_{i\in I}$. For each $j \in J$, let

$$I(j) = \{i \mid \varphi(i) = j\} \quad \text{and} \quad h_j = \textstyle\sum_{i\in I(j)} f_i.$$

By (ii), $h_j \in \mathscr{A}$ for all $j \in J$. Moreover, $S_j = \bigcup_{i\in I(j)} \operatorname{Supp}(f_i)$ is closed (see 2.9) for all $j \in J$; hence

$$\operatorname{Supp}(h_j) \subset S_j \subset \bigcup_{i\in I(j)} U_i \subset V_j.$$

We shall show now that $(S_j)_{j\in J}$ is locally finite. Let $t \in X$ and $V \in \mathscr{N}_X(t)$ such that $A = \{i \mid V \cap U_i \neq \varnothing\}$ is finite. If $j \in J$ and $V \cap S_j \neq \varnothing$, we have $V \cap U_i \neq \varnothing$ for some $i \in I(j)$, hence $j \in \varphi(A)$. We conclude that $\{j \mid V \cap S_j \neq \varnothing\}$ is finite, so that $(S_j)_{j\in J}$ is locally finite. Since

$$\textstyle\sum_{j\in J} h_j = \sum_{j\in J} (\sum_{i\in I(j)} f_i) = 1,$$

Corollary 21.4 is proved.

We shall denote by $(X, \mathscr{T})$ a separated space and by $\mathscr{B}$ an *algebra* of functions on X to $\mathbf{R}$ having the following properties:

(α) if $f \in \mathscr{B}$ and $f(x) \neq 0$ for all $x \in X$, then $1/f \in \mathscr{B}$;

(β) if $(f_i)_{i\in I}$ is a family of functions belonging to $\mathscr{B}$ such that $(\operatorname{Supp}(f_i))_{i\in I}$ is locally finite, then $\sum_{i\in I} f_i \in \mathscr{B}$.

---

* Recall that a paracompact space is normal (see 20.7).

We denote by $\mathscr{B}_0$ a *subalgebra* of $\mathscr{B}$ having the following properties:

(δ) if $f \in \mathscr{B}_0$ and $g \in \mathscr{B}$, then $fg \in \mathscr{B}_0$;

(ε) for every compact set $K \subset X$ and open set U containing K, there is $f \in \mathscr{B}_0$ such that:

(ε.1) $0 \leq f \leq 1$;

(ε.2) $f(x) = 1$ for $x \in K$;

(ε.3) Supp $(f) \subset U$.

Example 2.—If $(X, \mathscr{T})$ is locally compact, $\mathscr{B} = C_{\mathbf{R}}(X)$, and $\mathscr{B}_0 = \mathscr{K}(X)$ is the set of all $f \in \mathscr{B}$ such that Supp (f) is compact, then $\mathscr{B}$ is an algebra satisfying (α) and (β) and $\mathscr{B}_0$ is a subalgebra of $\mathscr{B}$ satisfying (δ) and (ε) (see 20.9).

21.5 Theorem.—*Let $(X, \mathscr{T})$ be a locally compact paracompact space and $(U_i)_{i\in I}$ be a locally finite open covering of X consisting of relatively compact sets. Then there is a $\mathscr{B}_0$-partition of unity of X subordinated to $(U_i)_{i\in I}$.*

Proof.—The theorem can be proved in the same way as Theorem 21.2. Notice (with the notations in the proof of Theorem 21.2) that each F_i $(i \in I)$ is compact and use (δ) and (ε).

21.6 Corollary.—*Let $(X, \mathscr{T})$ be a locally compact paracompact space, $F \subset X$ a closed set, and U an open set containing F. Then there is $f \in \mathscr{B}$ satisfying* (a), (b), *and* (c) *of* (iii).

Proof.—Let $(U_i)_{i\in I}$ be an open locally finite covering of X consisting of relatively compact sets such that for each $i \in I$ *either* $U_i \subset U$ or $U_i \subset \complement F$. Let $(f_i)_{i\in I}$ be a $\mathscr{B}_0$-partition of unity of X subordinated to $(U_i)_{i\in I}$. Let J be the set of all $i \in I$ such that $U_i \subset U$ and let $f = \sum_{i\in I} f_i$. Then $f \in \mathscr{B}$, and clearly, f satisfies (a) and (b) of (iii). Furthermore, $S = \bigcup_{j\in J}$ Supp (f_j) is closed (see 2.9) and contained in U. Since S is closed, Supp $(f) \subset S$. Hence, f satisfies (ε.3) also.

▼ *Example* 3.—Let $X = \mathbf{R}^n$ $(n \geq 1)$. For every $p \in \mathbf{N}$, let $\mathscr{B}^{(p)}$ be the set of all functions $f: X \to \mathbf{R}$ such that

$$\frac{\partial^{i_1+\dots+i_n}}{\partial x_1^{i_1} \dots \partial x_n^{i_n}} f$$

exists and is continuous on $\mathbf{R}^n$, for *every* system of n positive numbers $i_1, \dots, i_n$ such that $i_1 + \dots + i_n \leq p$. Let

$$\mathscr{B}^{(\infty)} = \bigcap_{p\in\mathbf{N}} \mathscr{B}^{(p)}.$$

If $\mathscr{B} = \mathscr{B}^{(p)}$ for some $1 \leq p \leq +\infty$, then $\mathscr{B}$ is an algebra satisfying (α) and (β). If $\mathscr{B}_0$ is the set of all $f \in \mathscr{B}$ having compact support, then $\mathscr{B}_0$ satisfies (δ) and (ε). The fact that $\mathscr{B}_0$ satisfies (δ) is obvious. Some indications of how to show that $\mathscr{B}_0$ also satisfies (ε) are given below (it is enough to consider the case $\mathscr{B} = \mathscr{B}^{(+\infty)}$).

For each $\varepsilon > 0$, let $\rho_\varepsilon : \mathbf{R} \to \mathbf{R}$ be defined by

$$\rho_\varepsilon(x) = \begin{cases} 0 & \text{if } |x| \geq \varepsilon \\ \exp\left(-\dfrac{1}{x^2 - \varepsilon^2}\right) & \text{if } |x| < \varepsilon. \end{cases}$$

We leave it to the reader to show that ρ_ε is *infinitely differentiable** on $\mathbf{R}$. Let $I = (a_1, b_1) \times \ldots \times (a_n, b_n)$ be a bounded open n-dimensional interval. For $j = 1, \ldots, n$, consider the function $\varphi_j : \mathbf{R} \to \mathbf{R}$ defined by

$$\varphi_j(x_j) = \rho_{(b_j - a_j)/2}(x_j - (a_j + b_j)/2)$$

for $x_j \in \mathbf{R}$; then φ_j is *infinitely differentiable* on $\mathbf{R}$. Since

$$x_j \in (a_j, b_j)_j \Leftrightarrow |x_j - (a_j + b_j)/2| < (b_j - a_j)/2,$$

we have

$$\varphi_j(x_j) = \begin{cases} 0 & \text{if } x_j \notin (a_j, b_j) \\ >0 & \text{if } x_j \in (a_j, b_j). \end{cases}$$

Define now $\varphi_I : \mathbf{R}^n \to \mathbf{R}$ by

$$\varphi_I(x) = \varphi_I(x_1, \ldots, x_n) = \varphi_1(x_1) \ldots \varphi_n(x_n)$$

for all $(x_1, \ldots, x_n) \in \mathbf{R}^n$. Then $\varphi \in \mathscr{B}$. Clearly, $\varphi_I(x) > 0$ if $x \in I$ and $\operatorname{Supp}(\varphi_I) \subset \bar{I}$.†

Now let $K \subset \mathbf{R}^n$ be a compact set and U an open set containing K. Let V be an open set such that $K \subset V \subset \bar{V} \subset U$. Using the fact that $\mathbf{R}^n$ is paracompact and Theorem 10.19, we deduce that there exists a locally finite covering of $\mathbf{R}^n$, $(I_\alpha)_{\alpha \in A}$, consisting of bounded open n-dimensional intervals such that for each $\alpha \in A$, either $I_\alpha \subset V$ or $I_\alpha \subset \complement K$. Now let

$$\varphi = \textstyle\sum_{\alpha \in A} \varphi_\alpha.$$

Then $\varphi \in \mathscr{B}$ and $\varphi(x) > 0$ for all $x \in \mathbf{R}^n$. If $B = \{\alpha \mid I_\alpha \cap K \neq \varnothing\}$, then B is finite (why?). Define, finally,

$$f = \frac{\sum_{\alpha \in B} \varphi_{I_\alpha}}{\varphi}.$$

---

* Use the following result: Let $I \subset \mathbf{R}$ be an interval, $f : I \to \mathbf{R}$ and $x_0 \in I$. Suppose that f is continuous on I, differentiable at every $x \in I - \{x_0\}$, and that $\lim_{x \to x_0} f'(x) = L$. Then f is differentiable at x_0 and $f'(x_0) = L$.

† In fact, $\bar{I} = [a_1, b_1] \times \ldots \times [a_n, b_n]$.

Then $f \in \mathscr{B}$ and $\operatorname{Supp} f \subset \bigcup_{\alpha \in B} \bar{I}_\alpha \subset \bar{V} \subset U$. Moreover, $0 \leq f \leq 1$. If $x \in K$, then

$$\varphi(x) = \sum_{\alpha \in B} \varphi_{I_\alpha}(x),$$

so that $f(x) = 1$ for $x \in K$. Hence $\mathscr{B}$ satisfies Condition (ε). ▲

Exercises for Chapter 21

1. Let $(X, \mathscr{T})$ be a separated space. Suppose that for every locally finite open cover $(U_i)_{i \in I}$ of X there is a partition of unity $(f_i)_{i \in I}$ subordinate to $(U_i)_{i \in I}$. Then $(X, \mathscr{T})$ is normal.

2. Suppose that $(U_i)_{i \in I}$ is a locally finite open covering of a paracompact space $(X, \mathscr{T})$ and that r is a mapping of I into $\mathbf{R}_+$. Then there is a continuous function $f: X \to \mathbf{R}$ such that for each x in X, $f(x) \leq \sup \{r(i) \mid x \in U_i\}$.

* 3. A real valued function f on a topological space $(X, \mathscr{T})$ is *lower semicontinuous* if $f^{-1}((-\infty, t])$ is closed for each t in $\mathbf{R}$ and *upper semicontinuous* if $f^{-1}((-\infty, t)) \in \mathscr{T}$ for each t in $\mathbf{R}$. If f is an upper semicontinuous function, and h is a lower semicontinuous function on a paracompact space $(X, \mathscr{T})$, and $f(x) < h(x)$ for each x in X, then there is a continuous mapping g of X into $\mathbf{R}$ such that $f(x) < g(x) < h(x)$ for each x in X.

Chapter 22

Baire Spaces

Let $(X, \mathscr{T})$ be a topological space. A set $A \subset X$ is *nowhere dense* if $\mathring{\bar{A}} = \varnothing$. A set $A \subset X$ is said to be *meager* (or of the *first category*) if it is the union of a *countable* family of nowhere dense sets.

Example 1.—If A is open or closed, then $Fr(A)$ is nowhere dense.

Exercise 1.—If $(A_i)_{i \in I}$ is a *finite* family of nowhere dense sets, then $\bigcup_{i \in I} A_i$ is nowhere dense.

22.1 *Definition.—A topological space $(X, \mathscr{T})$ is a Baire space if, for every family $(F_n)_{n \in \mathbf{N}}$ of closed nowhere dense parts of X, the union $\bigcup_{n \in \mathbf{N}} F_n$ has a void interior.*

When there is no ambiguity, we say that X is a Baire space, instead of saying that $(X, \mathscr{T})$ is a Baire space.

Notice that $(X, \mathscr{T})$ is a Baire space if and only if for every family $(F_n)_{n \in \mathbf{N}}$ of closed parts of X

$$\overbrace{\bigcup_{n \in \mathbf{N}} F_n}^{\circ} \neq \varnothing \Rightarrow \mathring{F}_n \neq \varnothing \quad \text{for some} \quad n \in \mathbf{N}.$$

22.2 Theorem.—*Let $(X, \mathscr{T})$ be a topological space. Then the following assertions are equivalent:*

(a) *the space $(X, \mathscr{T})$ is a Baire space;*

(b) *for every sequence $(U_n)_{n \in \mathbf{N}}$ of open parts of X dense in X, the set $\bigcap_{n \in \mathbf{N}} U_n$ is dense in X.*

Proof of (a) $\Rightarrow$ (b).—Let $(U_n)_{n \in \mathbf{N}}$ be a sequence of open parts of X dense in X. Then for each $n \in \mathbf{N}$, the set $\complement U_n$ is nowhere dense. Hence

$$\complement\left(\bigcap_{n \in \mathbf{N}} U_n\right) = \bigcup_{n \in \mathbf{N}} \complement U_n$$

 has no interior point, so that $\bigcap_{n \in \mathbf{N}} U_n$ is dense in X.

Proof of (b) $\Rightarrow$ (a).—Let $(F_n)_{n\in\mathbf{N}}$ be a sequence of closed nowhere dense parts of X. Then for each $n \in \mathbf{N}$, the set $\mathbf{C}F_n$ is open and dense in X. Hence

$$\mathbf{C}(\bigcup_{n\in\mathbf{N}} F_n) = \bigcap_{n\in\mathbf{N}} \mathbf{C}F_n$$

is dense in X so that $\bigcup_{n\in\mathbf{N}} F_n$ has a void interior. Since $(F_n)_{n\in\mathbf{N}}$ was arbitrary, $(X, \mathscr{T})$ is a Baire space.

Exercise 2.—(i) The space $(X, \mathscr{T})$ is a Baire space if and only if every open non-void part of X is not meager. (ii) If $(X, \mathscr{T})$ is a Baire space, and $U \subset X$ is open, then $(U, \mathscr{T}_U)$ is a Baire space.

Let $(X, \mathscr{T})$ be a topological space and let $\mathscr{T}^* = \{A \mid A \in \mathscr{T}, A \neq \varnothing\}$. We say that $(X, \mathscr{T})$ satisfies *Condition* $(\mathscr{I})$ if there is a function $f: \mathscr{T}^* \to \mathscr{T}^*$ such that:

(j) $f(U) \subset U$ for every $U \in \mathscr{T}^*$;

(jj) if $(U_n)_{n\in\mathbf{N}}$ is a sequence of elements of $\mathscr{T}^*$ such that $f(U_n) \supset U_{n+1}$ for every $n \in \mathbf{N}$, then $\bigcap_{n\in\mathbf{N}} U_n \neq \varnothing$.

Remark.—It is obvious that if $(X, \mathscr{T})$ satisfies $(\mathscr{I})$ and $U \subset X$ is open, then $(U, \mathscr{T}_U)$ satisfies $(\mathscr{I})$.

22.3 Theorem.—*A topological space* $(X, \mathscr{T})$ *that satisfies Condition* $(\mathscr{I})$ *is a Baire space.*

Proof.—Let $(X, \mathscr{T})$ be a topological space satisfying $(\mathscr{I})$ and assume that $(X, \mathscr{T})$ is not a Baire space. Then there is a sequence $(F_n)_{n\in\mathbf{N}}$ of nowhere dense closed sets such that

$$U = \overset{\circ}{\widehat{\bigcup_{n\in\mathbf{N}} F_n}} \neq \varnothing.$$

We construct now the sequence $(U_n)_{n\in\mathbf{N}}$ of open sets as follows: We write $U_1 = U$. Supposing that U_n $(n \in \mathbf{N})$ was defined, we write $U_{n+1} = f(U_n) - F_n$. It is easy to see that

$$f(U_n) \supset U_{n+1} \quad \text{and} \quad U_n \subset U$$

for all $n \in \mathbf{N}$. Moreover,

$$\bigcap_{n\in\mathbf{N}} U_n = U \cap (\bigcap_{n\in\mathbf{N}} f(U_n) \cap \mathbf{C}F_n) \subset U \cap \bigcap_{n\in\mathbf{N}} \mathbf{C}F_n$$
$$= U \cap \mathbf{C}(\bigcup_{n\in\mathbf{N}} F_n) \subset U \cap \mathbf{C}U = \varnothing.$$

Since $(X, \mathscr{T})$ satisfies Condition $(\mathscr{I})$, this is a contradiction. Hence, $(X, \mathscr{T})$ is a Baire space.

22.4 Theorem.—*A locally compact space satisfies Condition* $(\mathscr{I})$, *and hence is a Baire space.*

Proof.—For each $U \in \mathscr{T}^*$, we denote by $f(U)$ a non-void open relatively compact set, the closure of which is contained in U. We define this way a mapping $f: U \mapsto f(U)$ of $\mathscr{T}^*$ into $\mathscr{T}^*$ such that $\overline{f(U)} \subset U$ for all $U \in \mathscr{T}^*$. Clearly, f satisfies (j). Now let $(U_n)_{n\in\mathbf{N}}$ be a sequence of elements of $\mathscr{T}^*$ satisfying $f(U_n) \supset U_{n+1}$ for all $n \in \mathbf{N}$. Then

$$U_n \supset \overline{f(U_n)} \supset U_{n+1} \supset \overline{f(U_{n+1})}$$

for all $n \in \mathbf{N}$. Since $\overline{f(U_1)}$ is compact,

$$\bigcap_{n\in\mathbf{N}} \overline{f(U_n)} \neq \varnothing,$$

so that $\bigcap_{n\in\mathbf{N}} U_n \neq \varnothing$. Since $(U_n)_{n\in\mathbf{N}}$ was arbitrary, (jj) is satisfied. Hence, $(X, \mathscr{T})$ satisfies $(\mathscr{I})$. By 22.3, $(X, \mathscr{T})$ is a Baire space.

22.5 Theorem.—*A complete metric space* (X, d) *satisfies Condition* $(\mathscr{I})$, *and hence is a Baire space.*

Proof.—For each $U \in \mathscr{T}^*$, we denote by $f(U)$ a non-void open set such that $\overline{f(U)} \subset U$ and

$$\delta(f(U)) \leq \inf(\delta(U), 1)/2.$$

We define this way a mapping $f: U \mapsto f(U)$ of $\mathscr{T}^*$ into $\mathscr{T}^*$ such that $\overline{f(U)} \subset U$ for all $U \in \mathscr{T}^*$. Clearly, f satisfies (j).

Now let $(U_n)_{n\in\mathbf{N}}$ be a sequence of elements of $\mathscr{T}^*$ satisfying $f(U_n) \supset U_{n+1}$ for all $n \in \mathbf{N}$. Then

$$U_n \supset \overline{f(U_n)} \supset U_{n+1},$$

so that

$$U_n \supset \overline{U}_{n+1}$$

for all $n \in \mathbf{N}$. This implies that

$$\bigcap_{n\in\mathbf{N}} U_n = \bigcap_{n\in\mathbf{N}} \overline{U}_n.$$

By mathematical induction, we deduce† that $\delta(U_n) \leq 2^{-n+1}$ for all

† Suppose $\delta(U_p) \leq 2^{-p+1}$ for $p > 1$. Then

$$\begin{aligned}\delta(U_{p+1}) &\leq \delta(f(U_p)) \leq \inf(\delta(U_p), 1)/2 \\ &\leq 2^{-p+1}/2 = 2^{-(p+1)+1}.\end{aligned}$$

$n > 1$. Now let $x_n \in U_n$ for $n \in \mathbf{N}$; then for each $p > 1$,

$$d(x_n, x_m) \leq 2^{-p+1}$$

if $n \geq p$, $m \geq p$. Hence, $(x_n)_{n \in \mathbf{N}}$ is a Cauchy sequence. Since (X, d) is complete, $(x_n)_{n \in \mathbf{N}}$ converges to some $x_0 \in X$. If $p \in N$, then $x_n \in U_p$ for $n \geq p$, whence $x_0 \in \overline{U_p}$ (see 14.4). Hence, $x_0 \in \bigcap_{n \in \mathbf{N}} \overline{U_n}$, and hence $\bigcap_{n \in \mathbf{N}} U_n \neq \varnothing$. Since $(U_n)_{n \in \mathbf{N}}$ was arbitrary, (jj) is satisfied. Hence, (X, d) satisfies $(\mathscr{I})$. By 22.3, the space (X, d) is a Baire space.

22.6 Theorem.—*Let $(X, \mathscr{T})$ be a Baire space and let $(f_\alpha)_{\alpha \in A}$ be a family of functions on X to $\mathbf{R}$ such that:*

(a) *for each $\alpha \in A$ and $a \in \mathbf{R}$, the set $f_\alpha^{-1}((-\infty, a])$ is closed;*

(b) *for each $x \in X$, $\sup_{\alpha \in A} f_\alpha(x) < +\infty$.*

Then there is an open non-void set U and $L \in \mathbf{R}$ such that

$$f_\alpha(x) \leq L$$

for all $\alpha \in A$ and $x \in U$.

Proof.—For each $n \in \mathbf{N}$, let

$$F_n = \{x \mid f_\alpha(x) \leq n \quad \text{for all} \quad \alpha \in A\};$$

since

$$F_n = \bigcap_{\alpha \in A} f_\alpha^{-1}((-\infty, n]),$$

the set F_n is closed. Using (b), we deduce that $\bigcup_{n \in \mathbf{N}} F_n = X$. Since $(X, \mathscr{T})$ is a Baire space, we deduce that there is $n_0 \in \mathbf{N}$ such that $\mathring{F}_{n_0} \neq \varnothing$. If $L = n_0$ and $U = \mathring{F}_{n_0}$, then $f_\alpha(x) \leq L$ for all $\alpha \in A$ and $x \in U$.

Remarks.—(i) Hypothesis 22.6(a) is satisfied if $f_\alpha \in C_{\mathbf{R}}(X)$ for all $\alpha \in A$.

(ii) A function $f: X \to \mathbf{R}$ such that $f^{-1}((-\infty, a])$ is closed for every $a \in \mathbf{R}$ is said to be *inferior* (or *lower*) *semicontinuous.* A function $f: X \to \mathbf{R}$ such that $f^{-1}((-\infty, a))$ is open for every $a \in \mathbf{R}$ is said to be *superior* (or *upper*) *semicontinuous*(see also Exercise 3, Chapter 21).

Exercise 3.—A function f: $X \to \mathbf{R}$ is continuous if and only if f is both inferior semicontinuous and superior semicontinuous.

Exercises for Chapter 22

1. A countable connected topological space cannot be locally compact.

2. A Banach space cannot have a countably infinite linear dimension.

3. Let (X, p) be a Banach space and (Y, q) a normed space. Suppose $(f_n)_{n\in\mathbf{N}}$ is a sequence of continuous linear functions of X into Y such that for each x in X, $\lim_{n\in\mathbf{N}} f_n(x)$ exists. Show that the function f defined by $f(x) = \lim_{n\in\mathbf{N}} f_n(x)$ for each x in X is a continuous linear function of X into Y.

Chapter 23

Filters. Limits. Nets

In this chapter, we introduce the notion of filter bases and discuss certain related subjects. Using filter bases, we may present in a unified way the various notions of limit that are used in mathematics. Several examples are given later.

Filter Bases

Let X be a set. We shall introduce the following basic definition.

23.1 Definition.—*A set* $\mathscr{B} \subset \mathscr{P}(X)$ *is a filter basis on* X *if:*

$(\mathrm{B_I})$ $\mathscr{B} \neq \varnothing$ *and* $\mathscr{B} \not\ni \varnothing$;

$(\mathrm{B_{II}})$ *if* $A_1 \in \mathscr{B}$ *and* $A_2 \in \mathscr{B}$, *then there exists* $A_3 \in \mathscr{B}$ *such that* $A_3 \subset A_1 \cap A_2$.

Clearly, $(\mathrm{B_{II}})$ is satisfied if $B_1 \in \mathscr{B}$ and $B_2 \in \mathscr{B}$ implies that $B_1 \cap B_2 \in \mathscr{B}$. Notice also that if $\mathscr{B}$ is a filter basis, and if

$$B_1 \in \mathscr{B}, \ldots, B_n \in \mathscr{B},$$

then there exists $B \in \mathscr{B}$ such that

$$B \subset B_1 \cap \ldots \cap B_n.$$

We shall now give a series of examples of filter bases.

Example 1.—Let X be a set and $(x_n)_{n \in \mathbf{N}}$ a sequence of elements of X. For each $p \in \mathbf{N}$, let $S_p = \{x_n \mid n \geq p\}$. Then

$$\mathscr{B} = \{S_p \mid p \in \mathbf{N}\}$$

is a *filter basis* on X.

In fact, clearly, $S_p \neq \varnothing$ for all $p \in \mathbf{N}$, and hence $\mathscr{B} \not\ni \varnothing$; since $\mathbf{N} \neq \varnothing$, $\mathscr{B} \neq \varnothing$. Hence $\mathscr{B}$ satisfies $(\mathrm{B_I})$. Since

$$S_p \cap S_q = S_{\sup\{p,q\}}$$

for all p and q in $\mathbf{N}$, we deduce that $\mathscr{B}$ satisfies $(\mathrm{B_{II}})$, too. Hence, $\mathscr{B}$ is a filter basis on X.

The filter basis $\mathscr{B}$ defined above is called the *filter basis associated with the sequence* $(x_n)_{n\in\mathbf{N}}$.

Example 2.—Let $(X, \mathscr{T})$ be a topological space, $a \in X$, and $\mathscr{F}$ a fundamental system of a. Then $\mathscr{F}$ is a filter basis on X.

Since $\mathscr{N}(a) \neq \varnothing$, it follows that $\mathscr{F} \neq \varnothing$. Since every $V \in \mathscr{F}$ is a neighborhood of a, it follows that $V \ni a$, whence $V \neq \varnothing$. Hence $\mathscr{F}$ satisfies $(\mathrm{B_I})$. If $V_1 \in \mathscr{F}$ and $V_2 \in \mathscr{F}$, then V_1 and V_2 are neighborhoods of a. Hence (see 2.2(c)), $V_1 \cap V_2$ is a neighborhood of a. By the definition of a fundamental system, there is $V_3 \in \mathscr{F}$ such that $V_1 \cap V_2 \supset V_3$. Hence, $\mathscr{F}$ satisfies $(\mathrm{B_{II}})$, too. Hence, $\mathscr{F}$ is a filter basis on X.

It follows in particular that *$\mathscr{N}(a)$ is a filter basis on X.*

Example 3.—If $a \in \mathbf{R}$, then

$$\{(a - \varepsilon, a + \varepsilon) \mid \varepsilon > 0\} \quad \text{and} \quad \{[a - \varepsilon, a + \varepsilon] \mid \varepsilon > 0\}$$

are *filter bases* on $\mathbf{R}$.

By Example 2, Chapter 2, these sets are fundamental systems of a. Hence they are filter bases.

Example 4.—Let $(X, \mathscr{T})$ be a topological space, $A \subset X$, $a \in \bar{A}$, and $\mathscr{F}$ a fundamental system of a. Then

$$\mathscr{F}_A = \{V \cap A \mid V \in \mathscr{F}\}$$

is a *filter basis* on A.

Since $\mathscr{F} \neq \varnothing$, it follows that $\mathscr{F}_A \neq \varnothing$. If $V \in \mathscr{F}$, then $V \cap A \neq \varnothing$; whence $\mathscr{F}_A \not\ni \varnothing$. Hence $\mathscr{F}_A$ satisfies $(\mathrm{B_I})$. Now let $A_1 \in \mathscr{F}_A$ and $A_2 \in \mathscr{F}_A$. Then there are $V_1 \in \mathscr{F}$ and $V_2 \in \mathscr{F}$ such that

$$A_1 = V_1 \cap A \quad \text{and} \quad A_2 = V_2 \cap A.$$

Let V_3 in $\mathscr{F}$ be such that $V_3 \subset V_1 \cap V_2$. Let $A_3 = V_3 \cap A$. Then $A_3 \in \mathscr{F}_A$ and

$$A_1 \cap A_2 = (V_1 \cap A) \cap (V_2 \cap A) = (V_1 \cap V_2) \cap A \supset V_3 \cap A = A_3;$$

hence $A_1 \cap A_2 \supset A_3$. Hence $\mathscr{F}_A$ satisfies (B_{II}), too. Hence, $\mathscr{F}_A$ is a filter basis on A.

Example 5.—Let $S \subset \mathbf{R}$ be a set and let $c \in \bar{S}$. Then the sets

$$\{(c - \varepsilon, c + \varepsilon) \cap S \mid \varepsilon > 0\} \quad \text{and} \quad \{[c - \varepsilon, c + \varepsilon] \cap S \mid \varepsilon > 0\}$$

are *filter bases* on S.

Example 6.—For every $a \in X$, we denote by $\mathscr{B}_a$ the set of all parts of X containing a. Then $\mathscr{B}_a$ is a filter basis on X.

Example 7.—Let X and Y be two sets and $f: X \to Y$ a mapping. Let $\mathscr{B}$ be a filter basis on X and let

$$f(\mathscr{B}) = \{f(A) \mid A \in \mathscr{B}\}.$$

Then $f(\mathscr{B})$ is a *filter basis* on Y.

Since $\mathscr{B} \neq \varnothing$, $f(\mathscr{B}) \neq \varnothing$. Since $A \neq \varnothing \Rightarrow f(A) \neq \varnothing$, and since $\mathscr{B} \not\ni \varnothing$, it follows that $f(\mathscr{B}) \not\ni \varnothing$. Hence $f(\mathscr{B})$ satisfies (B_I). Now let $B_1 \in f(\mathscr{B})$ and $B_2 \in f(\mathscr{B})$. Then there are $A_1 \in \mathscr{B}$ and $A_2 \in \mathscr{B}$ such that

$$B_1 = f(A_1) \quad \text{and} \quad B_2 = f(A_2).$$

Let $A_3 \in \mathscr{B}$ be such that $A_3 \subset A_1 \cap A_2$ and let $B_3 = f(A_3)$. Then

$$B_1 \cap B_2 = f(A_1) \cap f(A_2) \supset f(A_1 \cap A_2) \supset f(A_3) = B_3;$$

that is, $B_1 \cap B_2 \supset B_3$. Hence $f(\mathscr{B})$ satisfies (B_{II}), and hence, $f(\mathscr{B})$ is a filter basis on Y.

Let X be a set and let $\mathscr{B}_1$ and $\mathscr{B}_2$ be two filter bases on X.

23.2 Definition.—*We say that $\mathscr{B}_1$ and $\mathscr{B}_2$ are equivalent if:*

(a) *every set $A_1 \in \mathscr{B}_1$ contains a set $A_2 \in \mathscr{B}_2$;*
(b) *every set $A_2 \in \mathscr{B}_2$ contains a set $A_1 \in \mathscr{B}_1$.*

Instead of saying that $\mathscr{B}_1$ and $\mathscr{B}_2$ are equivalent, we often write $\mathscr{B}_1 \sim \mathscr{B}_2$.

Example 8.—Let $(X, \mathscr{T})$ be a topological space, $a \in X$, and $\mathscr{F}'$ and $\mathscr{F}''$ two fundamental systems of a. Then $\mathscr{F}' \sim \mathscr{F}''$.

In fact, let $V' \in \mathscr{F}'$. Then V' is a neighborhood of a. Hence, there exists $V'' \in \mathscr{F}''$ such that $V' \supset V''$. In the same way, we see that if $W'' \in \mathscr{F}''$, then there is $W' \in \mathscr{F}'$ such that $W'' \supset W'$.

Example 9.—Let $(X, \mathscr{T})$ be a topological space, $A \subset X$, $a \in \bar{A}$ and $\mathscr{F}'$ and $\mathscr{F}''$ two fundamental systems of a. Then, with the notations of Example 4, $\mathscr{F}'_A \sim \mathscr{F}''_A$.

We leave to the reader the proof of this assertion.

Example 10.—Let X and Y be two sets and $f: X \to Y$ a mapping. Let $\mathscr{B}_1$ and $\mathscr{B}_2$ be two equivalent filter bases on X. Then

$$f(\mathscr{B}_1) \sim f(\mathscr{B}_2).$$

In fact, let $A_1 \in f(\mathscr{B}_1)$. Then there exists $B_1 \in \mathscr{B}_1$ such that $A_1 = f(B_1)$. Since $\mathscr{B}_1 \sim \mathscr{B}_2$, there exists $B_2 \in \mathscr{B}_2$ for which $B_2 \subset B_1$ whence $A_1 \supset A_2 = f(B_2)$ and $A_2 \in f(\mathscr{B}_2)$. Hence, $A_1 \in f(\mathscr{B}_1)$ contains $A_2 \in f(\mathscr{B}_2)$. In the same way, we see that every $A_2' \in f(\mathscr{B}_2)$ contains an $A_1' \in f(\mathscr{B}_1)$. Therefore $f(\mathscr{B}_1) \sim f(\mathscr{B}_2)$.

Example 11.—Let X be a set and let $a \in X$. Then $\mathscr{A}_a = \{a\}$ is a filter basis on X equivalent with $\mathscr{B}_a$ (see Example 6).

▼ *Example* 12.—Let $(X_i)_{i \in I}$ be a family of sets, $X = \Pi_{i \in I} X_i$, and for each $i \in I$, let $\mathscr{B}_i$ be a filter basis on X_i. Denote by $\Pi_{i \in I} \mathscr{B}_i$ the set of all parts of X of the form

$$\Pi_{i \in I} B_i$$

where for each $i \in I$, either $B_i \in \mathscr{B}_i$ or $B_i = X_i$ and

$$\{i \mid B_i \neq X_i\}$$

is finite. Then $\Pi_{i \in I} \mathscr{B}_i$ is a filter basis on X, called the *product of the family* $(\mathscr{B}_i)_{i \in I}$. If $I = \{1, \ldots, n\}$ then $\Pi_{i \in I} \mathscr{B}_i$ consists of all the parts of X of the form

$$B_1 \times \ldots \times B_n$$

where $B_i \in \mathscr{B}$ for $i = 1, \ldots, n$. ▲

Limits

Let $(X, \mathscr{T})$ be a topological space.

23.3 Definition.—*A filter basis* $\mathscr{B}$ *on* X *converges to* $a \in X$ *if for every* $V \in \mathscr{N}(a)$ *there exists* $B \in \mathscr{B}$ *contained in* V.

The element $a \in X$ is said to be *a limit* of $\mathscr{B}$. The notation

$$a = \lim \mathscr{B} \quad \text{(or, equivalently, } \lim \mathscr{B} = a\text{)}$$

means that $\mathscr{B}$ converges to a in the sense of Definition 23.3.

Instead of saying that $\mathscr{B}$ converges to a, we shall often say that $\mathscr{B}$ tends to a. Whenever we say that a filter basis $\mathscr{B}$ on X is *convergent*, we mean that $\mathscr{B}$ converges to some $a \in X$ in the sense of Definition 23.3.

Since $B \in \mathscr{B}$ in Definition 23.3 depends on $V \in \mathscr{N}(a)$, we shall sometimes use some notation to indicate this. For instance, we may write B_V instead of B.

We notice that a filter basis may converge to more than one element. For instance, if $(X, \mathscr{W})$ is the topological space introduced in Example 2, Chapter 1, and $\mathscr{B}$ is a filter basis on X, then $\mathscr{B}$ converges to *every* $a \in X$. However, as we shall see later, if X is separated, then a filter basis converges to at most one element. A filter basis may not converge to any element.

23.4 Theorem.—*Let $\mathscr{B}_1$ and $\mathscr{B}_2$ be two filter bases on X. Suppose $\mathscr{B}_1 \sim \mathscr{B}_2$. Then the following assertions are equivalent:*

(a) $\lim \mathscr{B}_1 = a$;
(b) $\lim \mathscr{B}_2 = a$.

Proof.—Suppose that $\lim \mathscr{B}_1 = a$. Let $V \in \mathscr{N}(a)$. Then there is $B_1 \in \mathscr{B}_1$ such that $B_1 \subset V$. Since $\mathscr{B}_1 \sim \mathscr{B}_2$, there exists $B_2 \in \mathscr{B}_2$ such that $B_2 \subset B_1$. Hence, $B_2 \subset V$. Since $V \in \mathscr{N}(a)$ was arbitrary, and since V contains B_2, which belongs to $\mathscr{B}_2$, we conclude that $\lim \mathscr{B}_2 = a$. Hence, (a) $\Rightarrow$ (b). In the same way, we prove the implication (b) $\Rightarrow$ (a).

23.5 Definition.—*Let $\mathscr{B}_1$ and $\mathscr{B}_2$ be two filter bases on a set X. We write $\mathscr{B}_1 \prec \mathscr{B}_2$ (or $\mathscr{B}_2 \succ \mathscr{B}_1$) if every set $A_1 \in \mathscr{B}_1$ contains a set $A_2 \in \mathscr{B}_2$.*

Notice that $\mathscr{B}_1 \prec \mathscr{B}_2$ *and* $\mathscr{B}_2 \prec \mathscr{B}_1$ if and only if $\mathscr{B}_1 \sim \mathscr{B}_2$.

23.6 Theorem.—*Let $(X, \mathscr{T})$ be a topological space. A filter basis $\mathscr{B}$ on X converges to $a \in X$ if and only if $\mathscr{B} \succ \mathscr{N}_X(a)$.*

Proof.—Obvious.

It is also obvious that if $\mathscr{B}_1 \prec \mathscr{B}_2$ are two filter bases on X, and if $\lim \mathscr{B}_1 = a$, then $\lim \mathscr{B}_2 = a$.

23.7 Theorem.—*Let $(X, \mathscr{T})$ be a separated topological space and $\mathscr{B}$ a filter basis on X that converges to a' and to a''. Then $a' = a''$.*

Proof.—If $a' \neq a''$, then (since $(X, \mathscr{T})$ is separated) there are $V' \in \mathscr{N}(a')$ and $V'' \in \mathscr{N}(a'')$ such that

$$V' \cap V'' = \varnothing.$$

Since $\lim \mathscr{B} = a'$, there exists $B' \in \mathscr{B}$ contained in V'. Since $\lim \mathscr{B} = a''$, there exists $B'' \in \mathscr{B}$ contained in V''. Since $\mathscr{B}$ is a filter basis, there exists $B''' \in \mathscr{B}$ such that $B''' \subset B' \cap B''$, whence

$$B''' \subset B' \cap B'' \subset V' \cap V''.$$

By (B_I), Definition 23.1, $B''' \neq \varnothing$. This implies that $V' \cap V'' \neq \varnothing$ and hence leads to a contradiction. Hence, $a' = a''$.

The next theorem shows that to prove that a filter basis on X converges to a, it is enough to verify the condition of Definition 23.3 for all V in some fundamental system of a.

23.8 Theorem.—*Let $a \in X$, $\mathscr{F}$ be a fundamental system of a, and $\mathscr{B}$ be a filter basis on X. Then the following assertions are equivalent:*

(a) *$\mathscr{B}$ converges to a;*
(b) *for every $V \in \mathscr{F}$, there exists $B \in \mathscr{B}$ contained in V.*

Proof.—Since $\mathscr{F} \subset \mathscr{N}(a)$, it is clear that (a) $\Rightarrow$ (b). Conversely, suppose that (b) is satisfied. Let $V \in \mathscr{N}(a)$; then there is $W \in \mathscr{F}$ satisfying $W \subset V$. By (b), there exists $B \in \mathscr{B}$ contained in W. Obviously, then, $B \subset V$. Since $V \in \mathscr{N}(a)$ was arbitrary, $\mathscr{B}$ converges to a. Hence, (b) $\Rightarrow$ (a).

Let $a \in R$. Then

$$\{(a - \varepsilon, a + \varepsilon) \mid \varepsilon > 0\}$$

is a fundamental system of a. Using Theorem 23.8, we deduce the following:

23.9 Corollary.—*Let $a \in \mathbf{R}$ and $\mathscr{B}$ be a filter basis on $\mathbf{R}$. Then $\mathscr{B}$ converges to a if and only if for every $\varepsilon > 0$ there exists $B_\varepsilon \in \mathscr{B}$ such that*

$$x \in B_\varepsilon \Rightarrow |x - a| < \varepsilon.$$

We obtain other equivalent formulations, using other fundamental systems of a.

Example 13.—Let $(X, \mathscr{T})$ be a topological space, $a \in X$, and $\mathscr{F}$ a fundamental system of a. Then

$$\lim \mathscr{F} = a.$$

Example 14.—Let $(X, \mathscr{T})$ be a topological space, $A \subset X$, $a \in \bar{A}$, and $\mathscr{F}$ a fundamental system of a. Then (see Example 4)

$$\mathscr{F} \prec \mathscr{F}_A \quad \text{and} \quad \lim \mathscr{F}_A = a.$$

Example 15.—Let $(X, \mathscr{T})$ be a topological space and $(x_n)_{n \in \mathrm{N}}$ a sequence of elements of X. Let $\mathscr{B}$ be the filter basis associated with the sequence $(x_n)_{n \in \mathrm{N}}$ (see Example 1).

By Definition 23.3, $\mathscr{B}$ converges to $a \in X$ if and only if for every $V \in \mathscr{N}(a)$ there exists $p \in \mathbf{N}$ such that

$$S_p = \{x_n \mid n \geq p\} \subset V.$$

This means that $\mathscr{B}$ converges to $a \in X$ if and only if for every $V \in \mathscr{N}(a)$ there exists $p \in \mathbf{N}$ such that

$$x_n \in V \quad \text{whenever} \quad n \geq p.$$

If we compare this with Definition 14.1, we see that $\mathscr{B}$ converges to a in the sense of Definition 23.3 if and only if $(x_n)_{n \in \mathbf{N}}$ converges to a in the sense of Definition 14.1.

It follows that Theorem 14.3 is a particular case of Theorem 23.7.

23.10 Theorem.—*Let $((X_i, \mathscr{T}_i))_{i \in I}$ be a family of topological spaces and $(X, \mathscr{T})$ its product. A filter basis $\mathscr{B}$ on X converges to $a = (a_i)_{i \in I}$ if and only if, for each $i \in I$, $pr_i(\mathscr{B})$ converges to a_i.*

Proof.—Suppose that $\mathscr{B}$ converges to a. Let $\alpha \in I$ and let $U \in \mathscr{N}_{X_\alpha}(a_\alpha)$. Let $V = \prod_{i \in I} U_i$, where $U_\alpha = U$ and $U_i = X_i$ if $i \neq \alpha$. Then $V \in \mathscr{N}_X(a)$. Since $\mathscr{B}$ converges to a, there is $A \in \mathscr{B}$ contained in V. Then

$$pr_\alpha(A) \subset pr_\alpha(V) = U_\alpha \subset U.$$

Since $pr_\alpha(A) \in pr_\alpha(\mathscr{B})$, and since $U \in \mathscr{N}_{X_\alpha}(a_\alpha)$ was arbitrary, we deduce that $pr_\alpha(A)$ converges to a_α.

Conversely, suppose that, for each $i \in I$, $pr_i(\mathscr{B})$ converges to a_i. Let $U = \prod_{i \in I} U_i$ be an *elementary part* of X containing a and let $J = \{i \mid U_i \neq X_i\}$; then J is finite. For each $i \in J$, let $A_i \in \mathscr{B}$ be such that $pr_i(A_i) \subset U_i$, and let $A \in \mathscr{B}$ be contained in $\bigcap_{i \in J} A_i$. Then

$$A \subset \bigcap_{i \in J} pr_i^{-1}(pr_i(A_i)) \subset \bigcap_{i \in J} pr_i^{-1}(U_i) = U.$$

Since the elementary parts of X containing a form a fundamental system of a (see Theorem 7.4), we deduce that $\mathscr{B}$ converges to a.

Limits of Functions

23.11 Definition.—*Let X be a set, $(Y, \mathscr{S})$ a topological space, $f: X \to Y$, and $\mathscr{B}$ a filter basis on X. We say that f converges (or tends) to $L \in Y$ with respect to $\mathscr{B}$ if*

$$L = \lim f(\mathscr{B}).$$

Whenever we write* $L = \lim_{\mathscr{B}} f$ (or $\lim_{\mathscr{B}} f = L$), we mean that f tends to L in the sense of Definition 23.11. Whenever we say that f has a limit with respect to $\mathscr{B}$, we mean that f converges to some L in the sense of Definition 23.11.

Let $\mathscr{F}$ be a fundamental system of $f(a)$. Comparing this with 23.3 and 23.8, we see that

$$L = \lim_{\mathscr{B}} f$$

if and only if for every $V \in \mathscr{F}$ *there is* $A \in \mathscr{B}$ *such that* $f(A) \subset V$.

If $\mathscr{B}_1$ and $\mathscr{B}_2$ are equivalent filter bases on X, then $f(\mathscr{B}_1)$ and $f(\mathscr{B}_2)$ are equivalent (see Example 10). We deduce then, from Theorem 23.4, that

$$L = \lim_{\mathscr{B}_1} f \quad \text{if and only if} \quad L = \lim_{\mathscr{B}_2} f.$$

Suppose that X is endowed with a topology, $S \subset X$, and $a \in \bar{S}$. Then

$$(\bigstar) \qquad \mathscr{B} = \{U \cap S \mid U \in \mathscr{N}_X(a)\}$$

is a filter basis on S. If $f: S \to Y$, we often write†

$$\lim_{x \to a} f(x) \quad \text{instead of} \quad \lim_{\mathscr{B}} f.$$

23.12 Theorem.—*Let* $(X, \mathscr{T})$ *and* $(Y, \mathscr{I})$ *be two topological spaces,* $S \subset X$, $a \in S$, *and* $f: S \to Y$. *Then the following assertions are equivalent:*

(a) f *is continuous at* a;
(b) $\lim_{x \to a} f(x) = f(a)$.

Proof.—We shall use Notation ($\bigstar$).

Proof of (a) $\Rightarrow$ (b).—Let $V \in \mathscr{N}_Y(f(a))$. Since f is continuous at a, there is $U \in \mathscr{N}_X(a)$ such that $f(U \cap S) \subset V$. Since $U \cap S \in \mathscr{B}$, and since $V \in \mathscr{N}_Y(f(a))$ was arbitrary, we deduce (b).

Proof of (b) $\Rightarrow$ (a).—Let $V \in \mathscr{N}_Y(f(a))$. Since $\lim_{x \to a} f(x) = f(a)$, there is $A \in \mathscr{B}$ such that $f(A) \subset V$. Since A is of the form $U \cap S$ with $U \in \mathscr{N}_X(a)$, and since $V \in \mathscr{N}_Y(f(a))$ was arbitrary, we deduce (a).

* The notation $\lim_{\mathscr{B}} f$ is read "limit of f with respect to $\mathscr{B}$."
† The notation $\lim_{x \to a} f(x)$ is read "limit of $f(x)$ when x tends to a."

▼ Let $((Y_i, \mathscr{I}_i))_{i\in I}$ be a family of separated* topological spaces, $(Y_\infty, \mathscr{I}_\infty)$ its product, $(Y, \mathscr{I})$ a separated topological space, $T \subset Y_\infty$, and $\varphi: T \to Y$ a continuous mapping. Then

23.13 Theorem.—*Let X be a set, $\mathscr{B}$ a filter basis on X, and for each $i \in I$,* let f_i *be a mapping of X into Y_i. Suppose that:*

(a) $a_i = \lim_{\mathscr{B}} f_i$, *for each* $i \in I$;
(b) $(f_i(x))_{i\in I} \in T$, *for each* $x \in X$;
(c) $(a_i)_{i\in I} \in T$.

Then, if h is the mapping $x \mapsto \varphi((f_i(x))_{i\in I})$ of X into Y, we have

$$\varphi((a_i)_{i\in I}) = \lim_{\mathscr{B}} h.$$

We leave the proof of 23.13 to the reader. From 23.13, 8.1, 8.2, and 8.6, we deduce the following corollary.

23.14 Corollary.—*Let X be a set, $\mathscr{B}$ a filter basis on X, and $f: X \to \mathbf{R}$ and $g: X \to \mathbf{R}$ two mappings such that*

$$A = \lim_{\mathscr{B}} f \quad \text{and} \quad B = \lim_{\mathscr{B}} g.$$

Then:

(a) $\lim_{\mathscr{B}} (\lambda f + \mu g) = \lambda \lim_{\mathscr{B}} f + \mu \lim_{\mathscr{B}} g$ *for all* λ and μ *in* $\mathbf{R}$;
(b) $\lim_{\mathscr{B}} (fg) = (\lim_{\mathscr{B}} f)(\lim_{\mathscr{B}} g)$;
(c) $\lim_{\mathscr{B}} (f/g) = A/B$ *if* $g(x) \neq 0$ *for* $x \in X$ *and* $B \neq 0$. ▲

Adherence of a Filter Basis

23.15 Definition.—*For any topological space $(X, \mathscr{T})$ and filter basis $\mathscr{B}$ on X, we write*

$$ad(\mathscr{B}) = \bigcap_{A\in\mathscr{B}} \bar{A}$$

and call $ad(\mathscr{B})$ the adherence of $\mathscr{B}$.

Notice that if $\mathscr{B}_1$ and $\mathscr{B}_2$ are two filter bases on X, then

$$\mathscr{B}_1 \sim \mathscr{B}_2 \Rightarrow ad(\mathscr{B}_1) = ad(\mathscr{B}_2).$$

If $\mathscr{B}$ is a filter basis on X and $a = \lim \mathscr{B}$, then

23.16 $a \in ad(\mathscr{B})$.

In fact, let $A \in \mathscr{B}$ and $V \in \mathscr{N}_X(a)$. By 23.3, there is $B \in \mathscr{B}$ contained

* The hypothesis that the spaces we consider here are separated is not necessary. However, the statement is easier to understand with this hypothesis.

in V. Since $A \cap B \neq \varnothing$, we deduce that $A \cap V \neq \varnothing$. Since $V \in \mathcal{N}_X(a)$ was arbitrary, $a \in \bar{A}$. Since $A \in \mathcal{B}$ was arbitrary, we obtain 23.16.

From the above remark, we deduce the following theorem.

23.17 Theorem.—*Let $(X, \mathcal{T})$ be a topological space, $A \subset X$, and $\mathcal{B} \subset \mathcal{P}(A)$ a filter basis on X. If $a = \lim \mathcal{B}$, then $a \in \bar{A}$.*

From the result in Example 15, it follows that Theorem 14.4 is a particular case of Theorem 23.17.

Comparing Theorem 9.15 with Definition 23.15, we deduce the following theorem.

23.18 Theorem.—*Let $(X, \mathcal{T})$ be a separated topological space. Then the following assertions are equivalent:*

(a) *the space $(X, \mathcal{T})$ is compact;*
(b) *for every filter basis $\mathcal{B}$ on X, we have $ad(\mathcal{B}) \neq \varnothing$.*

Filters

Let X be a set.

23.19 Definition.—*A set $\mathcal{F} \subset \mathcal{P}(X)$ is a filter if:*

(F_{I}) *$\mathcal{F} \neq \varnothing$ and $\mathcal{F} \not\ni \varnothing$;*
(F_{II}) *if $A_1 \in \mathcal{F}$ and $A_2 \in \mathcal{F}$, then $A_1 \cap A_2 \in \mathcal{F}$;*
(F_{III}) *if $A \in \mathcal{F}$ and $B \supset A$, then $B \in \mathcal{F}$.*

It is clear that a filter is a filter basis. Conversely, if $\mathcal{B}$ is a filter basis, then

$$\mathcal{F}_{\mathcal{B}} = \{A \mid A \in \mathcal{P}(X),\ A \supset \text{some set belonging to } \mathcal{B}\}$$

is a filter.

We leave it to the reader to show that:

23.20 *if $\mathcal{B}_1$ and $\mathcal{B}_2$ are two filter bases on X, then*

$$\mathcal{B}_1 \sim \mathcal{B}_2 \Leftrightarrow \mathcal{F}_{\mathcal{B}_1} = \mathcal{F}_{\mathcal{B}_2};$$

23.21 *if $\mathcal{B}_1$ and $\mathcal{B}_2$ are two filter bases on X, then*

$$\mathcal{B}_1 < \mathcal{B}_2 \Leftrightarrow \mathcal{F}_{\mathcal{B}_1} \subset \mathcal{F}_{\mathcal{B}_2}.$$

▼ Let X be a set and let $\mathcal{F}(X)$ be the set of all filters on X. We may order $\mathcal{F}(X)$ by writing

$$\mathcal{F}' \leq \mathcal{F}'' \quad \text{whenever} \quad \mathcal{F}' \subset \mathcal{F}''.$$

Exercise 1.—If $\mathscr{A} \in \mathscr{F}(X)$ and

$$\mathscr{G}(\mathscr{A}) = \{\mathscr{F} \mid \mathscr{F} \in \mathscr{F}(X),\ \mathscr{F} \supset \mathscr{A}\},$$

then $\mathscr{G}(\mathscr{A})$ is inductive.

In particular, $\mathscr{F}(X)$ is inductive (notice that $\{X\}$ is a filter).

The maximal elements of $\mathscr{F}(X)$ are called *ultrafilters*. An *ultrafilter basis* is a filter basis $\mathscr{B}$ such that $\mathscr{F}_{\mathscr{B}}$ is an ultrafilter.

Exercise 2.—A space $(X, \mathscr{T})$ is compact if and only if every ultrafilter basis on X is convergent. ▲

Nets

An ordered set A (with the order relation denoted $\leq$) is *directed* (or *filtering*) if for *any* a and b in A there is c in A satisfying $a \leq c$ *and* $b \leq c$.

Hence, A is directed if and only if every non-void finite part has an upper bound. Clearly, **N** (endowed with the usual order relation) is directed.

23.22 Definition.—*A net of elements of a set X is a family $(x_\alpha)_{\alpha \in A}$ of elements of X with A a directed set.*

Clearly, a sequence of elements of X is a net.

23.23 Definition.—*Let $(X, \mathscr{T})$ be a topological space and $(x_\alpha)_{\alpha \in A}$ a net of elements of X. We say that $(x_\alpha)_{\alpha \in A}$ converges to $a \in X$ if for every $V \in \mathscr{N}_X(a)$ there is $\beta \in A$ such that*

$$x_\alpha \in V \quad \text{whenever} \quad \alpha \geq \beta.$$

The element $a \in X$ is said to be *a limit* of $\mathscr{B}$. The notation

$$a = \lim_{\alpha \in A} x_\alpha \quad \text{(or, equivalently, } \lim_{\alpha \in A} x_\alpha = a)$$

means that $\mathscr{B}$ converges to a in the sense of Definition 23.23.

Instead of saying that $(x_\alpha)_{\alpha \in A}$ converges to a, we shall often say that $(x_\alpha)_{\alpha \in A}$ tends to a. Whenever we say that $(x_\alpha)_{\alpha \in A}$ *is convergent* in X we mean that the considered net converges to some $a \in X$ in the sense of Definition 23.22.

Example 16.—Let X be a set and $(x_\alpha)_{\alpha \in A}$ $(A \neq \varnothing)$ a net of elements of X. For each $\beta \in A$, let $S_\beta = \{x_\alpha \mid \alpha \geq \beta\}$. Then

$$\mathscr{B} = \{S_\beta \mid \beta \in A\}$$

is a *filter basis* on X.

In fact, clearly, $S_\beta \neq \varnothing$ for all $\beta \in A$, and hence $\mathscr{B} \not\ni \varnothing$; since $A \neq \varnothing$, $\mathscr{B} \neq \varnothing$. Hence, $\mathscr{B}$ satisfies (B_I). Now let α' and α'' be in A and let α''' be in A such that $\alpha''' \geq \alpha'$ and $\alpha''' \geq \alpha''$. Then

$$S_{\alpha'} \cap S_{\alpha''} \supset S_{\alpha'''}.$$

We deduce that $\mathscr{B}$ satisfies (B_{II}), too. Hence $\mathscr{B}$ is a filter basis on X.

The filter basis $\mathscr{B}$ defined above is called the *filter basis associated with the net* $(x_\alpha)_{\alpha \in A}$.

Example 17.—Let $(X, \mathscr{T})$ be a topological space and $(x_\alpha)_{\alpha \in A}$ a net of elements of X. Let $\mathscr{B}$ be the filter basis associated with the net $(x_\alpha)_{\alpha \in A}$ (see Example 16).

By Definition 23.3, $\mathscr{B}$ converges to $a \in X$ if and only if for every $V \in \mathscr{N}_X(a)$ there exists $\beta \in A$ such that

$$S_\beta = \{x_\alpha \mid \alpha \geq \beta\} \subset V.$$

This means that $\mathscr{B}$ converges to $a \in X$ if and only if for every $V \in \mathscr{N}_X(a)$ there exists $\beta \in A$ such that

$$x_\alpha \in V \quad \text{whenever} \quad \alpha \geq \beta.$$

If we compare this with Definition 23.23, we see that $\mathscr{B}$ converges to a in the sense of Definition 23.3 if and only if $(x_\alpha)_{\alpha \in A}$ converges to a in the sense of Definition 23.23.

Remarks.—(1) The results in Examples 1 and 15 are particular cases of the results in Examples 16 and 17, respectively.

(2) Comparing Theorem 23.7 with the result in Example 17, we deduce that if a net $(x_\alpha)_{\alpha \in A}$ converges to a' and a'', and if $(X, \mathscr{T})$ is separated, then $a' = a''$.

Exercise 3.—Let $((X_i, \mathscr{T}_i))_{i \in I}$ be a family of topological spaces and $(X, \mathscr{T})$ its product. A net $(x_\alpha)_{\alpha \in A}$ of elements of X converges to a if and only if $(pr_i(x_\alpha))_{\alpha \in A}$ converges to $pr_i(a)$ for every $i \in I$.

Exercises for Chapter 23

1. Let $\mathscr{B}_{+\infty}$ be the set of all parts of $\mathbf{R}$ containing some interval of the form $(a, +\infty)$, with $a \in \mathbf{R}$. Then $\mathscr{B}_{+\infty}$ is a filter on $\mathbf{R}$.

2. Let $\mathscr{B}_{-\infty}$ be the set of all parts of $\mathbf{R}$ containing some interval of the form $(-\infty, a)$, with $a \in \mathbf{R}$. Then $\mathscr{B}_{-\infty}$ is a filter on $\mathbf{R}$.

3. Let $t \in \mathbf{R}$ and let

$$\mathscr{B} = \{(t - \varepsilon, t) \cup (t, t + \varepsilon) \mid \varepsilon > 0\}.$$

Then $\mathscr{B}$ is a filter basis on $\mathbf{R}$.

4. Let $\mathscr{B}$ be the set of all parts of $\mathbf{R}^n$ $(n > 1)$ containing some set of the form

$$\{(x_1, \ldots, x_n) \mid x_1 > a_1, \ldots, x_n > a_n\}$$

with $(a_1, \ldots, a_n) \in \mathbf{R}^n$. Then $\mathscr{B}$ is a filter on $\mathbf{R}^n$.

5. Let $(X, \mathscr{T})$ be a topological space and $\mathscr{B}$ a filter basis on X. Then $x \in ad(\mathscr{B})$ if and only if there exists a filter basis $\mathscr{B}' \succ \mathscr{B}$ that converges to x.

6. Let $(X, \mathscr{T})$ be a topological space and $(\mathscr{F}_i)_{i \in I}$ a family of filters on X. If, for each $i \in I$, $\lim \mathscr{F}_i = x$, then

$$\lim \bigcap_{i \in I} \mathscr{F}_i = x.$$

7. Let X be a set and $\mathscr{F}$ a filter on X. Then $\mathscr{F}$ is the intersection of a set of filters of the form $\mathscr{F}_{\mathscr{B}}$, where $\mathscr{B}$ is a filter basis associated with a net.

8. Let X be a set and $\mathscr{U}$ an ultrafilter on X. If A and B are two parts of X such that $A \cup B \in \mathscr{U}$, then either $A \in \mathscr{U}$ or $B \in \mathscr{U}$.

9. Let X be a set and $\mathscr{F}$ a filter on X. Then $\mathscr{F}$ is the intersection of a set of ultrafilters on X.

10. Let X be a set, $\mathscr{B}$ an ultrafilter basis on X, Y a set, and $f: X \to Y$. Then $f(\mathscr{B})$ is an ultrafilter basis on Y.

11. Let E be the vector space of all bounded mappings $f: \mathbf{N} \to \mathbf{R}$. There is a linear mapping $\mathscr{L}: E \to \mathbf{R}$ such that:

(i) if $f = (f(n))_{n \in \mathbf{N}}$ is convergent, $\mathscr{L}(f) = \lim_{n \in \mathbf{N}} f(n)$;

(ii) if $f \in E$, $g \in E$, and $\{n \mid f(n) \neq g(n)\}$ is finite, then $\mathscr{L}(f) = \mathscr{L}(g)$;

(iii) if $f \in E$, $g \in E$, $g(n) \neq 0$ for $n \in \mathbf{N}$, $1/g \in E$ and $\mathscr{L}(g) \neq 0$, then $\mathscr{L}(f/g) = \mathscr{L}(f)/\mathscr{L}(g)$.

(Hint: Let $\mathscr{B}$ be the filter basis on N associated with the *sequence* N and let $\mathscr{U}$ be an ultrafilter basis on N such that $\mathscr{U} \succ \mathscr{B}$. For $f \in E$, define $\mathscr{L}(f) = \lim_{\mathscr{U}} f$.)

More Notations and Terminology

Let X be a set and $A \subset X \times X$, $B \subset X \times X$. We define the set $B \circ A \subset X \times X$ as follows:

$$(x, y) \in B \circ A \Leftrightarrow \text{there is } z \in X \text{ such that } (x, z) \in A \text{ and } (z, y) \in B.$$

If A, B, and C are three parts of $X \times X$, then

$$C \circ (B \circ A) = (C \circ B) \circ A.$$

Hence, we may define without ambiguity $C \circ B \circ A = (C \circ B) \circ A$. In the same way, if $A_1, \ldots, A_n$ are $n > 1$ parts of $X \times X$, we define $A_n \circ \ldots \circ A_1$. If $A_1 = \ldots = A_n = A$, we write

$$A_n \circ \ldots \circ A_1 = A^n.$$

If A, A', B, and B' are parts of $X \times X$, then

$$A \subset A' \quad \text{and} \quad B \subset B' \Rightarrow B \circ A \subset B' \circ A'.$$

If A is a part of $X \times X$ and $\Delta = \Delta_X$ $(= \{(x, x) \mid x \in X\})$, then

$$\Delta \circ A = A \circ \Delta = A.$$

If A is a part of $X \times X$ containing Δ, and B is a part of $X \times X$, then

$$A \circ B \supset B \quad \text{and} \quad B \circ A \supset B.$$

We notice that it is *not* necessarily true that, if A and B are parts of $X \times X$, then $A \circ B = B \circ A$.

For any $A \subset X \times X$, we write

$$A^{-1} = \{(x, y) \mid (y, x) \in A\}.$$

A set $A \subset X \times X$ is *symmetric* if $A = A^{-1}$. If $A \subset X \times X$ is arbitrary, then $A \cap A^{-1}$ is symmetric.

If $A \subset X \times X$ and $a \in X$, we write

$$A(a) = \{y \mid (a, y) \in A\}.$$

The set A is symmetric if and only if

$$\{y \mid (a, y) \in A\} = \{y \mid (y, a) \in A\}$$

for every $a \in X$.

More generally, if $A \subset X \times X$ and $K \subset X$, we define $A(K) = \{y \mid y \in X$, there is $x \in K$ such that $(x, y) \in A\}$.

Chapter 24

Uniform Structures. Uniform Spaces

We start with the following definition.

24.1 Definition.—*A uniform structure on* X *is a filter* $\mathscr{U}$ *on* $X \times X$ *having the following properties*:*

24.2 $U \supset \Delta$ *for every* $U \in \mathscr{U}$*;*

24.3 $U^{-1} \in \mathscr{U}$ *for every* $U \in \mathscr{U}$*;*

24.4 *for every* $U \in \mathscr{U}$ *there are* $V \in \mathscr{U}$ *and* $W \in \mathscr{U}$ *satisfying*† $W \circ V \subset U$.

Notice that if $\mathscr{U}$ is a uniform structure on X, and $\mathscr{B}$ is a *basis* of $\mathscr{U}$, then for every $U \in \mathscr{U}$ and integer $p > 1$ there is $V \in \mathscr{B}$ satisfying

24.5 $U \supset V^p$.

A set $U \subset X \times X$ is *symmetric* if $U = U^{-1}$. If $\mathscr{U}$ is a uniform structure on X, and if $\mathscr{U}^{(s)}$ is the set of all symmetric sets belonging to $\mathscr{U}$, then $\mathscr{U}^{(s)}$ *is a basis of* $\mathscr{U}$. It follows, in particular, that for every $U \in \mathscr{U}$ and integer $p > 1$ there is $V \in \mathscr{U}^{(s)}$ satisfying 24.5.

24.6 Theorem.—*Let* $\mathscr{B}$ *be a filter basis on* $X \times X$. *Then* $\mathscr{F}_{\mathscr{B}}$ *is a uniform structure if and only if the following conditions are satisfied:*

(a) $U \supset \Delta$ *for every* $U \in \mathscr{B}$*;*

(b) *for every* $U \in \mathscr{B}$, *there is* $V \in \mathscr{B}$ *such that* $V \subset U^{-1}$*;*

(c) *for every* $U \in \mathscr{B}$, *there are* $V \in \mathscr{B}$ *and* $W \in \mathscr{B}$ *satisfying* $W \circ V \subset U$.

---

* Recall that $\Delta = \Delta_X = \{(x, x) \mid x \in X\}$.

† If $V \in \mathscr{U}$ and $W \in \mathscr{U}$, then $W \circ V \in \mathscr{U}$.

Proof.—Assume that $\mathcal{F}_{\mathcal{B}}$ is a uniform structure and recall that $\mathcal{B} \subset \mathcal{F}_{\mathcal{B}}$. Clearly, (a) is satisfied. If $U \in \mathcal{B}$, then, by 24.3, $U^{-1} \in \mathcal{F}_{\mathcal{B}}$. Since $\mathcal{B}$ is a basis of $\mathcal{F}_{\mathcal{B}}$, there is $V \in \mathcal{B}$ such that $V \subset U^{-1}$. Hence, (b) is satisfied. Let $U \in \mathcal{B}$. By 24.4, there are $V' \in \mathcal{F}_{\mathcal{B}}$ and $W' \in \mathcal{F}_{\mathcal{B}}$ satisfying $W' \circ V' \subset U$. Since $\mathcal{B}$ is a basis of $\mathcal{F}_{\mathcal{B}}$, there are $V \in \mathcal{B}$ and $W \in \mathcal{B}$ such that $V \subset V'$ and $W \subset W'$. Then $W \circ V \subset W' \circ V' \subset U$. Hence, (c) is satisfied.

Assume now that $\mathcal{B}$ satisfies (a), (b), and (c). Let $U \in \mathcal{F}_{\mathcal{B}}$ and let $U' \in \mathcal{B}$ be such that $U \supset U'$. Since, by (a), U' contains Δ, we deduce that U contains Δ. Hence 24.2 is satisfied. Furthermore, $U^{-1} \supset (U')^{-1}$. By (b), there is $V \in \mathcal{B}$ such that $V \subset (U')^{-1}$. Then $V \subset U^{-1}$, so that $U^{-1} \in \mathcal{F}_{\mathcal{B}}$. Hence 24.3 is satisfied. Finally, let $V \in \mathcal{B}$ and $W \in \mathcal{B}$ satisfy $W \circ V \subset U'$. Then $W \circ V \subset U$, and hence 23.4 is also satisfied.

Let $\mathcal{U}$ be a *uniform structure* on X. For each $a \in X$, let

$$\mathcal{U}(a) = \{U(a) \mid U \in \mathcal{U}\};$$

then $\mathcal{U}(a)$ is a *filter* on X.

In fact, since $\mathcal{U} \neq \emptyset$ and $\mathcal{U} \not\ni \emptyset$, it follows that $\mathcal{U}(a) \neq \emptyset$ and $\mathcal{U}(a) \not\ni \emptyset$. Furthermore, if U and V belong to $\mathcal{U}$, then

$$U(a) \cap V(a) = (U \cap V)(a) \in \mathcal{U}(a).$$

Finally, let $U \in \mathcal{U}$ and $A \supset U(a)$. Then $V = U \cup (A \times A) \in \mathcal{U}$ and $V(a) = A$.

24.7 If $\mathcal{B}$ is a basis of $\mathcal{U}$, then for each $a \in X$,

$$\mathcal{B}(a) = \{U(a) \mid U \in \mathcal{B}\}$$

is a basis of $\mathcal{U}(a)$.

24.8 Recall that the set $U \subset X \times X$ is *symmetric* if and only if

$$U(a) = \{y \mid (a, y) \in U\} = \{y \mid (y, a) \in U\}$$

for every $a \in X$.

We shall show now that if $\mathcal{U}$ is a uniform structure on X, then the *filters* $\mathcal{U}(x)$, $x \in X$, satisfy the Conditions of Theorem 2.6. In fact, if $U \in \mathcal{U}$, then $U \supset \Delta$; hence $(x, x) \in U$, so that $U(x) \ni x$ for every $x \in X$. Hence, 2.6(a) is satisfied. Since, for every $x \in X$, $\mathcal{U}(x)$ is a filter, 2.6(b) and (c) are also satisfied. Now let $x \in X$ and $U \in \mathcal{U}$. Let $W \in \mathcal{U}$ be such that $W \circ W \subset U$. Then

$$y \in W(x) \Rightarrow W(y) \subset U(x).$$

In fact, let $z \in W(y)$. Then $(y, z) \in W$. Since $y \in W(x)$, we have $(x, y) \in W$. Hence, $(x, y) \in W$ and $(y, z) \in W$, so that $(x, z) \in W \circ W \subset U$. Hence $z \in U(x)$. Since $z \in W(y)$ was arbitrary, $W(y) \subset U(x)$.

We conclude that 2.6(d) is also satisfied. By Theorem 2.6, there is a *unique topology* $\mathcal{T}$ on X such that

$$\mathcal{N}_{(X,\mathcal{T})}(x) = \mathcal{U}(x)$$

for every $x \in X$.

24.9 Definition.—*The topology* $\mathcal{T}$ *is denoted* $\mathcal{T}_{\mathcal{U}}$ *and is said to be the topology defined by* $\mathcal{U}$.

If $\mathcal{B}$ is a *basis* of $\mathcal{U}$, then for every $x \in X$,

$$\mathcal{B}(x) = \{U(x) \mid U \in \mathcal{B}\}$$

is a *fundamental system* of $x \in X$. In particular,

$$\mathcal{U}^{(s)}(x) = \{V(x) \mid V \in \mathcal{U}, V = V^{-1}\}$$

is, for every $x \in X$, a fundamental system of $x \in X$.

24.10 Definition.—*Two uniform structures* $\mathcal{U}'$ *and* $\mathcal{U}''$ *on* X *are said to be equivalent if* $\mathcal{T}_{\mathcal{U}'} = \mathcal{T}_{\mathcal{U}''}$.

Notice that the uniform structures $\mathcal{U}'$ and $\mathcal{U}''$ are equivalent if and only if

$$\mathcal{U}'(x) = \mathcal{U}''(x)$$

for every $x \in X$ (see Theorem 2.5).

Example 1.—Let X be a set and let

$$\mathcal{U}' = \{U \subset X \times X \mid U \supset \Delta\}.$$

Then $\mathcal{U}'$ is a uniform structure on X and $\mathcal{T}_{\mathcal{U}'} = \mathcal{D} =$ the discrete topology on X.

Let $\mathcal{P}$ be the set of all *partitions* $(A_i)_{i \in I}$ of X and let

$$\mathcal{B} = \{\bigcup_{i \in I} (A_i \times A_i) \mid (A_i)_{i \in I} \in \mathcal{P}\}.$$

Then $\mathcal{U}'' = \mathcal{F}_{\mathcal{B}}$ is a uniform structure on X and $\mathcal{T}_{\mathcal{U}''} = \mathcal{D}$.

Hence, $\mathcal{U}'$ and $\mathcal{U}''$ are *equivalent*. However, $\mathcal{U}' = \mathcal{U}''$ *if and only if* X *is finite*.

24.11 Definition.—*A couple* $(X, \mathcal{U})$, *where* X *is a set and* $\mathcal{U}$ *is a uniform structure on* X, *is called a uniform space.*

When $(X, \mathscr{U})$ is a uniform space, we assume $X \times X$ to be endowed with the topology product of the topologies $\mathscr{T}_{\mathscr{U}}$ and $\mathscr{T}_{\mathscr{U}}$ (unless we explicitly indicate the contrary).

If $(X, \mathscr{U})$ is a uniform space, we say that $\mathscr{U}$ is the uniform structure of $(X, \mathscr{U})$. When there is no ambiguity as to what uniform structure we consider, we shall often say "the uniform space X" instead of "the uniform space $(X, \mathscr{U})$."

If $(X, \mathscr{U})$ is a uniform space, then we shall always suppose that X is endowed with the topology $\mathscr{T}_{\mathscr{U}}$. We shall often say that $\mathscr{T}_{\mathscr{U}}$ is the topology of the uniform space $(X, \mathscr{U})$.

Whenever we say, for instance, that a uniform space is separated, compact, or locally compact, we mean that $(X, \mathscr{T}_{\mathscr{U}})$ is separated, compact, or locally compact.

24.12 Theorem.—*If $(X, \mathscr{U})$ is a uniform space, then $\overset{\circ}{U} \in \mathscr{U}$ for every $U \in \mathscr{U}$.*

Proof.—Let $U \in \mathscr{U}$. Choose $V \in \mathscr{U}^{(s)}$ such that $V \circ V \circ V \subset U$. Now let $(a, b) \in V$ and let $(x, y) \in V(a) \times V(b)$. Then (since $V = V^{-1}$)

$$(x, a) \in V, \qquad (a, b) \in V, \qquad (b, y) \in V,$$

so that $(x, y) \in V \circ V \circ V \subset U$. Hence, $V(a) \times V(b) \subset U$, and hence $(a, b) \in \overset{\circ}{U}$. We deduce that $V \subset \overset{\circ}{U}$, so that $\overset{\circ}{U} \in \mathscr{U}$.

24.13 Theorem.—*Let $(X, \mathscr{U})$ be a uniform space and $\mathscr{B}$ a basis of $\mathscr{U}$. Then*

$$\overset{\circ}{\mathscr{B}} = \{\overset{\circ}{U} \mid U \in \mathscr{B}\}$$

is a basis of $\mathscr{U}$.

Proof.—By 24.12, $\overset{\circ}{\mathscr{B}} \subset \mathscr{U}$. Furthermore, if $V \in \mathscr{U}$, there is $U \in \mathscr{B}$ such that

$$\overset{\circ}{U} \subset U \subset V.$$

Since $V \in \mathscr{U}$ was arbitrary, $\overset{\circ}{\mathscr{B}}$ is a basis of $\mathscr{U}$.

24.14—Theorem.—*Let $(X, \mathscr{U})$ be a uniform space and $\mathscr{B}$ a basis of $\mathscr{U}$. Then*

$$(\bigstar) \qquad \bar{E} = \bigcap_{V \in \mathscr{B}} V \circ E \circ V$$

for every $E \subset X \times X$.

Proof.—Let $(a, b) \in \bar{E}$ and $V \in \mathscr{B}$. Let $U \in \mathscr{U}^{(s)}$ be such that $U \subset V$. Then $(U(a) \times U(b)) \cap E \neq \varnothing$, and hence there is

$$(x, y) \in (U(a) \times U(b)) \cap E.$$

Then $(a, x) \in U$, $(x, y) \in E$ and $(b, y) \in U$. Since U is symmetric, $(y, b) \in U$. We deduce that

$$(a, b) \in U \circ E \circ U \subset V \circ E \circ V.$$

Since $V \in \mathscr{B}$ was arbitrary, we conclude that

$$\text{(i)} \qquad \overline{E} \subset \bigcap_{V \in \mathscr{B}} V \circ E \circ V.$$

Conversely, let $(a, b) \in \bigcap_{V \in \mathscr{B}} V \circ E \circ V$ and $W \in \mathscr{N}_X(a, b)$. Let $V \in \mathscr{U}^{(s)}$ be such that $V(a) \times V(b) \subset W$ (see 24.7). Since* $(a, b) \in V \circ E \circ V$, there is $(x, y) \in E$ satisfying $(a, x) \in V$ and $(y, b) \in V$. Hence, $x \in V(a)$ and, since $V = V^{-1}$, $y \in V(b)$. Hence

$$W \cap E \supset (V(a) \times V(b)) \cap E \neq \varnothing.$$

Since $W \in \mathscr{N}(a, b)$ was arbitrary, $(a, b) \in \overline{E}$. Since (a, b) was arbitrary,

$$\text{(ii)} \qquad \bigcap_{V \in \mathscr{B}} V \circ E \circ V \subset \overline{E}.$$

Comparing (i) and (ii), we deduce (★).

24.15 Theorem.—*If $(X, \mathscr{U})$ is a uniform space and $\mathscr{B}$ a basis of $\mathscr{U}$, then*

$$\overline{\mathscr{B}} = \{\overline{U} \mid U \in \mathscr{B}\}$$

is a basis of $\mathscr{U}$.

Proof.—Let $V \in \mathscr{U}$ and let $U \in \mathscr{B}$ (see 24.5) be such that

$$U \circ U \circ U \subset V.$$

Then (see Theorem 24.14)

$$\overline{U} \subset U \circ U \circ U \subset V.$$

Since $V \in \mathscr{U}$ was arbitrary and since, clearly, $\overline{\mathscr{B}} \subset \mathscr{U}$, we deduce that $\overline{\mathscr{B}}$ is a basis of $\mathscr{U}$.

It follows in particular that $\overline{\mathscr{U}} = \{\overline{U} \mid U \in \mathscr{U}\}$ is a basis of $\mathscr{U}$.

Remarks.—Let $(X, \mathscr{U})$ be a uniform space and $U \in \mathscr{U}$. Then:

(i) if U is closed, $U(x)$ is closed for every $x \in X$;
(ii) if U is open, $U(K)$ is open for every $K \subset X$;
(iii) if $U \in \mathscr{U}$ and $A \subset X$, there is $B \subset X$ open such that $A \subset B \subset U(A)$.

* Since $\mathscr{B}$ is a basis, there is $U \in \mathscr{B}$ contained in V, whence $U \circ E \circ U \subset V \circ E \circ V$.

To prove (i), we notice that, for every $x \in X$, $k: y \mapsto (x, y)$ is continuous and

$$U(x) = \{y \mid (x, y) \in U\} = k^{-1}(U).$$

To prove (ii), we use the continuity of k and the relation

$$U(K) = \bigcup_{x \in K} U(x).$$

To prove (iii), we use 23.13 and (ii).

24.16 Corollary.—*A uniform space $(X, \mathscr{U})$ is regular.*

Proof.—Let $x \in X$ and $V \in \mathscr{N}_X(x)$. Then there is $U \in \mathscr{U}$ such that $U(x) = V$. By 24.15, there is $W \in \mathscr{U}$ satisfying $\overline{W} \subset U$. We deduce that $\overline{W}(x) \subset U(x) \subset V$. By (i), $\overline{W}(x)$ is *closed*. Since $x \in X$ and $V \in \mathscr{N}_X(x)$ were arbitrary, Corollary 24.16 is proved.

24.17 Corollary.—*A uniform space $(X, \mathscr{U})$ is separated if and only if*

$$\bigcap_{U \in \mathscr{U}} U = \Delta.$$

Proof.—By Theorem 24.15,

$$\bigcap_{U \in \mathscr{U}} U = \bigcap_{U \in \mathscr{U}} \overline{U},$$

so that $\bigcap_{U \in \mathscr{U}} U$ is closed. Hence, if $\bigcap_{U \in \mathscr{U}} U = \Delta$, then Δ is closed, and then (by Theorem 7.9) $(X, \mathscr{U})$ is separated.

Conversely, suppose that $(X, \mathscr{U})$ is separated and let $(x, y) \notin \Delta$. Then $x \neq y$, and hence there is $U \in \mathscr{U}$ such that $y \notin U(x)$. Since this means that $(x, y) \notin U$, we deduce that $\Delta = \bigcap_{U \in \mathscr{U}} U$.

Let X be an arbitrary set and let $\mathscr{U}(X)$ be the set of all uniform structures on X. We may order $\mathscr{U}(X)$ by writing

$$\mathscr{U}' \leq \mathscr{U}'' \quad \text{whenever} \quad \mathscr{U}' \subset \mathscr{U}''.$$

If $(\mathscr{U}_i)_{i \in I}$ is a family of elements of $\mathscr{U}(X)$, then $\bigcap_{i \in I} \mathscr{U}_i$ is a uniform structure on X, and it is easy to see that

$$\bigcap_{i \in I} \mathscr{U}_i = \inf_{i \in I} \mathscr{U}_i.$$

We leave it to the reader to show that if $(\mathscr{U}_i)_{i \in I}$ is a family of elements of $\mathscr{U}(X)$, then $\sup_{i \in I} \mathscr{U}_i$ exists (observe that $\mathscr{U} = \mathscr{F}_{\{\Delta\}} \in \mathscr{U}(X)$ and $\mathscr{U} \geq \mathscr{U}'$ for every $\mathscr{U}' \in \mathscr{U}(X)$).

24.18 Definition.—(a) *Let X be a set and $\mathcal{T}$ a topology on X. We say that $\mathcal{T}$ is uniformizable if there is a uniform structure $\mathcal{U}$ on X such that $\mathcal{T} = \mathcal{T}_{\mathcal{U}}$.*

(b) *We say that a topological space $(X, \mathcal{T})$ is uniformizable if $\mathcal{T}$ is uniformizable.*

Subspaces

Let $(X, \mathcal{U})$ be a uniform space and let $A \subset X$. It is easily seen that

$$\mathcal{U}_A = \{U \cap (A \times A) \mid U \in \mathcal{U}\}$$

is a *filter* on $A \times A$ satisfying 24.2, 24.3, and 24.4. Hence, $\mathcal{U}_A$ is a *uniform structure* on A. Thus, we define the uniform space $(A, \mathcal{U}_A)$. Any uniform space of this form will be called a *subspace* of $(X, \mathcal{U})$. We shall often say "the *subspace* A of X" instead of "the subspace $(A, \mathcal{U}_A)$ of $(X, \mathcal{U})$."

Exercise 1.—If $\mathcal{B}$ is a basis of $\mathcal{U}$, then

$$\mathcal{B}_A = \{U \cap (A \times A) \mid U \in \mathcal{B}\}$$

is a basis of $\mathcal{U}_A$.

Let $a \in A$ and $U \in \mathcal{U}$. Then

$$U(a) \cap A = \{x \mid (a, x) \in U\} \cap A = \{x \mid x \in A, (a, x) \in U\}$$
$$= \{x \mid (a, x) \in U \cap (A \times A)\} = (U \cap (A \times A))(a).$$

Hence (see 4.3),

$$\mathcal{N}_{(A, (\mathcal{T}_{\mathcal{U}})_A)}(a) = \mathcal{N}_{(A, \mathcal{T}_{\mathcal{U}_A})}(a)$$

for every $a \in A$, and hence

$$(\mathcal{T}_{\mathcal{U}})_A = \mathcal{T}_{\mathcal{U}_A}.$$

24.19 Theorem.—*Assume that $\bar{A} = X$. Then*

$$\mathcal{B} = \{\overline{U \cap (A \times A)} \mid U \in \mathcal{U}\}$$

is a basis of $\mathcal{U}$.

(*) *Proof.*—Let $W = U \cap (A \times A)$ with $U \in \mathcal{U}$; then

$$\mathring{U} \subset \overline{W}.$$

In fact, let $(x, y) \in \mathring{U}$ and let $B \in \mathcal{N}_{X \times X}(x, y)$. Then

$$B \cap W = B \cap (U \cap (A \times A)) = (B \cap U) \cap (A \times A) \neq \varnothing,$$

since $B \cap U \in \mathcal{N}_{X \times X}(x, y)$ and $(x, y) \in \overline{A \times A}$. Since $(x, y) \in \mathring{U}$ was arbitrary (*), is proved. We deduce that $\mathcal{B} \subset \mathcal{U}$.

Now let $U \in \mathcal{U}$. By 24.15, there is $V \in \mathcal{U}$ such that $\overline{V} \subset U$. Then

$$\overline{V \cap (A \times A)} \subset \overline{V} \subset U.$$

Hence, every set in $\mathcal{U}$ contains a set belonging to $\mathcal{B}$.

We conclude that $\mathcal{B}$ is a basis of $\mathcal{U}$.

24.20 Corollary.—*Let $\mathcal{U}'$ and $\mathcal{U}''$ be two uniform structures on X defining the same topology and A a set dense in X. If $\mathcal{U}'_A = \mathcal{U}''_A$ then*

$$\mathcal{U}' = \mathcal{U}''.$$

Proof.—Let

$$\mathcal{B}' = \{\overline{U \cap (A \times A)} \mid U \in \mathcal{U}'\}$$

and

$$\mathcal{B}'' = \{\overline{U \cap (A \times A)} \mid U \in \mathcal{U}''\}.$$

By Theorem 24.19, $\mathcal{B}'$ is a basis of $\mathcal{U}'$ and $\mathcal{B}''$ a basis of $\mathcal{U}''$. By hypothesis, $\mathcal{B}' = \mathcal{B}''$, whence $\mathcal{U}' = \mathcal{U}''$.

Remark.—Notice that $\mathcal{U}_A$ is the uniform structure on A, *inverse image of $\mathcal{U}$* by the mapping $j_{A,X}: x \mapsto x$ of A into X (see later discussion and also Example 2 of Chapter 6).

Initial Uniform Structures

Let X be a set, $(Y, \mathcal{U})$ a uniform space, and f a mapping of X into Y. Let $\mathcal{B}$ be the set of all parts of $X \times X$ of the form

$$W(U) = \{(x, y) \mid (f(x), f(y)) \in U\},$$

for some $U \in \mathcal{U}$. Then $\mathcal{B}$ is a filter basis on $X \times X$ satisfying the conditions of Theorem 24.6.* Hence $\mathcal{W} = \mathcal{F}_{\mathcal{B}}$ is a uniform structure on X. This uniform structure is called *the inverse image of $\mathcal{U}$ by f.*

We observe now that (see Chapter 6)

24.21 $\mathcal{T}_{\mathcal{W}} = f^{-1}(\mathcal{T}_{\mathcal{U}})$.

Proof.—Let $A \in \mathcal{T}_{\mathcal{W}}$ and $x \in A$. Then there is $U_x \in \mathcal{U}$, $U_x = \mathring{U}_x$, such that

$$A \supset W(U_x)(x).$$

* Let $\Delta_X = \{(x, x) \mid x \in X\}$ and notice that $W(U) \supset \Delta_X$ and $W(U)^{-1} = W(U^{-1})$ for every $U \in \mathcal{U}$, and that if $U \in \mathcal{U}$ and $V \in \mathcal{U}$ satisfy $V \circ V \subset U$, then $W(V) \circ W(V) \subset W(U)$.

But* $W(U_x)(x) = f^{-1}(U_x(f(x)))$ and $U_x(f(x)) \in \mathcal{T}_{\mathcal{U}}$. We deduce that

$$A = \bigcup_{x\in A} f^{-1}(U_x(f(x))) = f^{-1}(\bigcup_{x\in A} U_x(f(x))) \in f^{-1}(\mathcal{T}_{\mathcal{U}}).$$

Since $A \in \mathcal{T}_{\mathcal{W}}$ was arbitrary, $\mathcal{T}_{\mathcal{W}} \subset f^{-1}(\mathcal{T}_{\mathcal{U}})$.

Now let $B = f^{-1}(\mathcal{T}_{\mathcal{U}})$. Then $B = f^{-1}(C)$ for some $C \in \mathcal{T}_{\mathcal{U}}$. Let $x \in B$ and $U \in \mathcal{U}$ be such that $U(f(x)) \subset C$. Then

$$W(U)(x) = f^{-1}(U(f(x))) \subset f^{-1}(C) = B.$$

Since $W(U)(x) \in \mathcal{N}_{(X,\mathcal{T}_{\mathcal{W}})}(x)$, and since $x \in B$ was arbitrary, $B \in \mathcal{T}_{\mathcal{W}}$. Hence $f^{-1}(\mathcal{T}_{\mathcal{U}}) \subset \mathcal{T}_{\mathcal{W}}$.

We conclude that $\mathcal{T}_{\mathcal{W}} = f^{-1}(\mathcal{T}_{\mathcal{U}})$.

We shall now generalize the above construction.

Let X be a set, $((Y_i, \mathcal{U}_i))_{i\in I}$ a family of uniform spaces, and for each $i \in I$ let f_i be a mapping of X into Y_i.

For every family $(U_j)_{j\in J}$, where J is a finite part of I and $U_j \in \mathcal{U}_j$ for each $j \in J$, we define†

24.22 $W((U_j)_{j\in J}) = \bigcap_{j\in J} \{(x,y) \mid (f_j(x), f_j(y)) \in U_j\}$.

Denote by $\mathcal{B}$ the set of all parts of $X \times X$ of the form 24.22. Then *$\mathcal{B}$ is a filter basis on $X \times X$.*

Note that

$$W((U'_j)_{j\in J}) \cap W((U''_j)_{j\in J}) = W((U'_j \cap U''_j)_{j\in J})$$

for any families $(U'_j)_{j\in J}$ and $(U''_j)_{j\in J}$, and that $\mathcal{B} \neq \varnothing$ and $\mathcal{B} \not\ni \varnothing$.

Moreover, $\mathcal{B}$ satisfies the conditions of Theorem 24.6. In fact, it is clear that every set in $\mathcal{B}$ contains Δ_x. To prove that $\mathcal{B}$ satisfies 24.6(b), we notice that

$$W((U_j)_{j\in J})^{-1} = W((U_j^{-1})_{j\in J})$$

---

* We have

$$y \in W(U_x)(x) \Leftrightarrow (x,y) \in W(U_x) \Leftrightarrow (f(x), f(y)) \in U_x \Leftrightarrow f(y) \in U_x(f(x))$$
$$\Leftrightarrow y \in f^{-1}(U_x(f(x))).$$

Hence $W(U_x)(x) = f^{-1}(U_x(f(x)))$.

† If $H \subset J$ and $U_j = Y_j$ for $j \in J - H$, then

$$W((U_j)_{j\in J}) = W((U_j)_{j\in H}).$$

for every family $(U_j)_{j \in J}$. To prove that $\mathscr{B}$ satisfies 24.6(c), we observe that if $(V_j)_{j \in J}$ and $(U_j)_{j \in J}$ are such that $V_j \circ V_j \subset U_j$ for every $j \in J$, then

$$W((V_j)_{j \in J}) \circ W((V_j)_{j \in J}) \subset W((U_j)_{j \in J}).$$

Hence, $\mathscr{W} = \mathscr{F}_{\mathscr{B}}$ is a uniform structure on X. The uniform structure $\mathscr{W}$ *is called the initial uniform structure on* X *associated with the families* $((Y_i, \mathscr{U}_i))_{i \in I}$ *and* $(f_i)_{i \in I}$.

24.23 Theorem.— *The initial topology* $\mathscr{T}$ *on* X *associated with the families* $((Y_i, \mathscr{T}_{\mathscr{U}_i}))_{i \in I}$ *and* $(f_i)_{i \in I}$ *coincides with* $\mathscr{T}_{\mathscr{W}}$.*

Proof.—Let $A \in \mathscr{T}_{\mathscr{W}}$ and $x \in A$. Then there is a finite family $(U_{j,x})_{j \in J}$, where $U_{j,x} = \overset{\circ}{U}_{j,x} \in \mathscr{U}_j$ for every $j \in J$, such that

$$A \supset W((U_{j,x})_{j \in J})(x).$$

But

$$W((U_{j,x})_{j \in J})(x) = \bigcap_{j \in J} f_j^{-1}(U_{j,x}(x))$$

and $f_j^{-1}(U_{j,x}(x)) \in \mathscr{T}$ for every $j \in J$ (see 6.2); hence $W(U_{j,x})(x) \in \mathscr{T}$. Since $x \in A$ was arbitrary, we deduce that $A \in \mathscr{T}$. Since $A \in \mathscr{T}_{\mathscr{W}}$ was arbitrary, $\mathscr{T}_{\mathscr{W}} \subset \mathscr{T}$.

Now let $j \in J$, $C \in \mathscr{T}_{\mathscr{U}_j}$, and $B = f_j^{-1}(C)$. Then $B \in \mathscr{T}_{\mathscr{W}}$ (see the proof of the inclusion $f^{-1}(\mathscr{T}_{\mathscr{U}}) \subset \mathscr{T}_{\mathscr{W}}$ in 24.21). Since $C \in \mathscr{T}_{\mathscr{U}_j}$ was arbitrary, f_j is continuous when X is endowed with the topology $\mathscr{T}_{\mathscr{W}}$ and Y_j with $\mathscr{T}_{\mathscr{U}_j}$. From Theorem 6.2, we deduce that $\mathscr{T} \subset \mathscr{T}_{\mathscr{W}}$.

We conclude that $\mathscr{T} = \mathscr{T}_{\mathscr{W}}$.

Product of Uniform Spaces

Let $((X_i, \mathscr{U}_i))_{i \in I}$ be a family of uniform spaces and let

$$X = \prod_{i \in I} X_i.$$

For each k in I, let pr_k be the *projection* $(x_i)_{i \in I} \mapsto x_k$ of X onto X_k.

The *initial uniform structure* $\mathscr{U}$ on X, associated with the families $((X_i, \mathscr{U}_i))_{i \in I}$ and $(pr_i)_{i \in I}$, is *called the product of the family* $(\mathscr{U}_i)_{i \in I}$ *of uniform structures. The uniform space* $(X, \mathscr{U})$ *is called the product of the family* $((X_i, \mathscr{U}_i))_{i \in I}$.

From 7.1 and 24.23, it follows that $\mathscr{T}_{\mathscr{U}}$ is *the product of the family* $(\mathscr{T}_{\mathscr{U}_i})_{i \in I}$.

Remarks.—If $(U_j)_{j \in J}$ is a finite family such that $U_j \in \mathscr{U}_j$ for every $j \in J$, then $W((U_j)_{j \in J})$ is the set of all (x, y) ($x = (x_i)_{i \in I}$ and

* A particular form of this result is 24.21.

$y = (y_i)_{i \in I})$ such that

$$(x_j, y_j) \in U_j$$

for every $j \in J$.

If $I = \{1, \ldots, n\}$, and if $U_i \in \mathscr{U}_i$ for all $i \in I$, then we often write $W(U_1, \ldots, U_n)$ instead of $W((U_j)_{j \in I})$. Notice that $W(U_1, \ldots, U_n)$ is the set

$$\{((x_j)_{1 \leq j \leq n}, (y_j)_{1 \leq j \leq n}) \mid (x_j, y_j) \in U_j \text{ for } j = 1, \ldots, n\}.$$

Notice also that if $\mathscr{B}_1, \ldots, \mathscr{B}_n$ are bases of $\mathscr{U}_1, \ldots, \mathscr{U}_n$, respectively, then

$$\{W(U_1, \ldots, U_n) \mid U_1 \in \mathscr{B}_1, \ldots, U_n \in \mathscr{B}_n\}$$

is a basis of the uniform structure "product of $\mathscr{U}_1, \ldots, \mathscr{U}_n$."

Exercises for Chapter 24

1. Give an example of a topological space that is not uniformizable.

2. Let $(X, \mathscr{U})$ be a separated uniform space, V a closed subset of $X \times X$, and K a compact subset of X. Then $V(K)$ is closed.

3. Let G be a topological group (see Exercise 7 at the end of Chapter 8). If $\mathscr{F}$ is a fundamental system of e, then

$$\mathscr{V} = \{\{(x, y) \mid yx^{-1} \in V\} \mid V \in \mathscr{F}\}$$

is a filter basis on $G \times G$ satisfying the conditions of Theorem 24.6. Moreover, if $\mathscr{U} = \mathscr{F}_{\mathscr{V}}$, then $\mathscr{T}_{\mathscr{U}}$ is the topology of G.

4. Show that the topological space $(X, \mathscr{T})$ in Appendix II, Chapter 13, is unformizable.

* 5. Let $(X, \mathscr{T})$ be paracompact and let $\mathscr{U}$ be the set of all parts of $X \times X$, the interior of which contain Δ_X. Then $\mathscr{U}$ is a uniform structure on X and* $\mathscr{T}_{\mathscr{U}} = \mathscr{T}$.

6. Let $\mathscr{U}$ be a uniform structure on a set X and let $U \in \mathscr{U}$. Then there is a sequence $(V_n)_{n \in \mathbf{N}}$ of elements of $\mathscr{U}$ such that

$$V_1 \supset V_2 \supset \ldots \supset V_n \supset \ldots$$

and

$$V_n \circ \ldots \circ V_1 \subset U$$

for all $n \in \mathbf{N}$.

---

* Hence, a paracompact space is unformizable.

7. Let $((X_i, \mathscr{U}_i))_{i \in I}$ be a family of uniform spaces and $(X, \mathscr{U})$ its product. Let $A_i \subset X_i$ for each $i \in I$ and $A = \prod_{i \in I} A_i$. Then $\mathscr{U}_A$ is the product of the family $(\mathscr{U}_{A_i})_{i \in I}$.

8. Suppose that $(X, \mathscr{T})$ is a topological space such that for each $x \in X$ there is a neighborhood V_x of x such that the subspace $(V_x, \mathscr{T}_{V_x})$ is uniformizable. If $(X, \mathscr{T})$ is regular, $(X, \mathscr{T})$ is uniformizable.

Chapter 25

Semi-Metrics and Uniform Spaces

Let X be a set. A *semi-metric* on X is a mapping $d: (x, y) \mapsto d(x, y)$ of $X \times X$ into $\mathbf{R}$ having the following properties:

25.1 $d(x, x) = 0$ for all $x \in X$;

25.2 $d(x, y) = d(y, x)$ for all $x \in X, y \in X$;

25.3 $d(x, y) \leq d(x, z) + d(z, y)$ for all $x \in X, y \in X, z \in X$.

Clearly, a metric is a semi-metric. The converse is not necessarily true (see Example 1). Notice that the only difference between metrics and semi-metrics is the following: If d is a metric, then $d(x, y) = 0$ if and only if $x = y$. If d is a semi-metric, we have $d(x, y) = 0$ if $x = y$; however, $d(x, y) = 0$ *does not imply* that $x = y$.

Example 1.—Let X be a set and $f: X \to \mathbf{R}$. Define $d: X \times X \to \mathbf{R}$ by

$$d(x, y) = |f(x) - f(y)|$$

for $(x, y) \in X \times X$. Then d is a semi-metric on X. Moreover, d is a metric if and only if f is *injective*.

If X is a set and d a semi-metric on X, then

25.4 $d(x, y) \geq 0$ for all $x \in X, y \in X$;

25.5 $|d(x, z) - d(y, z)| \leq d(x, y)$ for all $x \in X, y \in X, z \in X$.

The proofs of 24.4 and 24.5 are similar to those of 13.4 and 13.6.

Example 2.—Denote by X a set. Then:

(i) If d is a semi-metric on X, and $\delta: X \times X \to \mathbf{R}$ is defined by

$$\delta(x, y) = \inf \{d(x, y), 1\}$$

for all $(x, y) \in X \times X$, then δ is a semi-metric on X.

(ii) If $\lambda \geq 0$ and d is a semi-metric on X, then λd is a semi-metric on X.

(iii) If $(d_i)_{i \in I}$ is a non-void finite family of semi-metrics on X, then

$$\textstyle\sum_{i \in I} d_i \quad \text{and} \quad \sup_{i \in I} d_i$$

are semi-metrics on X.

(iv) Let d be a semi-metric on X and $A \subset X$. Then $d \mid A \times A \to \mathbf{R}$ is a semi-metric on A. This semi-metric is usually denoted by d_A.

Let X be a set and d a *semi-metric* on X. We define*

$$V_{r,d} = \{(x, y) \mid d(x, y) < r\} \quad \text{for} \quad r > 0$$

and

$$W_{r,d} = \{(x, y) \mid d(x, y) \leq r\} \quad \text{for} \quad r \geq 0.$$

Notice that $V_{r,d}$ and $W_{r,d}$ are *symmetric* sets for all r and d.

If $\lambda > 0$, then

$$V_{\lambda r, d} = \{(x, y) \mid d(x, y) < \lambda r\} = V_{r, d/\lambda};$$

in the same way we see that $W_{\lambda r, d} = W_{r, d/\lambda}$.

If $(d_i)_{i \in I}$ is a non-void finite family of semi-metrics on X, and

$$d = \sup_{i \in I} d_i,$$

then

$$V_{\varepsilon, d} = \bigcap_{i \in I} V_{\varepsilon, d_i} \quad \text{and} \quad W_{\varepsilon, d} = \bigcap_{i \in I} W_{\varepsilon, d_i}$$

for all $\varepsilon > 0$.

For each $a \in X$, we have

$$V_{r,d}(a) = \{x \mid d(a, x) < r\} \quad \text{if} \quad r > 0$$

---

* When there is no ambiguity, we write V_r and W_r instead of $V_{r,d}$ and $W_{r,d}$, respectively.

and

$$W_{r,d}(a) = \{x \mid d(a, x) \leq r\} \quad \text{if} \quad r \geq 0.$$

Let $(X, \mathscr{U})$ be a uniform space and d a semi-metric on X such that $V_{\varepsilon,d} \in \mathscr{U}$ for every $\varepsilon > 0$. Then d is *continuous** on $X \times X$. In fact, let $(a, b) \in X \times X$. Then for every $(x, y) \in X \times X$, we have (see the proof of 13.21)

25.6 $|d(a, b) - d(x, y)| \leq d(a, x) + d(b, y).$

If $(x, y) \in V_{\varepsilon/2,d}(a) \times V_{\varepsilon/2,d}(b)$, then

$$|d(a, b) - d(x, y)| < \varepsilon.$$

Since $V_{\varepsilon/2,d}(a) \times V_{\varepsilon/2,d}(b) \in \mathscr{N}(a, b)$, and since $\varepsilon > 0$ was arbitrary, we deduce that d is continuous at (a, b). Since $(a, b) \in X \times X$ was arbitrary, d is continuous on $X \times X$.

In the same way, we may prove that if d is a semi-metric on X such that $W_{\varepsilon,d} \in \mathscr{U}$ for every $\varepsilon > 0$, then d is continuous on X.

Let X be a set. A family $(d_i)_{i \in A}$ of semi-metrics on X is *directed* (or *filtering*) if for every α and β in A there is γ in A such that

$$d_\gamma \geq d_\alpha \quad \text{and} \quad d_\gamma \geq d_\beta.$$

Clearly, if $(d_i)_{i \in I}$ is a directed family of semi-metrics on X, then for every non-void finite set $I \subset A$ there is $\varphi \in A$ satisfying $d_\varphi \geq d_i$ for all $i \in I$.

Let $\mathscr{A} = (d_i)_{i \in I}$ be an arbitrary family of semi-metrics on X. Let $\mathscr{F}(I)$ be the set of all finite parts of I. For every $J \in \mathscr{F}(I)$, let

$$d_J = \sup_{i \in J} d_i$$

(we write $d_J = 0$ if $J = \varnothing$). Then

$$\mathscr{A}^* = (d_J)_{J \in \mathscr{F}(I)}$$

is a *directed* family of semi-metrics on X.

Now let X be a set and $\mathscr{A} = (d_i)_{i \in A}$ a directed family of semi-metrics on X. Let

$$\mathscr{V} = \{V_{r,d_i} \mid r > 0, i \in A\}.$$

* As we shall see in Chapter 27, 25.6 implies that d is uniformly continuous on $X \times X$.

Then $\mathscr{V}$ is a filter basis on X satisfying the conditions of Theorem 24.6 (we leave it to the reader to establish these assertions). Then by Theorem 24.6, $\mathscr{F}_{\mathscr{V}}$ is a *uniform structure* on X.

25.7 Definition. *The uniform structure* $\mathscr{F}_{\mathscr{V}}$ *is denoted by* $\mathscr{U}^{(\mathscr{A})}$ *and is said to be the uniform structure defined by the directed family* $\mathscr{A}$ *of semi-metrics.*

The topology $\mathscr{T}_{\mathscr{U}^{(\mathscr{A})}}$ is often denoted by $\mathscr{T}^{(\mathscr{A})}$. Notice that for every $a \in X$,

$$\mathscr{V}(a) = \{V_{r,d_i}(a) \mid r > 0, i \in A\}$$

is a fundamental system of a.

If

$$\mathscr{W} = \{W_{r,d_i} \mid r > 0, i \in A\},$$

then $\mathscr{W}$ is a filter basis on $X \times X$ equivalent with $\mathscr{V}$. We deduce that, for every $a \in X$,

$$\mathscr{W}(a) = \{W_{r,d_i}(a) \mid r > 0, i \in A\}$$

is a fundamental system of a.

Remarks.—(i) If $\mathscr{A}$ is as above, then $V_{\varepsilon,d_i} \in \mathscr{U}^{(\mathscr{A})}$ for every $i \in I$ and $\varepsilon > 0$. It follows that for every $i \in I$, $d_i : X \times X \to \mathbf{R}$ *is continuous* (in fact, we shall see in Chapter 27 that this mapping is uniformly continuous).

(ii) Let d be a semi-metric on X, $I = \{i_0\}$ and $d_{i_0} = d$. Then $\mathscr{A} = (d_i)_{i \in I}$ is a directed family of semi-metrics on X. In this case, we often write $\mathscr{U}_d$ instead of $\mathscr{U}^{(\mathscr{A})}$ and $\mathscr{T}_d$ instead of $\mathscr{T}^{(\mathscr{A})}$.

It is obvious that the notation $\mathscr{T}_d$ is consistent with that introduced in Chapter 13.

A uniform structure $\mathscr{U}$ on a set X is said to be *metrizable* if there is a metric d on X satisfying $\mathscr{U} = \mathscr{U}_d$.

If (X, d) is a metric space, we always assume that X is endowed with the uniform structure $\mathscr{U}_d$.

We have shown that to any directed family of semi-metrics on a set X, we may associate a uniform structure on X (and a topology). We shall show that any uniform structure can be defined by a directed family of semi-metrics. For the proof of this result, we need several auxiliary propositions, which we shall establish first.

25.8 Let $(W_n)_{0 \leq n < +\infty}$ be a sequence of symmetric sets such that

$$W_n \circ W_n \circ W_n \subset W_{n-1}$$

for all $n \in \mathbf{N}$. Define $f: X \times X \to \mathbf{R}$ by

$$\textbf{25.9} \quad f(x,y) = \begin{cases} 1 & \text{if} \quad (x,y) \in CW_o \\ 2^{-n} & \text{if} \quad (x,y) \in W_{n-1} - W_n \ (n \in \mathbf{N}) \\ 0 & \text{if} \quad (x,y) \in \bigcap_{n \in \mathbf{N}} W_n. \end{cases}$$

Then the following properties hold:

25.10 $f(x, x) = 0$ *for every* $x \in X$;

25.11 $0 \leq f(x,y) \leq 1$ *for every* $(x,y) \in X \times X$;

25.12 $f(x,y) = f(y, x)$ *for every* $(x,y) \in X \times X$;

25.13 $\frac{1}{2} f(x,y) \leq \sum_{0 \leq i \leq p-1} f(z_i, z_{i+1})$

for every $(x,y) \in X \times X$, $p \in \mathbf{N}$, *and sequence* $(z_i)_{0 \leq i \leq p}$ *of elements of* X *such that* $z_0 = x$ *and* $z_p = y$.

Assertions 25.10, 25.11, and 25.12 are easily established. We shall prove 25.13 by induction.

Let A be the set of all $p \in \mathbf{N}$ such that for every $(x,y) \in X \times X$, $p \in \mathbf{N}$, and sequence $(z_i)_{0 \leq i \leq p}$ of elements of X such that $z_0 = x$ and $z_p = y$, the relation 25.13 holds.

Clearly, $1 \in A$. It is also clear that if $p \in A$, then $q \in \mathbf{N}$ and $1 \leq q \leq p$ imply $q \in A$.

Now let $p \in A$ and let $(x,y) \in X \times X$ and $(z_i)_{0 \leq i \leq p+1}$ be a sequence of elements of X such that $z_0 = x$ and $z_{p+1} = y$. We shall show that

$$(\bigstar) \qquad \tfrac{1}{2} f(x,y) \leq \sum_{0 \leq i \leq p} f(z_i, z_{i+1}).$$

Let $a = \sum_{0 \leq i \leq p} f(z_i, z_{i+1})$. If $a \geq \frac{1}{2}$, then ($\bigstar$) is satisfied (use 25.11). Suppose now that $a < \frac{1}{2}$. Let h $(0 < h \leq p)$ be the largest index such that*

$$\sum_{0 \leq i < h} f(z_i, z_{i+1}) \leq a/2.$$

Then

$$\sum_{0 \leq i \leq h} f(z_i, z_{i+1}) > a/2 \quad \text{and} \quad \sum_{h < i \leq p} f(z_i, z_{i+1}) \leq a/2.$$

---

* We have $\{ i \mid 0 \leq i < h\} = \varnothing$ if $h = 1$ and $\{i \mid h < i \leq p\} = \varnothing$ if $h = p$. We recall that $\sum_{i \in \phi} x_i = 0$.

Now, clearly, $f(z_h, z_{h+1}) \leq a$. Since $p \in A$, we also have

$$\tfrac{1}{2}f(x, z_h) \leq \textstyle\sum_{0 \leq i < h} f(z_i, z_{i+1}) \leq a/2$$

and

$$\tfrac{1}{2}f(z_{h+1}, y) \leq \textstyle\sum_{h < i \leq p} f(z_i, z_{i+1}) \leq a/2;$$

hence $f(x, z_h) \leq a$ and $f(z_{h+1}, y) \leq a$.

Now let m be the smallest element of $\mathbf{N}$ such that $2^{-m} \leq a$. Since $a < \frac{1}{2}$, we have $m \geq 2$. Clearly,* $(x, z_h) \in W_{m-1}$, $(z_h, z_{h+1}) \in W_{m-1}$, and $(z_{h+1}, y) \in W_{m-1}$ whence $(x, y) \in W_{m-2}$. We deduce that

$$f(x, y) \leq 2^{-(m-1)} = 2 \cdot 2^{-m} \leq 2a.$$

Hence, (★) is satisfied. Since (x, y) and $(z_i)_{0 \leq i \leq p+1}$ were arbitrary, we deduce that $p + 1 \in A$. By induction, $A = \mathbf{N}$, and hence 25.13 holds.

25.14 Let $f: X \times X \to \mathbf{R}$ be a *positive* function satisfying 25.10 and 25.12. Define $d_f: X \times X \to \mathbf{R}$ by the following: For every $(x, y) \in X \times X$,

$$d_f(x, y) = \inf \textstyle\sum_{0 \leq i \leq p-1} f(z_i, z_{i+1}),$$

the inf being taken over all $p \in \mathbf{N}$ and the sequence $(z_i)_{0 \leq i \leq p-1}$ such that $z_0 = x$ and $z_p = y$. Then *d_f is a semi-metric on X.* It is clear that 25.1 and 25.2 are satisfied. To prove 25.3, we reason as follows: Let x, y, and z be in X and let $\varepsilon > 0$. Let $(s_j)_{0 \leq j \leq n}$ be a sequence of elements of X such that $s_0 = x$, $s_n = y$, and

$$\textstyle\sum_{0 \leq k \leq n-1} f(s_k, s_{k+1}) \leq d_f(x, y) + \varepsilon/2;$$

let $(t_j)_{0 \leq j \leq m}$ be a sequence of elements of X such that $t_0 = y$, $t_m = z$ and

$$\textstyle\sum_{0 \leq k \leq m-1} f(t_k, t_{k+1}) \leq d_f(y, z) + \varepsilon/2.$$

Let $r_i = s_i$ for $0 \leq i \leq n$ and $r_{i+n} = t_i$ for $0 \leq i \leq m$. Then

$$d_f(x, z) \leq \textstyle\sum_{0 \leq k \leq m+n-1} f(r_k, r_{k+1}) \leq d_f(x, y) + d_f(y, z) + \varepsilon.$$

Since $\varepsilon > 0$ was arbitrary,

$$d_f(x, y) \leq d_f(x, y) + d_f(y, z).$$

Since x, y, and z were arbitrary, 24.3 is satisfied. Hence, d_f is a semi-metric on X.

---

* If $(u, v) \in X \times X$ and $f(u, v) \leq a$, then $f(u, v) \leq 2^{-m}$, whence $(u, v) \in W_{m-1}$.

25.15 Let $(X, \mathscr{U})$ be a uniform space and $U \in \mathscr{U}$. Then *there is a semi-metric d on X such that:*

(i) $V_{\varepsilon,d} \in \mathscr{U}$ *for all* $\varepsilon > 0$;

(ii) $V_{1,d} \subset U$.

Let $W_0 = U \cap U^{-1}$. By induction, we define a sequence $(W_n)_{0 \leq n < +\infty}$ of symmetric elements of $\mathscr{U}$ such that (see 24.4 and 24.5)

$$W_n \circ W_n \circ W_n \subset W_{n-1}$$

for all $n \in \mathbf{N}$. Let $f: X \times X \to \mathbf{R}$ be defined by 25.9 and let d_f be the corresponding semi-metric defined in 25.14. Using 25.13, we deduce that

$$\tfrac{1}{2}f(x,y) \leq d_f(x,y) \leq f(x,y)$$

for all $(x,y) \in X \times \mathbf{N}$.

Let $n \in \mathbf{N}$ and $(x,y) \in W_n$. Then $f(x,y) \leq 2^{-(n+1)}$ and hence, $d_f(x,y) \leq 2^{-(n+1)}$. Hence

$$W_n \subset W_{2^{-(n+1)},d_f} \subset V_{2^{-n},d_f}.$$

We deduce that $V_{2^{-n},d_f} \in \mathscr{U}$ for all $n \in \mathbf{N}$. Now let $\varepsilon > 0$ and let $n \in \mathbf{N}$ be such that $2^{-n} \leq \varepsilon$; then $V_{2^{-n},d_f} \subset V_{\varepsilon,d_f}$ so that $V_{\varepsilon,d_f} \in \mathscr{U}$.

Furthermore, if $(x,y) \in X \times X$ is such that $d_f(x,y) \leq \frac{1}{3}$, then $f(x,y) < 1$, so that $(x,y) \in W_0$. Hence $V_{1/3},d_f \subset W_0 \subset U$. If $d = 3d_f$, then d is a semi-metric on X, $V_{\varepsilon,d} = V_{\varepsilon/3,d_f} \in \mathscr{U}$, and $V_{1,d} = V_{1/3,d_f} \subset U$. Hence 25.15 is proved.

25.16 Theorem.—*Let* $(X, \mathscr{U})$ *be a uniform space. Then there is a directed family* $\mathscr{A}$ *of semi-metrics on X such that* $\mathscr{U} = \mathscr{U}^{(\mathscr{A})}$.

Proof.—By 25.15, for each $U \in \mathscr{U}$ there is a semi-metric d_U on X such that $V_{\varepsilon,d_U} \in \mathscr{U}$ for every $\varepsilon > 0$ and $V_{1,d_U} \subset U$.

Let A be the set of all elements of the form $(U_j)_{1 \leq j \leq n}$, where $n \in \mathbf{N}$ and $U_j \in \mathscr{U}$ for all $j = 1, \ldots, n$. For every $\alpha = (U_j)_{1 \leq j \leq n} \in A$, let

$$(\cdot) \qquad d_\alpha = \sup\{d_{U_1}, \ldots, d_{U_n}\}.$$

Clearly, $\mathscr{A} = (d_\alpha)_{\alpha \in A}$ is a *directed* family of semi-metrics on X. If $U \in \mathscr{U}$, then $V_{1,d_U} \subset U$, so that $U \in \mathscr{U}^{(\mathscr{A})}$. *Hence* $\mathscr{U} \subset \mathscr{U}^{(\mathscr{A})}$. If $\alpha \in A$, and if d_α is defined by $(\cdot)$ then

$$V_{\varepsilon,d_\alpha} = \bigcap_{1 \leq j \leq n} V_{\varepsilon,d_{U_j}} \in \mathscr{U}.$$

Hence $\mathscr{U}^{(\mathscr{A})} \subset \mathscr{U}$. We conclude that $\mathscr{U} = \mathscr{U}^{(\mathscr{A})}$.

25.17 Theorem.—*Let $(X, \mathscr{U})$ be a uniform space and $\mathscr{A} = (d_i)_{i\in A}$ a directed set of semi-metrics on X such that $\mathscr{U} = \mathscr{U}^{(\mathscr{A})}$. Then $(X, \mathscr{U})$ is separated if and only if for every $x \in X$, $y \in X$, $x \neq y$ there is $i \in A$ such that $d_i(x,y) \neq 0$.*

Proof.—The set

$$\mathscr{V} = \{V_{\varepsilon,d_i} \mid \varepsilon > 0, i \in A\}$$

is a basis of $\mathscr{U}$. By 24.17, $(X, \mathscr{U})$ is separated if and only if

$$\bigcap_{\varepsilon>0, i\in A} V_{\varepsilon,d_i} = \Delta.$$

Hence, $(X, \mathscr{U})$ is separated if and only if for every $x \in X, y \in X, x \neq y$ there are $\varepsilon > 0$ and $i \in A$ such that $(x,y) \notin V_{\varepsilon,d_i}$. Hence, $(X, \mathscr{U})$ is separated if and only if for every $x \in X$, $y \in X$, $x \neq y$ there is $i \in A$ such that $d_i(x,y) \neq 0$.

25.18 Definition.—*Two directed families $\mathscr{A}'$ and $\mathscr{A}''$ of semi-metrics on a set X are said to be equivalent if $\mathscr{U}^{(\mathscr{A}')} = \mathscr{U}^{(\mathscr{A}'')}$.*

Notice that if $\mathscr{A}'$ and $\mathscr{A}''$ are equivalent, then $\mathscr{T}^{(\mathscr{A}')} = \mathscr{T}^{(\mathscr{A}'')}$. However, the converse is not true (see Chapter 28).

Example 3.—Let $(X, \mathscr{U})$ be a uniform space and $\mathscr{A} = (d_i)_{i\in A}$ a directed family of semi-metrics on X such that $\mathscr{U} = \mathscr{U}^{(\mathscr{A})}$. If $Y \subset X$, then $\mathscr{B} = ((d_i)_Y)_{i\in A}$ is a directed family of semi-metrics on Y and

$$\mathscr{U}^{(\mathscr{B})} = (\mathscr{U}^{(\mathscr{A})})_Y.$$

▼ *Example* 4.—Let X be a set and $\mathscr{A} = (d_i)_{i\in A}$ a *countable* directed family of semi-metrics on X. Let $(a_i)_{i\in A}$ be a sequence of strictly positive numbers such that $\sum_{i\in A} a_i < +\infty$. Define $d: X \times X \to \mathbf{R}$ by

$$d = \sum_{i\in A} a_i \frac{d_i}{1 + d_i}$$

We leave it to the reader to show that $\mathscr{U}_d = \mathscr{U}^{(\mathscr{A})}$. Using this result and 25.15, we easily deduce that *a uniform structure on X is metrizable if and only if it is separated and has a countable basis.* ▲

Exercises for Chapter 25

1. Let X be a set and $(d_i)_{i\in I}$ an arbitrary family of semi-metrics such that

$$\sup_{i\in I} d_i(x,y) < +\infty$$

for every $(x, y) \in X \times X$. Then the mapping $d: X \times X \rightarrow \mathbf{R}$ defined by

$$d(x, y) = \sup_{i \in I} d_i(x, y)$$

for $(x, y) \in X \times X$ is a semi-metric.

2. Prove Relation 25.6.

3. Prove the assertions in Example 3.

4. Prove the assertions in Example 4.

* 5. Let $(X, \mathscr{T})$ be a topological space such that for each $x \in X$ there is a neighborhood V_x of x such that $(V_x, \mathscr{T}_{V_x})$ is metrizable. Show that $(X, \mathscr{T})$ is metrizable if and only if $(X, \mathscr{T})$ is paracompact.

6. Let $(X, \mathscr{T})$ be the topological space introduced in Appendix II, Chapter 13. Determine a directed family of semi-metrics on X defining $\mathscr{T}$.

7. Let $(X, \mathscr{T})$ be a separated topological space and $\mathscr{K}$ the set of all compact parts of X. For each $K \in \mathscr{K}$, define $p_K: C_{\mathbf{R}}(X) \rightarrow \mathbf{R}$ by

$$p_K(f) = \sup_{x \in K} |f(x)| \quad \text{for } f \text{ in } \quad C_{\mathbf{R}}(X).$$

For each $K \in \mathscr{K}$, define $d^K: C_{\mathbf{R}}(X) \times C_{\mathbf{R}}(X) \rightarrow \mathbf{R}$ by

$$d^K(f, g) = p_K(f - g) \quad \text{for } f \text{ and } g \text{ in } \quad C_{\mathbf{R}}(X).$$

Then $\mathscr{A} = (d^K)_{K \in \mathscr{K}}$ is a directed family of semi-metrics on X ($\mathscr{K}$ is directed by inclusion). The uniform space* $(X, \mathscr{U}^{(\mathscr{A})})$ is separated.

8. † Let $\mathscr{A}$ be a subalgebra of $C_{\mathbf{R}}(X)$ that separates the points of X and such that for every $x \in X$ there is $f \in \mathscr{A}$ satisfying $f(x) \neq 0$. Then $\bar{\mathscr{A}} = C_{\mathbf{R}}(X)$ (hint: Use the Stone-Weierstrass theorem).

9. If $(X, \mathscr{T})$ is compact and d is the distance on $C_{\mathbf{R}}(X)$ corresponding to the usual norm (see Chapter 18), then $\mathscr{U}_c = \mathscr{U}_d$.

* 10. Assume $(X, \mathscr{T})$ to be locally compact. Then $\mathscr{U}_c$ is metrizable if and only if X is σ-compact (hint: See Example 4, Chapter 25).

---

* The uniform structure $\mathscr{U}^{(\mathscr{A})}$ is often denoted $\mathscr{U}_c$. The topology $\mathscr{T}_{\mathscr{U}_c}$ is often denoted $\mathscr{T}_c$ and called *the topology of uniform convergence on the compact parts of X.*

† In Exercises 8, 9, and 10, we use the notations introduced in Exercise 7.

Chapter 26

Complete Uniform Spaces

Let $(X, \mathscr{U})$ be a uniform space. Then

26.1 Definition.—*A filter basis $\mathscr{B}$ on X is said to be a Cauchy filter basis (or a Cauchy basis) if for every $U \in \mathscr{U}$ there is $A \in \mathscr{B}$ such that*

$$A \times A \subset U.$$

Since A depends on U, we shall often use some notation to indicate this. For instance, instead of writing A, we may write A_U.

Notice that if $\mathscr{B}$ is a Cauchy filter basis on X and $\mathscr{D} \succ \mathscr{B}$ is a filter basis, then $\mathscr{D}$ is also a Cauchy filter basis.

26.2 Theorem.—*Let $(X, \mathscr{U})$ be a uniform space and $\mathscr{B}$ a convergent filter basis on X. Then $\mathscr{B}$ is a Cauchy filter basis.*

Proof.—Let $U \in \mathscr{U}$ and $W \in \mathscr{U}$ symmetric and such that $W \circ W \subset U$. Let $a \in X$ be such that $\mathscr{B}$ converges to a. Then there is $A \in \mathscr{B}$ contained in $W(a)$. Hence $(x, y) \in A \times A \Rightarrow (x, y) \in W(a) \times W(a) \Rightarrow x \in W(a)$ and $y \in W(a) \Rightarrow$ (recall that W is symmetric) $(x, a) \in W$ and $(a, y) \in W \Rightarrow (x, y) \in W \circ W \subset U$. Hence $A \times A \subset U$. Since $U \in \mathscr{U}$ was arbitrary, we deduce that $\mathscr{B}$ is a Cauchy filter basis.

26.3 Theorem.—*Let $(X, \mathscr{U})$ be a uniform space and $\mathscr{B}$ a Cauchy filter basis on X such that $ad(\mathscr{B}) \neq \varnothing$. Then $\mathscr{B}$ converges to every $a \in ad(\mathscr{B})$.*

Proof.—Let $a \in ad(\mathscr{B})$, and take $U \in \mathscr{U}$ and $W \in \mathscr{U}$ symmetric and such that $W \circ W \subset U$. Since $\mathscr{B}$ is a Cauchy filter basis, there is $A \in \mathscr{B}$ satisfying $A \times A \subset W$. Let $y \in A \cap W(a)$, so that $(a, y) \in W$. If $x \in A$, then $(y, x) \in A \times A \subset W$. Hence, $(a, x) \in W \circ W \subset U$; that is, $x \in U(a)$. We deduce that $A \subset U(a)$. Since $U \in \mathscr{U}$ was arbitrary, $\mathscr{B}$ converges to a.

Remark.—We deduce that if $(X, \mathscr{U})$ is a separated uniform space and $\mathscr{B}$ is a Cauchy filter basis on X, then $ad(\mathscr{B})$ contains *at most one* element.

26.4 Definition.—*A uniform space $(X, \mathscr{U})$ is complete if every Cauchy filter basis on X is convergent.*

When there is no ambiguity, instead of saying that $(X, \mathscr{U})$ is complete, we say that X is complete.

In Chapter 28 (Theorem 28.4), we shall show that *any separated uniform space can be embedded in a separated complete uniform space.*

Let $(X, \mathscr{U})$ be a uniform space. We say that $A \subset X$ *is complete if $(A, \mathscr{U}_A)$ is complete.* It is obvious that A is complete if and only if every Cauchy filter basis $\mathscr{B}$ consisting of sets *contained in* A (that is, such that $\mathscr{B} \subset \mathscr{P}(A)$) converges to some element belonging to A.

26.5 Theorem.—*Let $(X, \mathscr{U})$ be a separated complete uniform space and let $A \subset X$. Then the following assertions are equivalent:*

(a) $A = \bar{A}$;
(b) A *is complete.*

Proof of (a) $\Rightarrow$ (b).—Let $\mathscr{B} \subset \mathscr{P}(A)$ be a Cauchy filter basis. Since $(X, \mathscr{U})$ is complete, $\mathscr{B}$ converges to some $a \in X$. By 23.17, $a \in \bar{A} = A$. Hence, A is complete.

Proof of (b) $\Rightarrow$ (a).—Suppose A to be complete and let $a \in \bar{A}$. By Examples 4 and 14 of Chapter 23,

$$\mathscr{B} = \{V \cap A \mid V \in \mathscr{N}_X(x)\}$$

is a filter basis that converges to x. Hence $\mathscr{B} \subset \mathscr{P}(A)$ and $\mathscr{B}$ is a Cauchy filter basis. Since A is complete, $\mathscr{B}$ converges to some $x' \in A$. Since $(X, \mathscr{U})$ is separated, $a = x' \in A$. Hence, $A = \bar{A}$.

26.6 Theorem.—*Let $(X, \mathscr{U})$ be a separated complete uniform space and $A \subset X$ a compact part.* Then A is complete.*

Proof.—Let $\mathscr{B} \subset \mathscr{P}(A)$ be a Cauchy filter basis. Then (see 23.18)

$$A \cap ad(\mathscr{B}) \neq \varnothing,$$

and hence $\mathscr{B}$ (see 26.3) converges to some $a \in A$. Hence A is complete.

---

* It is enough to assume that A is quasi-compact.

Example 1.—Let $((X_i, \mathscr{U}_i)_{i \in I}$ be a family of uniform spaces and $(X, \mathscr{U})$ its product. Then

26.7 *If, for each $i \in I$, $(X_i, \mathscr{U}_i)$ is complete, then $(X, \mathscr{U})$ is complete.*

In fact, let $\mathscr{B}$ be a Cauchy filter basis on X and let $i \in I$. Then $pr_i(\mathscr{B})$ is a Cauchy filter basis on X_i (we shall not prove this assertion here, since a more general result is established in Theorem 27.11). Since $(X_i, \mathscr{U}_i)$ is complete, $pr_i(\mathscr{B})$ converges to some $a_i \in X_i$. Since $i \in I$ was arbitrary, $\mathscr{B}$ converges to $(a_i)_{i \in I}$. Since $\mathscr{B}$ was arbitrary, $(X, \mathscr{U})$ is complete.

The converse is also true. Namely, if $(X, \mathscr{U})$ is complete and $X \neq \varnothing$, then $(X_i, \mathscr{U}_i)$ is complete for each $i \in I$. We leave the proof of this assertion to the reader (hint: Use Cauchy bases of the form

$$\prod_{i \in I} \mathscr{B}_i,$$

where for all indices j, with the exception of an *arbitrary one*, $\mathscr{B}_j = \{\{a_j\}\}$ for some $a_j \in X_j$).

Sequences

Let (X, d) be a metric space and $(x_n)_{n \in \mathbf{N}}$ a sequence of elements of X. Let

$$S_p = \{x_n \mid n \geq p\}$$

for all $p \in \mathbf{N}$ and let $\mathscr{B}$ be the filter basis

$$\{S_p \mid p \in \mathbf{N}\}$$

(hence $\mathscr{B}$ is the filter basis associated with the sequence $(x_n)_{n \in \mathbf{N}}$).

Then the following assertions are equivalent:

26.8 $(x_n)_{n \in \mathbf{N}}$ *is a Cauchy sequence* (in the sense of Definition 15.1);

26.9 $\mathscr{B}$ *is a Cauchy basis* (in the uniform space $(X, \mathscr{U}_d)$).

Proof of 26.8 $\Rightarrow$ 26.9—Let $U \in \mathscr{U}_d$ and let $\varepsilon > 0$ be such that $V_{\varepsilon,d} \subset U$. Since $(x_n)_{n \in \mathbf{N}}$ is a Cauchy sequence, there is $p \in \mathbf{N}$ such that $d(x_n, x_m) < \varepsilon$ if $n \geq p$, $m \geq p$. Hence, $(x_n, x_m) \in V_{\varepsilon,d}$ if $n \geq p$, $m \geq p$; that is,

$$S_p \times S_p \subset V_{\varepsilon,d} \subset U.$$

Since $U \in \mathscr{U}_d$ was arbitrary, $\mathscr{B}$ is a Cauchy basis.

Proof of 26.9 $\Rightarrow$ 26.8.—Let $\varepsilon > 0$. Since $\mathscr{B}$ is a Cauchy basis, there is $p \in \mathbf{N}$ such that $S_p \times S_p \subset V_{\varepsilon,d}$. This means that $(x_n, x_m) \in V_{\varepsilon,d}$ if $n \geq p$, $m \geq p$; that is,

$$d(x_n, x_m) < \varepsilon$$

if $n \geq p$, $m \geq p$. Since $\varepsilon > 0$ was arbitrary, $(x_n)_{n \in \mathbf{N}}$ is a Cauchy sequence.

A sequence $(x_n)_{n \in \mathbf{N}}$ of elements of a uniform space X is said to be a Cauchy sequence if the associated filter basis is a Cauchy basis.

Let X be a set and d a metric on X. Then

26.10 Theorem.—*The uniform space $(X, \mathscr{U}_d)$ is complete if and only if every Cauchy sequence $(x_n)_{n \in \mathbf{N}}$ of elements of X is convergent.*

Proof.—Suppose $(X, \mathscr{U}_d)$ to be complete and let $(x_n)_{n \in \mathbf{N}}$ be a Cauchy sequence of elements of X. Then the filter basis associated with $(x_n)_{n \in \mathbf{N}}$ is a Cauchy basis. Since $(X, \mathscr{U}_d)$ is complete, $\mathscr{B}$ is convergent. This means that $(x_n)_{n \in \mathbf{N}}$ is convergent.

Conversely, suppose that every Cauchy sequence of elements of X is convergent. Let $\mathscr{B}$ be a Cauchy basis on X. For each $n \in \mathbf{N}$, let $A_n \in \mathscr{B}$ satisfy

$$A_n \times A_n \subset V_{1/n,d}$$

(we may and shall suppose that $A_1 \supset A_2 \supset \ldots \supset A_n \supset \ldots$). For each $n \in \mathbf{N}$, let $x_n \in A_n$. If $p \in N$ and $n \geq p$, $m \geq p$, then

$$(x_n, x_m) \in A_p \times A_p \subset V_{1/p,d};$$

that is,

$$d(x_n, x_m) < 1/p.$$

Hence $(x_n)_{n \in \mathbf{N}}$ is a Cauchy sequence. By hypothesis, $(x_n)_{n \in \mathbf{N}}$ converges to some $a \in X$.

Now let $U \in \mathscr{U}$ and $p \in \mathbf{N}$ be such that $V_{1/p,d} \circ V_{1/p,d} \subset U$. Let $q \in \mathbf{N}$, $q \geq p$, such that $d(x_q, a) < 1/p$. Then $y \in A_q$ implies that $(y, x_q) \in V_{1/p,d}$. Since $(x_q, a) \in V_{1/p,d}$, we obtain

$$(y, a) \in V_{1/p,d} \circ V_{1/p,d} \subset U.$$

We deduce that $A_q \subset U(a)$. Since $U \in \mathscr{U}$ was arbitrary, and $A_q \in \mathscr{B}$, we conclude that $\mathscr{B}$ converges to a. Since $\mathscr{B}$ was arbitrary, $(X, \mathscr{U})$ is complete.

It follows that *Definitions* 15.3 *and* 26.4 *are equivalent in the case of metric spaces.*

Remark.—Let d and δ be the metrics introduced in Example 6, Chapter 15. Then $\mathscr{U}_d \neq \mathscr{U}_\delta$ (since, for instance, $(n)_{n\in N}$ is a Cauchy sequence in $(R, \mathscr{U}_\delta)$, but not in (R, d)). However, $\mathscr{T}_{\mathscr{U}_d} = \mathscr{T}_d = \mathscr{T}_\delta = \mathscr{T}_{\mathscr{U}_\delta}$.

Nets

Let $(X, \mathscr{U})$ be a uniform space and $(x_\alpha)_{\alpha\in A}$ a net of elements of X, We say that $(x_\alpha)_{\alpha\in A}$ is a *Cauchy net* if for every $U \in \mathscr{U}$ there is $\alpha_U \in A$ such that

$$(x_{\alpha'}, x_{\alpha''}) \in U$$

whenever $\alpha' \geq \alpha_U$, $\alpha'' \geq \alpha_U$.

Let $(x_\alpha)_{\alpha\in A}$ be a net of elements of X. Let

$$S_\beta = \{x_\alpha \mid \alpha \geq \beta\}$$

for all $\beta \in A$ and let $\mathscr{B}$ be the filter basis

$$\{S_\beta \mid \beta \in A\}$$

(hence, $\mathscr{B}$ is the filter basis associated with the net $(x_\alpha)_{\alpha\in A}$). By the same method as we used to prove that 26.8 and 26.9 are equivalent, we may show that *$(x_\alpha)_{\alpha\in A}$ is a Cauchy net if and only if $\mathscr{B}$ is a Cauchy basis.*

We leave it to the reader to show that *a uniform space $(X, \mathscr{U})$ is complete if and only if every Cauchy net of elements of X is convergent.*

▼ Let $((X_i, \mathscr{U}_i))_{i\in I}$ be a family of uniform spaces and $(X, \mathscr{U})$ its product.

Exercise 1.—A net $(x_\alpha)_{\alpha\in A}$ of elements of X is a Cauchy net if and only if, for every $i \in I$, $(pr_i(x_\alpha))_{\alpha\in A}$ is a Cauchy net.

Exercise 2.—Using the result in Exercise 1, show that if $X \neq \varnothing$, then $(X, \mathscr{U})$ is complete if and only if every $(X_i, \mathscr{U}_i)$ $(i \in I)$ is complete. ▲

Exercises for Chapter 26

1. Let $(X, \mathscr{U})$ be a uniform space and $\mathscr{A} = (d_i)_{i\in I}$ a directed family of semi-metrics on X such that $\mathscr{U} = \mathscr{U}^{(\mathscr{A})}$. Let $\mathscr{B}$ be a filter basis on X. Then $\mathscr{B}$ is a Cauchy filter basis in $(X, \mathscr{U})$ if and only if $\mathscr{B}$ is a Cauchy filter basis in $(X, \mathscr{U}_{d_i})$, for each i in I.

2. Prove the assertion in Exercise 1 in the text.

3. Prove the assertion in Exercise 2 in the text.

4. (Here, we use the notations of Exercise 7 at the end of Chapter 25.) Show that $(C_{\mathbf{R}}(X), \mathscr{U}_c)$ is not necessarily complete.

5. (Here, we use the notations of Exercise 7 at the end of Chapter 25.) Show that if $(X, \mathscr{T})$ is locally compact, then $(C_{\mathbf{R}}(X), \mathscr{U}_c)$ is complete.

Chapter 27

Uniformly Continuous Functions

In this chapter, we shall discuss uniformly continuous functions having uniform spaces for domain and range.

We denote below by $(X_1, \mathscr{U}_1)$ and $(X_2, \mathscr{U}_2)$ two uniform spaces.

27.1 Definition.—*We say that $f: X_1 \to X_2$ is uniformly continuous on X_1 if for every $U_2 \in \mathscr{U}_2$ there exists $U_1 \in \mathscr{U}_1$ such that*

$$(f(x), f(y)) \in U_2$$

whenever $(x, y) \in U_1$.

Let $\mathscr{B}_1$ be a *basis* of $\mathscr{U}$ and $\mathscr{B}_2$ a *basis* of $\mathscr{U}_2$. It is clear that $f: X_1 \to X_2$ is uniformly continuous on X_1 *if and only if* for every $B_2 \in \mathscr{B}_2$ there exists $B_1 \in \mathscr{B}_1$ such that

$$(f(x), f(y)) \in B_2$$

whenever $(x, y) \in B_1$.

Then the following results are immediate consequences.

27.2 Assume that there is a semi-metric d_1 on X_1 satisfying $\mathscr{U}_{d_1} = \mathscr{U}_1$ and a semi-metric d_2 on X_2 satisfying $\mathscr{U}_{d_2} = \mathscr{U}_2$. Then

$$\mathscr{V}_1 = \{V_{\varepsilon, d_1} \mid \varepsilon > 0\} \quad \text{and} \quad \mathscr{V}_2 = \{V_{\varepsilon, d_2} \mid \varepsilon > 0\}$$

are bases of $\mathscr{U}_1$ and $\mathscr{U}_2$, respectively. We deduce that $f: X_1 \to X_2$ is uniformly continuous if and only if for every $\varepsilon > 0$ there exists $\delta > 0$ such that

$$(f(x), f(y)) \in V_{\varepsilon, d_2}$$

whenever $(x, y) \in V_{\delta, d_1}$.

It follows that if (X_1, d_1) and (X_2, d_2) are metric spaces, then Definitions 17.1 and 27.1 (of uniformly continuous functions) are equivalent.

27.3 A function $f: X_1 \to \mathbf{R}$ is uniformly continuous if and only if for every $\varepsilon > 0$ there is $U \in \mathscr{U}_1$ such that

$$|f(x) - f(y)| < \varepsilon$$

whenever $(x, y) \in U$.

Exercise 1.—Let $(X, \mathscr{U})$ be a uniform space and $C^u_{\mathbf{R}}(X)$ the set of all uniformly continuous mappings of X into $\mathbf{R}$. Then:

(i) $C^u_{\mathbf{R}}(X)$ is a vector space (when endowed with the usual multiplication by scalars and the usual addition);
(ii) if $f \in C^u_{\mathbf{R}}(X)$, then $|f| \in C^u_{\mathbf{R}}(X)$;
(iii) if $f_1, \ldots, f_p$ belong to $C^u_{\mathbf{R}}(X)$, then $\inf\{f_1, \ldots, f_p\}$ and $\sup\{f_1, \ldots, f_p\}$ belong to $C^u_{\mathbf{R}}(X)$.

If $A \subset X_1$, we say that $f: A \to X_2$ is *uniformly continuous on* A, when considered as a mapping of the uniform space $(A, (\mathscr{U}_1)_A)$ (see Chapter 24) into $(X_2, \mathscr{U}_2)$, f is uniformly continuous in the sense of Definition 27.1.

Hence, if $\mathscr{B}_1$ and $\mathscr{B}_2$ are bases of $\mathscr{U}_1$ and $\mathscr{U}_2$, respectively, then $f: A \to X_2$ is uniformly continuous on A if and only if for every $B_2 \in \mathscr{U}_2$ there is $B_1 \in \mathscr{U}_1$ such that

$$(f(x), f(y)) \in B_2$$

whenever $(x, y) \in B_1 \cap (A \times A)$.

27.4 Theorem.—*Let $A \subset X_1$ and let $f: A \to X_2$ be uniformly continuous. Then f is continuous on A.*

Proof.—Let $a \in A$ and $V \in \mathscr{N}_{X_2}(f(a))$. Let $U_2 \in \mathscr{U}_2$ satisfy $U_2(f(a)) \subset V$. Since f is uniformly continuous on A, there is $U_1 \in \mathscr{U}$ such that

$$(f(u), f(v)) \in U_2$$

whenever u and v belong to A and $(u, v) \in U_1$. Hence $x \in A$ and $x \in U_1(a)$ imply that $(a, x) \in U_1$, so that $(f(a), f(x)) \in U_2$. Hence, $x \in A \cap U_1(a)$ implies $f(x) \in U_2(f(a)) \subset V$. Since $V \in \mathscr{N}_{X_2}(f(a))$ was arbitrary, f is continuous at a. Since $a \in X_1$ was arbitrary, f is continuous on X_1.

27.5 Theorem.—*Let* $(X_1, \mathscr{U}_1)$ *be separated,** $A \subset X_1$ *compact, and* $f: A \to X_2$ *a continuous function. Then* f *is uniformly continuous.*

Proof.—Let $V \in \mathscr{U}_2$ and $W \in \mathscr{U}_2$ be *symmetric* and satisfying $W \circ W \subset V$. Since $f: A \to X_2$ is continuous on A, for each $t \in A$ there is $B_t \in \mathscr{U}_1$ such that

$$(\bigstar) \qquad (t, x) \in B_t \Rightarrow (f(t), f(x)) \in W.$$

For each $t \in A$, let $U_t \in \mathscr{U}_1$ satisfy $U_t \circ U_t \subset B_t$. Since A is compact, there is a finite family $(t_j)_{1 \le j \le n}$ of elements of A such that $(U_{t_j}(t_j))_{1 \le j \le n}$ is a covering of A_n.

Let $U = \bigcap_{j=1} U_{t_j}$ and let x and y be elements of A satisfying $(x, y) \in U$. Since $(U_{t_j}(t_j))_{1 \le j \le n}$ is a covering of A, there is j_0 such that $x \in U_{t_{j_0}}(t_{j_0})$. Then

$$(t_{j_0}, x) \in U_{t_{j_0}} \subset B_{t_{j_0}};$$

since $(x, y) \in U \subset U_{t_{j_0}}$,

$$(t_{j_0}, y) \in U_{t_{j_0}} \circ U_{j_0} \subset B_{t_{j_0}}.$$

By $(\bigstar)$,

$$(f(t_{j_0}), f(x)) \in W \quad \text{and} \quad (f(t_{j_0}), f(x)) \in W.$$

Since W is symmetric, we obtain $(f(x), f(y)) \in W \circ W \subset V$. Since $V \in \mathscr{U}_2$ was arbitrary we conclude that f is uniformly continuous on X.

Let $(X', \mathscr{U}')$, $(X'', \mathscr{U}'')$, and $(X''', \mathscr{U}''')$ be three uniform spaces. Then:

27.6 Theorem.—*Let* $S \subset X'$ *and* $f: S \to X''$ *and let* $T \subset X''$ *and* $g: T \to X'''$. *Suppose that:*

(a) $x \in S \Rightarrow f(x) \in T$;

(b) f *is uniformly continuous on* S *and* g *is uniformly continuous on* T.

Then $g \circ f$ *is uniformly continuous on* S.

The proof is left to the reader.

Example 1.—Let $(X, \mathscr{U})$ be a locally compact† uniform space. Then every function‡ $f \in C_{\mathbf{R},\infty}(X)$ is uniformly continuous on X.

---

* The proof shows that it is enough to assume that A is *quasi-compact*.

† As we shall see in Chapter 28, for every locally compact topological space $(X, \mathscr{T})$ there is at least one uniform structure $\mathscr{U}$ on X satisfying $\mathscr{T}_{\mathscr{U}} = \mathscr{T}$.

‡ A function $f: X \to \mathbf{R}$ belongs to $C_{\mathbf{R},\infty}(X)$ if and only if it is continuous and if for every $\varepsilon > 0$ there is a compact $K_\varepsilon \subset X$ such that $|f(x)| < \varepsilon$ for $x \in X - K_\varepsilon$.

Let $\varepsilon > 0$ and let $K \subset X$ be a compact set such that

$$|f(x)| < \varepsilon/2 \quad \text{if} \quad x \in X - K.$$

Since f is continuous on X, for each $t \in K$, there is $B_t \in \mathscr{U}$ such that

$$(\bigstar\bigstar) \qquad (t, x) \in B_t \Rightarrow |f(x) - f(t)| < \varepsilon/2$$

For each $t \in K$, let $U_t \in \mathscr{U}$ satisfy $U_t \circ U_t \subset B_t$. Since K is compact, there is a finite family $(t_j)_{1 \leq j \leq n}$ of elements of K such that $(U_{t_j}(t_j))_{1 \leq j \leq n}$ is a covering of K.

Let $U = \bigcap_{j=1}^n U_{t_j}$ and let $(x, y) \in U$. Suppose that at least one of the elements, x and y, belongs to K (*for instance*, x). Since $(U_{t_j}(t_j))_{1 \leq j \leq n}$ is a covering of K, there is j_0 such that $x \in U_{t_{j_0}}(t_{j_0})$. Then

$$(t_{j_0}, x) \in U_{t_{j_0}} \subset B_{t_{j_0}};$$

since $(x, y) \in U \subset U_{t_{j_0}}$,

$$(t_{j_0}, y) \in U_{t_{j_0}} \circ U_{t_{j_0}} \subset B_{t_{j_0}}.$$

By $(\bigstar\bigstar)$,

$$|f(t_{j_0}) - f(y)| < \varepsilon/2 \quad \text{and} \quad |f(t_{j_0}) - f(x)| < \varepsilon/2;$$

hence $|f(x) - f(y)| < \varepsilon$. If $x \notin K$ *and* $y \notin K$, then

$$|f(x) - f(y)| \leq |f(x)| + |f(y)| < 2(\varepsilon/2) = \varepsilon.$$

Hence, for any elements x and y in X, $(x, y) \in U$ implies that $|f(x) - f(y)| < \varepsilon$. Since $\varepsilon > 0$ was arbitrary, we deduce that f is uniformly continuous on X.

Before discussing the next example, we notice that the following result is an immediate consequence of the definition of uniformly continuous functions.

27.7 *Let $(X_1, \mathscr{U}_1)$ and $(X_2, \mathscr{U}_2)$ be two uniform spaces and $\mathscr{B}_2$ a basis of $\mathscr{U}_2$. Let $A \subset X_1$ and $f: A \to X_2$. Then f is uniformly continuous on A if and only if*

$$\{(x, y) \mid (f(x), f(y)) \in U\} \in (\mathscr{U}_1)_A$$

for every $U \in \mathscr{B}_2$.

Example 2.—Let X be a set, $((Y_i, \mathscr{U}_i))_{i \in I}$ a family of uniform spaces, and for each $i \in I$ let f_i be a mapping of X into Y_i. Let $\mathscr{W}$ be the initial uniform structure on X associated with the families $((Y_i, \mathscr{U}_i))_{i \in I}$ and $(f_i)_{i \in I}$. Then:

27.8 *The uniform structure $\mathscr{W}$ is the smallest uniform structure on X such that all the functions f_i $(i \in I)$ are uniformly continuous on X.*

In fact, let $i \in I$. Then

$$\{(x, y) \mid (f_i(x), f_i(y)) \in U_i\} = W(U_i) \in \mathscr{W}$$

for all $U_i \in \mathscr{U}_i$, and hence f_i is uniformly continuous on X.

Conversely, let $\mathscr{U}'$ be a uniform structure on X such that for each $i \in I, f_i$, considered as a mapping of $(X, \mathscr{U}')$ into $(Y_i, \mathscr{U}_i)$, is uniformly continuous on X. Then for every family $(U_j)_{j \in J}$, where J is a finite part of I and $U_j \in \mathscr{U}_j$, for each $j \in J$, we have

$$W((U_j)_{j \in J}) = \bigcap_{j \in J} \{(x, y) \mid (f_j(x), f_j(y)) \in U_j\} \in \mathscr{U}'.$$

We conclude that $\mathscr{U}' \supset \mathscr{W}$. Hence, 27.8 is proved.

27.9 *Let $(Z, \mathscr{L})$ be a uniform space and $g: Z \to X$. Then g is uniformly continuous if and only if $f_i \circ g$ is uniformly continuous for every $i \in I$.*

The proof is left to the reader.

Now let $((X_i, \mathscr{U}_i))_{i \in I}$ be a family of uniform spaces and $(X, \mathscr{U})$ its *product*. From 27.8, we deduce that

27.8′ *The uniform structure $\mathscr{U}$ is the smallest uniform structure on X such that all the mappings pr_i $(i \in I)$ are uniformly continuous.*

From 27.9, we deduce

27.9′ *Let $(Z, \mathscr{L})$ be a uniform space and $g: Z \to X$. Then g is uniformly continuous if and only if $pr_i \circ g$ is uniformly continuous for every $i \in I$.*

Example 3.—Let $(X, \mathscr{U})$ be a uniform space and d a semi-metric on X. Then the following assertions are equivalent:

(i) *$V_{\varepsilon,d} \in \mathscr{U}$ for $\varepsilon > 0$;*
(ii) *d is uniformly continuous on $X \times X$.**

Proof of (i) ⇒ (ii).—For every (x_1, x_2) and (y_1, y_2) in $X \times X$, we have (see 25.6)

$$|d(x_1, x_2) - d(y_1, y_2)| \leq d(x_1, y_1) + d(x_2, y_2).$$

* We assume $X \times X$ to be endowed with the uniform structure $\mathscr{W}$, "product of $\mathscr{U}$ with itself." Recall that if $U_1 \in \mathscr{U}$ and $U_2 \in \mathscr{U}$, then

$$W(U_1, U_2) = \{((x_1, x_2), (y_1, y_2)) \mid (x_1, y_1) \in U_1, (x_2, y_2) \in U_2\}.$$

Now let $\varepsilon > 0$ and let

$$W = W(V_{\varepsilon/2,d}, V_{\varepsilon/2,d}).$$

If $((x_1, x_2), (y_1, y_2)) \in W$, then $(x_1, y_1) \in V_{\varepsilon/2,d}$ and $(x_2, y_2) \in V_{\varepsilon/2,d}$ (so that $d(x_1, y_1) < \varepsilon/2$ and $d(x_2, y_2) < \varepsilon/2$); hence

$$|d(x_1, x_2) - d(y_1, y_2)| < 2(\varepsilon/2) = \varepsilon.$$

Since $\varepsilon > 0$ was arbitrary, we deduce that d is uniformly continuous $X \times X$.

Proof of (ii) $\Rightarrow$ (i).—Let $\varepsilon > 0$. Since d is uniformly continuous on $X \times X$,

$$W = \{((x_1, x_2), (y_1, y_2)) \mid |d(x_1, x_2) - d(y_1, y_2)| < \varepsilon\} \in \mathscr{W}.$$

Hence, there are $U \in \mathscr{U}$ and $V \in \mathscr{U}$ such that $W(U, V) \subset W$. *Now let* $(x_1, x_2) \in U$. Then

$$(x_1, x_2) \in U \quad \text{and} \quad (x_2, x_2) \in V \Rightarrow ((x_1, x_2), (x_2, x_2)) \in W,$$

so that

$$(x_1, x_2) \in U \Rightarrow d(x_1, x_2) = |d(x_1, x_2) - d(x_2, x_2)| < \varepsilon.$$

Hence $U \subset V_{\varepsilon,d}$, that is, $V_{\varepsilon,d} \in \mathscr{U}$. Since $\varepsilon > 0$ was arbitrary, (i) is proved.

Example 4.—Let $(X, \mathscr{U})$ be a uniform space and let $\mathscr{C} = (d_i)_{i \in K}$ be the *directed* family of all semi-metrics on X that are uniformly continuous on $X \times X$. Then

27.10 $\mathscr{U}^{(\mathscr{C})} = \mathscr{U}.$

By Theorem 25.16, there is a directed family $\mathscr{A}$ of semi-metrics on X such that $\mathscr{U} = \mathscr{U}^{(\mathscr{A})}$. Since, by the results in Example 3, each element of the family $\mathscr{A}$ is uniformly continuous on $X \times X$, we deduce that $\mathscr{U} = \mathscr{U}^{(\mathscr{A})} \subset \mathscr{U}^{(\mathscr{C})}$. If d is a semi-metric on X, uniformly continuous on $X \times X$, we have $V_{\varepsilon,d} \in \mathscr{U}$ for all $\varepsilon > 0$. We deduce that $\mathscr{U}^{(\mathscr{C})} \subset \mathscr{U}$. Hence, we conclude that $\mathscr{U}^{(\mathscr{C})} = \mathscr{U}$.

Example 5.—Let $(X, \mathscr{U})$ be a uniform space and d a semi-metric on X, uniformly continuous on $X \times X$. Let

$$R_d = \{(x, y) \mid d(x, y) = 0\}.$$

Using the properties of a semi-metric, we deduce that R_d is an *equivalence relation* in X. Denote by $\varphi_d : x \mapsto \dot{x}_d$ the canonical mapping of X onto $X_d = X/R_d$. Let

$$\dot{d} : X_d \times X_d \to \mathbf{R}$$

be defined by

$$\dot{d}(\dot{x}_d, \dot{y}_d) = d(x, y)$$

for $(\dot{x}_d, \dot{y}_d) \in X_d \times X_d$. It is easy to show that $\dot{d}$ is well-defined on $X_d \times X_d$ and that $\dot{d}$ is a *metric* on X_d. Moreover, $\varphi_d : X \to X_d$ *is uniformly continuous as a mapping of* $(X, \mathscr{U})$ *onto* $(X_d, \dot{d})$. In fact, if $\varepsilon > 0$, then

$$\{(x, y) \mid (\varphi_d(x), \varphi_d(y)) \in V_{\varepsilon, \dot{d}}\} = V_{\varepsilon, d} \in \mathscr{U};$$

hence, since $\varepsilon > 0$ was arbitrary, we deduce that φ_d is uniformly continuous on X (see Example 3).

Example 6.—Let X be a set and $(g_i)_{i \in I}$ a family of mappings of X into $\mathbf{R}$. For each $i \in I$, let d_i be the semi-metric $(x, y) \mapsto |g_i(x) - g_i(y)|$. Let $\mathscr{A} = (d_i)_{i \in I}$. Then $\mathscr{A}^* = (d_J)_{J \in \mathscr{F}(I)}$† is a directed family of semi-metrics on X and $\mathscr{U} = \mathscr{U}^{(\mathscr{A}^*)}$ *is the smallest uniform structure on* X *such that all the mappings* g_i $(i \in I)$ *are uniformly continuous (with respect to* $\mathscr{U}$*).*

In fact, if $i \in I$ and $\varepsilon > 0$, then

$$\{(x, y) \mid |g_i(x) - g_i(y)| < \varepsilon\} = V_{\varepsilon, d_i} = V_{\varepsilon, d_{\{i\}}} \in \mathscr{U};$$

since $\varepsilon > 0$ was arbitrary, g_i is uniformly continuous. Conversely, let $\mathscr{U}'$ be a uniform structure on X such that all g_i $(i \in I)$ are uniformly continuous on X with respect to $\mathscr{U}'$. Then $V_{\varepsilon, d_i} \in \mathscr{U}'$ for all $\varepsilon > 0$ and $i \in I$, whence $V_{\varepsilon, d_J} \in \mathscr{U}'$ for all $\varepsilon > 0$ and $J \in \mathscr{F}(I)$. We deduce‡ that $\mathscr{U} \subset \mathscr{U}'$.

For each $i \in I$, let $Y_i = \mathbf{R}_i$ and let $\mathscr{U}_i$ be the uniform structure on $\mathbf{R}$ corresponding to the usual distance. We see then that the uniform structure $\mathscr{U}$ on X we defined above is the initial uniform structure corresponding to the families $((Y_i, \mathscr{U}_i))_{i \in I}$ and $(g_i)_{i \in I}$. The fact that $\mathscr{U}$ is the smallest uniform structure on X such that all the functions g_i $(i \in I)$ are uniformly continuous on X follows from 27.8. Due to the importance of this example, we found it instructive to study it more directly.

Notice that, by Theorem 24.33, the topology $\mathscr{T}_{\mathscr{U}}$ is the initial topology on X associated with the families $((Y_i, \mathscr{T}_{\mathscr{U}_i}))_{i \in I}$ and $(g_i)_{i \in I}$. From Theorem 6.2, we deduce that $\mathscr{T}_{\mathscr{U}}$ *is the smallest topology on* X *such that all the functions* g_i $(i \in I)$ *are continuous.*

† Recall that $\mathscr{F}(I)$ is the set of all finite parts of I and that for $J \in \mathscr{F}(I)$, $d_J = \sup_{i \in J} d_i$.

‡ Notice that $(X, \mathscr{U})$ is separated if and only if $(g_i)_{i \in I}$ separates the points of X.

Extentions of Uniformly Continuous Mappings

Denote by $(X_1, \mathscr{U}_1)$ and $(X_2, \mathscr{U}_2)$ two uniform spaces.

27.11 Theorem.—*Let $A \subset X_1$, $\mathscr{B} \subset \mathscr{P}(A)$ be a Cauchy filter basis and $f: A \to X_2$ be a uniformly continuous mapping. Then $f(\mathscr{B})$ is a Cauchy filter basis.*

Proof.—Let $V \in \mathscr{U}_2$ and

$$U = \{(x, y) \mid (f(x), f(y)) \in V\};$$

then (see 27.7) $U \in \mathscr{U}_1$. Since $\mathscr{B}$ is a Cauchy filter basis, there is $A \in \mathscr{B}$ satisfying $A \times A \subset U$. Let $(s, t) \in f(A) \times f(A)$; then there is $(x, y) \in A \times A$ such that $s = f(x)$ and $t = f(y)$. We obtain $(s, t) \in V$. Since (s, t) was arbitrary, we deduce that $f(A) \times f(A) \subset V$. Since V was arbitrary, we conclude that $f(\mathscr{B})$ is a Cauchy basis.

27.12 Theorem.—*Assume $(X_2, \mathscr{U}_2)$ to be separated and complete and let $A \subset X_1$ and $f: A \to X_2$. Then the following assertions are equivalent:*

(a) *f is uniformly continuous on A;*

(b) *there is a unique uniformly continuous mapping $\bar{f}: \bar{A} \to X_2$ such that $\bar{f} \mid A = f$.*

Proof of (a) $\Rightarrow$ (b).—Let $y \in \bar{A}$ and let

$$\mathscr{B}_y = \{A \cap V \mid V \in \mathscr{N}_X(y)\}.$$

Then $\mathscr{B}_y$ is a filter basis, contained in $\mathscr{P}(A)$, that converges to the element $y \in \bar{A}$. Hence, $\mathscr{B}_y$ is a Cauchy basis, and then, by Theorem 27.11, $f(\mathscr{B}_y)$ is a Cauchy basis. Since $(X_2, \mathscr{U}_2)$ is complete, $f(\mathscr{B}_y)$ is convergent. Define

$$\bar{f}(y) = \lim f(\mathscr{B}_y)$$

for $y \in \bar{A}$. Since $(X_2, \mathscr{U}_2)$ is separated, $\bar{f}$ is well-defined. Since f is uniformly continuous, and hence continuous on A, $\bar{f}(y) = f(y)$ for $y \in A$.

It remains to be shown that $\bar{f}$ is uniformly continuous and unique.

Let $V \in \mathscr{U}_2$ and let W be symmetric in $\mathscr{U}_2$ and satisfy $W \circ W \circ W \subset V$. Let $U \in \mathscr{U}_1$ be such that

$$(x, y) \in U \cap (A \times A) \Rightarrow (f(x), f(y)) \in W,$$

and let B be symmetric in $\mathscr{U}_1$ and satisfy $\overline{B} \circ \overline{B} \circ \overline{B} \subset U$.

Now let $(s, t) \in \overline{B \cap (A \times A)}$. Then there are $A_s \in \mathscr{B}_s$ and $A_t \in \mathscr{B}_t$ such that

$$A_s \subset B(s), \qquad f(A_s) \subset W(\bar{f}(s))$$

and

$$A_t \subset B(t), \qquad f(A_t) \subset W(\bar{f}(t)).$$

If $x \in A_s$ and $y \in A_t$, then $(x, s) \in B$ and $(t, y) \in B$; since $(s, t) \in \bar{B}$, we deduce that $(x, y) \in B \circ \bar{B} \circ B \subset U$. Hence $(f(x), f(y)) \in W$. Now $(\bar{f}(s), f(x)) \in W$ and $(f(y), \bar{f}(t)) \in W$. We deduce that

$$(\bar{f}(s), \bar{f}(t)) \in W \circ W \circ W \subset V.$$

Since $\overline{B \cap (A \times A)} \in (\mathscr{U}_1)_{\bar{A}}$ (see 24.19*), and since (s, t) and V were arbitrary, $\bar{f}$ is uniformly continuous on $\bar{A}$.

Now let $f_1 : \bar{A} \to X_2$ and $f_2 : \bar{A} \to X_2$ be two uniformly continuous mappings such that $f_1 \mid A = f_2 \mid A$. By 7.21, the set

$$C = \{x \in \bar{A} \mid f_1(x) = f_2(x)\}$$

is closed. Since $C \supset A$, we conclude that $C = \bar{A}$; that is, $f_1 = f_2$.

Proof of (b) $\Rightarrow$ (a).—Since $\bar{f}$ is uniformly continuous, we deduce that $f = \bar{f} \mid A$ is uniformly continuous.

Example 7.—Let $(X, \mathscr{U})$ be a uniform space and A a set dense in X. Let $\mathscr{A} = (d_i)_{i \in I}$ be a directed family of semi-norms on A such that $\mathscr{U}_A = \mathscr{U}^{(\mathscr{A})}$. By the result in Example 3, for each $i \in I$, the semi-metric d_i is uniformly continuous on $A \times A$; by Theorem 27.12, d_i can be extended to a function $\bar{d}_i$ uniformly continuous on $X \times X$. It is easy to see that $\bar{\mathscr{A}} = (\bar{d}_i)_{i \in I}$ is a *directed family of semi-metrics* on X. From 27.10 we obtain $\mathscr{U}^{(\bar{\mathscr{A}})} \subset \mathscr{U}$. Now for every i in I and $\varepsilon > 0$ we have

$$W_{\varepsilon/2, \bar{d}_i} \subset \bar{W}_{\varepsilon, d_i}.$$

By 24.19, $\{\bar{W}_{\varepsilon, d_i} \mid \varepsilon > 0, i \text{ in } I\}$ is a basis of $\mathscr{U}$; we deduce that $\mathscr{U} \subset \mathscr{U}^{(\bar{\mathscr{A}})}$. Hence $\mathscr{U} = \mathscr{U}^{(\bar{\mathscr{A}})}$.

Exercises for Chapter 27

1. Prove Theorem 27.6.

2. Prove 27.9.

---

* Notice that $((\mathscr{U}_1)_{\bar{A}})_A = (\mathscr{U}_1)_A$.

3. Let $(X, \mathscr{U})$ be a uniform space, $\mathscr{A} = (d_j)_{j \in J}$ a directed family of semi-metrics satisfying $\mathscr{U} = \mathscr{U}^{(\mathscr{A})}$, and $f: X \to \mathbf{R}$. If there is $j \in J$ and $M \in \mathbf{R}$ such that

$$|f(x) - f(y)| \leq M d_j(x, y)$$

for all $(x, y) \in X \times X$, then f is uniformly continuous.

4. (Here we use the notations of Exercise 1 in the text.) If f and g belong to $C^u_{\mathbf{R}}(X)$ and are bounded, then $fg \in C^u_{\mathbf{R}}(X)$.

Chapter 28

Isomorphisms. Completion of Uniform Spaces

Let $(X_1, \mathscr{U}_1)$ and $(X_2, \mathscr{U}_2)$ be two uniform spaces.

28.1 Definition.—*We call an isomorphism of* $(X_1, \mathscr{U}_1)$ *onto* $(X_2, \mathscr{U}_2)$ *any bijection* $f: X_1 \to X_2$ *such that both* f *and* f^{-1} *are uniformly continuous.*

When there is no ambiguity, we shall say simply "*isomorphism of* X_1 *onto* X_2," instead of "isomorphism of $(X_1, \mathscr{U}_1)$ onto $(X_2, \mathscr{U}_2)$."

Notice that f is an isomorphism of X_1 onto X_2 if and only if f^{-1} is an isomorphism of X_2 onto X_1 (observe that $(f^{-1})^{-1} = f$).

Two uniform spaces are said to be *isomorphic* if there exists an isomorphism of one onto the other.

If f is an isomorphism of $(X_1, \mathscr{U}_1)$ onto $(X_2, \mathscr{U}_2)$ then, by definition, f and f^{-1} are uniformly continuous. By 27.4, f and f^{-1} are continuous, whence f *is a homeomorphism of* $(X_1, \mathscr{T}_{\mathscr{U}_1})$ *onto* $(X_2, \mathscr{T}_{\mathscr{U}_2})$.

Using Theorem 27.11, we easily see that if $(X_1, \mathscr{U}_1)$ and $(X_2, \mathscr{U}_2)$ are isomorphic, then $(X_1, \mathscr{U}_1)$ *is complete if and only if* $(X_2, \mathscr{U}_2)$ *is complete.*

In Example 6 of Chapter 15, we considered the *metrics* d and δ on $\mathbf{R}$, and we noticed that $\mathscr{T}_d = \mathscr{T}_\delta$, $(\mathbf{R}, d)$ is complete and $(\mathbf{R}, \delta)$ is *not* complete. This shows that $(\mathbf{R}, \mathscr{U}_d)$ and $(\mathbf{R}, \mathscr{U}_\delta)$ are not isomorphic, although $\mathscr{T}_{\mathscr{U}_d} = \mathscr{T}_{\mathscr{U}_\delta}$. Hence, there are non-isomorphic uniform spaces $(X_1, \mathscr{U}_1)$ and $(X_2, \mathscr{U}_2)$ such that $(X_1, \mathscr{T}_{\mathscr{U}_1})$ and $(X_2, \mathscr{T}_{\mathscr{U}_2})$ are homeomorphic.

Let $(X_1, \mathscr{U}_1)$, $(X_2, \mathscr{U}_2)$, and $(X_3, \mathscr{U}_3)$ be three uniform spaces. Then

28.2 Theorem.—*If* f *is an isomorphism of* X_1 *onto* X_2 *and* g *an isomorphism of* X_2 *onto* X_3, *then* $g \circ f$ *is an isomorphism of* X_1 *onto* X_3.

The proof is left to the reader.

Denote by $(X_1, \mathscr{U}_1)$ and $(X_2, \mathscr{U}_2)$ two complete uniform spaces by Y_1 a set dense in X_1, and by Y_2 a set dense in X_2. Then:

28.3 Theorem.—*Let f be an isomorphism of $(Y_1, (\mathscr{U}_1)_{Y_1})$ onto $(Y_2, (\mathscr{U}_2)_{Y_2})$. Then there is a unique isomorphism $\bar{f}$ of $(X_1, \mathscr{U}_1)$ onto $(X_2, \mathscr{U}_2)$ such that $\bar{f} \mid Y_1 = f$.*

Proof.—By Theorem 27.12, there are uniformly continuous mappings $\bar{f}_1 : X_1 \to X_2$ and $\bar{f}_2 : X_2 \to X_1$ such that

$$\bar{f}_1 \mid Y_1 = f \quad \text{and} \quad \bar{f}_2 \mid Y_2 = f^{-1}.$$

Now, $\bar{f}_2 \circ \bar{f}_1(x) = f^{-1} \circ f(x) = x$ for every $x \in Y_1$. Since $\bar{f}_2 \circ \bar{f}_1$ is continuous, we deduce that $\bar{f}_2 \circ \bar{f}_1(x) = x$ for every $x \in X_1$ (use 7.21). In the same way, we prove that $\bar{f}_1 \circ \bar{f}_2(x) = x$ for every $x \in X_2$. Hence, $\bar{f}_1$ is a bijection of X_1 onto X_2 and $(\bar{f}_1)^{-1} = \bar{f}_2$. Hence, $\bar{f} = \bar{f}_1$ is an isomorphism of X_1 onto X_2 such that $\bar{f} \mid Y_1 = f$.

The uniqueness of $\bar{f}$ follows from Theorem 27.12.

Theorem 28.3 is similar to Theorem 17.12.

28.4 Theorem.—*Let $(X, \mathscr{U})$ be a separated uniform space and let:*

(a) *$\mathscr{A} = (d_i)_{i\in I}$, a family of semi-metrics on X satisfying $\mathscr{U} = \mathscr{U}^{(\mathscr{A}^*)}$;*
(b) *$(X_i, \delta_i)_{i\in I}$, a family of metric spaces;*
(c) *for each $i \in I$, a mapping $\varphi_i : X \to X_i$ satisfying*

$$\delta_i(\varphi_i(x), \varphi_i(y)) = d_i(x, y)$$

for all x and y in X.

Let $(X_\infty, \mathscr{U}_\infty)$ be the uniform space product of the family $((X_i, \mathscr{U}_{\delta_i}))_{i\in I}$ φ the mapping $x \mapsto ((\varphi_i(x))_{i\in I}$ of X into X_∞, and $X' = \varphi(X)$. Then φ (considered as a mapping into X') is an isomorphism of $(X, \mathscr{U})$ onto $(X', (\mathscr{U}_\infty)_{X'})$.

Proof.—From (c), we deduce that for each $i \in I$, φ_i is uniformly continuous. Since, for each $i \in I$, $pr_i \circ \varphi = \varphi_i$, we deduce from 27.9′ that $\varphi : X \to X_\infty$ is uniformly continuous.

Then φ, considered as a mapping into X', is clearly *uniformly continuous*. Moreover, if x and y belong to X and $x \neq y$, there is $i \in I$ such that $d_i(x, y) \neq 0$, since

$$\delta_i(\varphi_i(x), \varphi_i(y)) = d_i(x, y) \neq 0.$$

We deduce that φ is a *bijection* of X onto X'.

It remains to be shown that $\varphi^{-1} : X' \to X$ is uniformly continuous. For this, we proceed as follows: Let $U \in \mathscr{U}$ and let J be a finite part of I and $\varepsilon > 0$ such that*

$$V_{\varepsilon,d_J} = \bigcap_{i\in J} V_{\varepsilon,d_i} \subset U.$$

Let

$$V = W((V_{\varepsilon,\delta_i})_{i\in J}) \cap (X' \times X') \in (\mathcal{U}_\infty)_{X'}.$$

Then $(u, v) \in V \Rightarrow (u, v) \in X' \times X' \Rightarrow$ there are x and y in X such that $u = \varphi(x)$ and $v = \varphi(y)$. We deduce that

$$\begin{aligned}(u, v) \in V &\Rightarrow (pr_i(\varphi(x)), pr_i(y))) \in V_{\varepsilon,\delta_i} \quad \text{for} \quad i \in J \\ &\Rightarrow (\varphi_i(x), \varphi_i(y)) \in V_{\varepsilon,\delta_i} \quad \text{for} \quad i \in J \\ &\Rightarrow \delta_i(\varphi_i(x), \varphi_i(y)) < \varepsilon \quad \text{for} \quad i \in J \\ &\Rightarrow d_i(x,y) < \varepsilon \quad \text{for} \quad i \in J \\ &\Rightarrow d_J(x,y) < \varepsilon \Rightarrow (x,y) \in V_{\varepsilon,\delta_J} \subset U.\end{aligned}$$

Hence, $(u, v) \in V \Rightarrow (\varphi^{-1}(u), \varphi^{-1}(v)) \in U$. Since $U \in \mathcal{U}$ was arbitrary, φ^{-1} is uniformly continuous.

Hence, Theorem 28.4 is proved.

28.5 Theorem.—*Any separated uniform space is isomorphic to a subspace of a product of a family of complete metric spaces.*

Proof.—Let $(X, \mathcal{U})$ be a uniform space and $\mathcal{A} = (d_i)_{i\in I}$ a directed family of semi-metrics on X such that $\mathcal{U} = \mathcal{U}^{(\mathcal{A})}$.

For each $i \in I$, consider (see Example 5, Chapter 27) the quotient space $X_i = X_{d_i}$, the canonical mapping $\varphi_i = \varphi_{d_i}$ of X onto X_i, and the metric $\dot{d}_i$ on X_i defined by

$$\dot{d}_i(\varphi_i(x), \varphi_i(y)) = d_i(x,y)$$

for $(x,y) \in X \times X$. For each $i \in I$, let $(\hat{X}_i, \hat{d}_i)$ be the completion of $(X_i, \dot{d}_i)$; we assume that $X_i \subset \hat{X}_i$ and $(\hat{d}_i)_{X_i} = \dot{d}_i$ (see the remarks following Theorem 17.15).

Let $(X_\infty, \mathcal{U}_\infty)$ be the uniform space product of the family $((\hat{X}_i, \hat{d}_i))_{i\in I}$, let φ be the mapping $x \mapsto (\varphi_i(x))_{i\in I}$ of X into X_∞, and let $X' = \varphi(X)$. Theorem 28.5 follows, then, from 28.4.

28.6 Theorem.—*Let $(X, \mathcal{U})$ be a separated uniform space. Then there is a complete separated uniform space $(\hat{X}, \hat{\mathcal{U}})$, $X' \subset \hat{X}$ satisfying $\overline{X'} = \hat{X}$, and an isomorphism φ of $(X, \mathcal{U})$ onto $(X', \hat{\mathcal{U}}_{X'})$.*

Proof.—By 26.7, a uniform space, product of a family of complete uniform spaces, is complete. From Theorem 28.5, we deduce that

* $d_J = \sup_{i\in J} d_i$.

there is a complete separated uniform space $(X_\infty, \mathscr{U}_\infty)$, a subset $X' \subset X_\infty$, and an isomorphism φ of $(X, \mathscr{U})$ onto $(X', (\mathscr{U}_\infty)_{X'})$. If we take $\hat{X} = X'$ and $\hat{\mathscr{U}} = (\mathscr{U}_\infty)_{\bar{X}'}$, and consider φ as a mapping of X into $\hat{X}$, then the conclusions of Theorem 28.6 are satisfied.

Theorem 28.6 is, in fact, a generalization of Theorem 17.15. However, in Theorem 17.15, φ is an isometry.

Remarks.—The uniform space $(\hat{X}, \hat{\mathscr{U}})$ is called the *completion* of $(X, \mathscr{U})$. If we *identify* X with $\varphi(X)$, then:

(i) $X \subset \hat{X}$;

(ii) $\hat{\mathscr{U}}_X = \mathscr{U}$;

(iii) $\bar{X} = \hat{X}$.

Now let $\mathscr{B} = (d_j)_{i \in J}$ be a directed family of semi-metrics on X such that $\mathscr{U} = \mathscr{U}^{(\mathscr{B})}$. For each $j \in J$, d_j can be extended to a semi-metric $\hat{d}_j$ on $\hat{X}$ (see Example 7, Chapter 27). Then $\hat{\mathscr{B}} = (\hat{d}_j)_{j \in J}$ is a directed family of semi-metrics and $\mathscr{U}^{(\hat{\mathscr{B}})} = \hat{\mathscr{U}}$ (see Example 7, Chapter 27).

We shall show that, in a certain sense, $(\hat{X}, \hat{\mathscr{U}})$ is "unique."

28.7 Theorem.—*Let $(X, \mathscr{U})$ be a separated uniform space, and, for each $i = 1.2$, let $(X_i, \mathscr{U}_i)$ be a complete separated uniform space, $X'_i \subset X_i$, satisfying $\overline{X'_i} = X_i$, and let φ_i be an isomorphism of $(X, \mathscr{U})$ onto the space $(X', (\mathscr{U}_i)_{X_i'})$. Then $(X_1, \mathscr{U}_1)$ and $(X_2, \mathscr{U}_2)$ are isomorphic.*

Proof.—The conclusion follows from 28.3. The details are left to the reader.

Exercises for Chapter 28

1. Prove Theorem 28.2.

2. Prove Theorem 28.7.

3. Let $(X, \mathscr{U})$ be a uniform space. We say that X is *totally bounded* if for each U in $\mathscr{U}$ there is a finite set F such that $U(F) \supset X$. A subset A of X is totally bounded if A is totally bounded when considered as a uniform subspace of X. Show that a subset A of X is totally bounded if and only if for each U in $\mathscr{U}$ there is a finite sequence of sets $A_1, \ldots, A_n$ such that $A \subset \bigcup_{i=1}^n A_i$ and $A_i \times A_i \subset U$ for each $1 \leq i \leq n$. Prove that X is compact if and only if $(X, \mathscr{U})$ is complete and X is totally bounded.

Chapter 29

Uniformizable Spaces. Completely Regular Spaces

Let X be a set and $(g_i)_{i \in I}$ a family of mappings of X into $\mathbf{R}$. Then (see Example 6, Chapter 28):

(1) there is on X a smallest topology $\mathscr{T}$ such that all g_i $(i \in I)$ are continuous;

(2) there is on X a smallest uniform structure $\mathscr{U}$ such that all g_i $(i \in I)$ are uniformly continuous;

(3) $\mathscr{T}_{\mathscr{U}} = \mathscr{T}$.

Hence, the topology $\mathscr{T}$ is *uniformizable* (see Definition 24.18).

29.1 Theorem.—*Let* $(X, \mathscr{T})$ *be a topological space satisfying the following condition*:

(★) *For every* $t \in X$ *and* $V \in \mathscr{N}(t)$, *there is* $f \in C_{\mathbf{R}}(X)$ *such that:*

(a) $0 \leq f \leq 1$;
(b) $f(t) = 1$;
(c) $f(x) = 0$ *for* $x \in \mathbf{C}V$.

Then $\mathscr{T}$ *is the smallest topology on* X *such that all functions*† $f \in C^b_{\mathbf{R}}(X)$ *are continuous.*

Proof.—Let $\mathscr{T}'$ be the smallest topology on X such that all $f \in C^b_{\mathbf{R}}(X)$ are continuous with respect to $\mathscr{T}'$. Clearly, $\mathscr{T}' \subset \mathscr{T}$.

† Here, $C^b_{\mathbf{R}}(X)$ is the set of all bounded functions on X to $\mathbf{R}$, continuous when X is endowed with $\mathscr{T}$.

Now let $t \in X$ and $V \in \mathscr{N}_{(X,\mathscr{T})}(t)$. By hypothesis, there is $f \in C^b_{\mathbf{R}}(X)$ satisfying 29.1. Since f is continuous with respect to $\mathscr{T}'$, $f^{-1}((0, 2))$ belongs to $\mathscr{N}_{(X,\mathscr{T}')}(t)$. Since $f^{-1}((0, 2)) \subset V$, we deduce that $V \in \mathscr{N}_{(X,\mathscr{T}')}(t)$. Hence, for *every* $t \in X$,

$$\mathscr{N}_{(X,\mathscr{T})}(t) \subset \mathscr{N}_{(X,\mathscr{T}')}(t),$$

and hence $\mathscr{T} \subset \mathscr{T}'$. We conclude that $\mathscr{T} = \mathscr{T}'$.

29.2 Theorem.—*Let* $(X, \mathscr{T})$ *be a topological space. Then the following assertions are equivalent:*

(a) *the space* $(X, \mathscr{T})$ *satisfies* $(\bigstar)$;
(b) *the space* $(X, \mathscr{T})$ *is uniformizable.*

Proof of (a) $\Rightarrow$ (b).—Let $\mathscr{U}$ be the smallest uniform structure on X such that all the functions $f \in C^b_{\mathbf{R}}(X)$ are uniformly continuous (with respect to $\mathscr{U}$). By the remarks at the beginning of this chapter, $\mathscr{T}_{\mathscr{U}}$ is the smallest topology on X such that all the functions $f \in C^b_{\mathbf{R}}(X)$ are continuous (with respect to $\mathscr{T}_{\mathscr{U}}$). Comparing 29.2 with 29.1, we deduce that $\mathscr{T} = \mathscr{T}_{\mathscr{U}}$, so that $(X, \mathscr{T})$ is uniformizable.

Proof of (b) $\Rightarrow$ (a).—Let $\mathscr{U}$ be a uniform structure on X such that $\mathscr{T}_{\mathscr{U}} = \mathscr{T}$, and let $\mathscr{A} = (d_i)_{i \in I}$ be a directed family of semi-metrics on X satisfying $\mathscr{U}^{(\mathscr{A})} = \mathscr{U}$. Let $t \in X$ and $V \in \mathscr{N}(t)$. Then there are $i \in I$ and $\varepsilon > 0$ such that

$$V_{\varepsilon, d_i}(t) \subset V.$$

Define

$$f(x) = 1 - \inf\{1, (1/\varepsilon) d_i(x, t)\}$$

for $x \in X$. Then $f \in C_{\mathbf{R}}(X)$, and f satisfies 29.1; hence $(X, \mathscr{T})$ satisfies $(\bigstar)$.

Remarks.—(i) A uniform space is regular (see 24.16). Hence, a topological space satisfying $(\bigstar)$ is regular. (ii) It follows from (i) that $(\bigstar)$ is equivalent with the condition obtained by replacing 29.1(c) with

(c′) $\operatorname{Supp}(f) \subset V$.

29.3 Definition.—*A topological space is completely regular if it is separated and satisfies* $(\bigstar)$.

Hence, a topological space is completely regular if and only if it is separated and uniformizable.

Remarks.—(i) Every subspace of a completely regular space is completely regular (every subspace of a uniformizable space is uniformizable). (ii) Let $(X', \mathscr{T}')$ and $(X'', \mathscr{T}'')$ be two homeomorphic topological spaces. Then $(X', \mathscr{T}')$ is completely regular if and only if $(X'', \mathscr{T}'')$ is. (iii) A normal space (see 20.1) is obviously completely regular. Hence, a compact space is completely regular. (iv) From 20.9, it follows that a locally compact space is completely regular.

We have already noticed that a compact space is completely regular, and therefore uniformizable. A stronger result, however, is the following.

29.4 Theorem.—*Let $(X, \mathscr{T})$ be a compact space and let*†

$$\mathscr{U}^* = \{U \mid \overset{\circ}{U} \supset \Delta_X\}.$$

Then $\mathscr{U}^*$ is the unique uniform structure on X such that $\mathscr{T}_{\mathscr{U}^*} = \mathscr{T}$.

Proof.—Since $(X, \mathscr{T})$ is compact, it is uniformizable, and hence there is a uniform structure $\mathscr{U}$ on X such that $\mathscr{T}_{\mathscr{U}} = \mathscr{T}$. We shall show that $\mathscr{U} = \mathscr{U}^*$.

If $U \in \mathscr{U}$, then by 24.12, $\overset{\circ}{U} \in \mathscr{U}$, so that $\overset{\circ}{U} \supset \Delta_X$. Hence, $U \in \mathscr{U}^*$. Since U was arbitrary,

$$\mathscr{U} \subset \mathscr{U}^*.$$

Conversely, let $U \in \mathscr{U}^*$. Suppose that $U \notin \mathscr{U}$. Then $\overline{V} \cap \mathbf{C}\overset{\circ}{U} \neq \varnothing$ for every $V \in \mathscr{U}$. Since

$$\mathscr{B} = \{\overline{V} \cap \mathbf{C}\overset{\circ}{U} \mid V \in \mathscr{U}\}$$

is a filter basis on $X \times X$ consisting of closed sets, we have

$$\bigcap_{V \in \mathscr{U}} \overline{V} \cap \mathbf{C}\overset{\circ}{U} \neq \varnothing.$$

But $\bigcap_{V \in \mathscr{U}} \overline{V} = \Delta_X$ and $\Delta_X \cap \mathbf{C}\overset{\circ}{U} = \varnothing$. Hence, the hypothesis $U \notin \mathscr{U}$ leads to a contradiction. Hence $U \in \mathscr{U}$. Since U was arbitrary, $\mathscr{U}^* \subset \mathscr{U}$. We conclude that $\mathscr{U} = \mathscr{U}^*$.

Let $I \neq \varnothing$ be a set and for each $i \in I$ let $X_i = [0, 1]$ and δ_i be the usual metric on $[0, 1]$. We denote by $\mathscr{T}^I$ the topology on $[0, 1]^I$, product of the family $(\mathscr{T}_{\delta_i})_{i \in I}$. The space $([0, 1]^I, \mathscr{T}^I)$ is *compact.* We denote by $\mathscr{U}^I$ the uniform structure on $[0, 1]^I$, product of the family $(\mathscr{U}_{\delta_i})_{i \in I}$. By Theorem 29.4, $\mathscr{U}^I$ is the unique uniform structure on X that defines the topology $\mathscr{T}^I$.

† $\Delta_X = \{(x, x) \mid x \in X\}$.

Now let $(X, \mathcal{T})$ be a completely regular space and let $(g_i)_{i \in I}$ be the family of all *continuous* mappings of X into $[0, 1] \subset \mathbf{R}$. Let $\mathcal{U}$ be the smallest uniform structure on X such that all the functions g_i, $i \in I$, are uniformly continuous with respect to $\mathcal{U}$. It is easy to see that $\mathcal{U}$ coincides with the smallest uniform structure on X such that all the functions $g \in C^b_{\mathbf{R}}(X)$ are uniformly continuous with respect to $\mathcal{U}$.*

For each $i \in I$, let d_i be the semi-metric on X defined by

$$d_i(x, y) = |g_i(x) - g_i(y)|$$

for $(x, y) \in X \times X$, and let $\mathcal{A} = (d_i)_{i \in I}$. Then $\mathcal{U} = \mathcal{U}^{(\mathcal{A}^*)}$ (see Example 6, Chapter 27). By Theorem 28.3, the uniform space $(X, \mathcal{U})$ is isomorphic to a subspace of the uniform space $([0, 1], \mathcal{U}^I)$. Hence:

29.5 Theorem.—*If $(X, \mathcal{T})$ is completely regular and if $\mathcal{U}$ is the smallest uniform structure on X such that all the functions $f \in C^b_{\mathbf{R}}(X)$ are uniformly continuous with respect to $\mathcal{U}$, then $(X, \mathcal{U})$ is isomorphic to a subspace of a space of the form $([0, 1]^I, \mathcal{U}^I)$.*

From 29.5, we deduce the:

29.6 Corollary.—*A topological space is completely regular if and only if it is homeomorphic with a subspace of a space of the form $([0, 1]^I, \mathcal{T}^I)$.*

29.7 Theorem.—*Let $(X, \mathcal{T})$ be a completely regular space. There is then a compact space $(\tilde{X}, \tilde{\mathcal{T}})$ and a homeomorphism φ of X onto a subspace X' of $\tilde{X}$ such that:*

(a) $\overline{\varphi(X)} = \tilde{X}$;

(b) *for every $f \in C^b_{\mathbf{R}}(X)$ there is $\tilde{f} \in C^b_{\mathbf{R}}(\tilde{X})$ satisfying*† $\tilde{f} \circ \varphi = f$.

Proof.—Let $\mathcal{U}$ be the smallest uniform structure on X such that all $f \in C^b_{\mathbf{R}}(X)$ are uniformly continuous (with respect to $\mathcal{U}$). By Theorem 29.5, there are a set I, a subspace X' of the uniform space $([0, 1]^I, \mathcal{U}^I)$ and an isomorphism φ of X onto X'.

Let $\tilde{X} = \bar{X}'$, $\tilde{\mathcal{T}} = (\mathcal{T}^I)_{\tilde{X}}$, and consider φ as a mapping into $\tilde{X}$. Clearly, φ is a homeomorphism of X onto X' and $\overline{\varphi(X)} = \tilde{X}$. Let $f \in C_b^{\mathbf{R}}(X)$. Then $f \circ \varphi^{-1}: X' \to \mathbf{R}$ is uniformly continuous. By Theorem 27.12, there is a uniformly continuous function $\tilde{f}: \tilde{X} \to \boldsymbol{R}$

* In fact, notice that if $f \in C^b_{\mathbf{R}}(X)$, then $f = \lambda u + \mu V$, where λ, μ are real numbers and u and v are continuous mappings of X into $[0, 1]$.

† Clearly, $\tilde{f}$ is unique.

such that $\tilde{f} \mid X' = f \circ \varphi^{-1}$. Then, if $x \in X$, $\tilde{f} \circ \varphi(x) = (f \circ \varphi^{-1}) \circ \varphi(x) = f(x)$. Hence the theorem is proved.

Remarks.—(i) Notice that if $(X, \mathscr{T})$, $(\tilde{X}, \tilde{\mathscr{T}})$ and φ satisfy the conditions of Theorem 29.7, if $\mathscr{U}$ is the uniform structure on X introduced above, and if $\tilde{\mathscr{U}}$ is the unique uniform structure on $\tilde{X}$ such that $\mathscr{T}_{\tilde{\mathscr{U}}} = \mathscr{T}$, then *$\varphi$ is an isomorphism of X onto X'.*

(ii) The space $(\tilde{X}, \tilde{\mathscr{T}})$ and the mapping φ, in Theorem 29.7, are "unique." In fact, suppose that $(\tilde{X}_1, \tilde{\mathscr{T}}_1)$, φ_1 and $(\hat{X}_2, \hat{\mathscr{T}}_2)$, φ_2 satisfy the conditions of Theorem 29.7. Then $\varphi_2 \circ \varphi_1^{-1}$ is an isomorphism of $\varphi_1(X)$ onto $\varphi_2(X)$. By Theorem 28.2, $\varphi_2 \circ \varphi_1^{-1}$ can be extended to an isomorphism h of $\tilde{X}_1$ onto $\tilde{X}_2$. Hence $\tilde{X}_1$ and $\tilde{X}_2$ are isomorphic and $\varphi_2 = h \circ \varphi_1$.

(iii) The compact space $\tilde{X}$ is called the Stone-Čech *compactification* of X. If we *identify* X with $\varphi(X)$, then:

(a) $X \subset \tilde{X}$;

(b) $\tilde{\mathscr{T}}_X = \mathscr{T}$;

(c) $\bar{X} = \tilde{X}$;

(d) every $f \in C^b_{\mathbf{R}}(X)$ can be extended to a function $\tilde{f} \in C^b_{\mathbf{R}}(\tilde{X})$.

▼ *Exercise.*—(Here, we use the notations of Theorem 29.7.) Let Z be a *compact* space and $g: X \to Z$ continuous. Then there is a unique $\tilde{g}: \tilde{X} \to Z$ continuous such that $\tilde{g} \circ \varphi = g$ (Hint: Use Theorems 29.7 and 29.5). ▲

Exercises for Chapter 29

1. Let $\mathbf{Z}$ be endowed with the discrete topology $\mathscr{P}(\mathbf{Z})$ and consider the Stone-Čech compactification $\tilde{\mathbf{Z}}$ of $\mathbf{Z}$. Then:

(a) the closure of every open set in $\tilde{\mathbf{Z}}$ is open ($\tilde{\mathbf{Z}}$ is "extremely disconnected");

(b) if $(x_n)_{n \in \mathbf{N}}$ is a sequence in $\tilde{\mathbf{Z}}$ such that $n \neq m \Rightarrow x_n \neq x_m$, then $(x_n)_{n \in \mathbf{N}}$ does not converge;

(c) $\tilde{\mathbf{Z}}$ is not metrizable.

2. Any product of completely regular spaces is completely regular.

3. Any subspace of a completely regular space is completely regular.

4. Let $(X, \mathscr{U})$ be a uniform space. A subset A of X is said to be *bounded* if for each U in $\mathscr{U}$ there is a finite set F contained in X and n in $\mathbf{N}$ such that $A \subset U^n(F)$. Show that every totally bounded subset of X is bounded.

5. Let $(X, \mathscr{U})$ be a uniform space such that $(X, \mathscr{T}_{\mathscr{U}})$ is locally compact. Suppose that:

(i) if B is bounded in $(X, \mathscr{U})$, B is relatively compact in $(X, \mathscr{T}_{\mathscr{U}})$;

(ii) there is W in $\mathscr{U}$ such that $W = W^{-1}$, and if B is bounded, $W(B)$ is bounded;

(a) $W_\infty = \bigcup \{W^n \mid n \in \mathbf{N}\}$ is an equivalence relation on X;

(b) for each x in X, $W_\infty(x)$ is both open and closed. (by 10.9, $W_\infty(x)$ is locally compact);

(c) for each x in X, $W_\infty(x)$ is σ-compact;

(d) $(X, \mathscr{T}_{\mathscr{U}})$ is paracompact;

(e) if there is U in $\mathscr{U}$ such that $U(x)$ is compact for each x in X, then $(X, \mathscr{U})$ is said to be uniformly locally compact. If $(X, \mathscr{U})$ is uniformly locally compact, then $(X, \mathscr{U})$ satisfies (i) and (ii).

Bibliography

1. Bartle, R. G., and Ionescu Tulcea, C.: Honors Calculus. Scott, Foresman and Co., 1970.
2. Bourbaki, N.: Topologie Générale. Hermann (Paris), Chapters 1–10.
3. Choquet, G.: Topology. Academic Press, 1966.
4. Dugundji, J.: Topology. Allyn and Bacon, 1966.
5. Fairchild, W. W., and Ionescu Tulcea, C.: Sets. W. B. Saunders, 1970.
6. Kelley, J. L.: General Topology, Van Nostrand, 1955.
7. Schwartz, L.: Théorie des Distributions. Hermann (Paris), 1966.

INDEX OF NOTATIONS

Symbol	Page
$\leq$	xi
$\mapsto$	xi
$\mathscr{P}$	xi
$\mathbf{C}$	xi
pr_J	xi
$\mathbf{N}$	xii
$\mathbf{Z}$	xii
$\mathbf{Q}$	xii
$\mathbf{R}$	xii
$\mathbb{C}$	xii
sup, inf	xii
$\mathbf{R}^*$	xii, 33
$\mathscr{D}$	2
$\mathscr{E}'\mathscr{E}''$	3
$\mathscr{T}$	3
$(X,\mathscr{T})$	4
$\mathscr{U}$	5
$\mathscr{N}(x)$, $\mathscr{N}_X(x)$, $\mathscr{N}_{\mathscr{T}}(x)$	8
Å	15
Ā	16
$Fr(A)$	20
$\mathscr{A}(t)$	21, 30
$\mathscr{T}_S$	22
$j_{S,X}$	27
$C_{\mathbf{R}}(S)$	32
$\triangle_X$	47
$\overset{\lim}{\leftarrow}$	56
$\precsim$	56
$\mathscr{U}_I$	77
$\mathbf{T}^2$	84
$\mathbf{T}^1$	84
$\odot$	91
$C(x)$	97
$d(a, A)$	109
$\delta(A)$	110
$V_r(a)$	110
$W_r(a)$	110
d_n	111
$\mathscr{T}_d$	114
(X, d)	114
d_A	134
$C^u_{\mathbf{R}}A)$	149
$(\hat{X}, \hat{d})$	153
(X, p)	157
$((X_n)_{n\in\mathbf{N}}, (S_n)_{n\in\mathbf{N}})$	159
$\Vert\ \Vert$	159, 162

Symbol	Page
$C^b_{\mathbf{R}}(X)$	161
$C^b_{\mathbf{R},\,a}(X)$	164
$C_{\mathbf{R},\,\infty}(X)$	164
$L(X, Y)$	168
$\otimes$	175
$C_{\mathbf{C}}(X)$	177
$C_{\mathbf{C},\,\infty}(X)$	177
Supp (f)	180
$\mathscr{T}^*$	199
$\mathscr{F}' \sim \mathscr{F}''$	205
$f(\mathscr{B})$	205
$\lim \mathscr{B}$	206
$\mathscr{B}_2 > \mathscr{B}_1$	207
$ad(\mathscr{B})$	211
$\lim_{x \in A} X_x$	213
$B \circ A$	216
A^{-1}	216
A^n	216
$A(K)$	217
$\mathscr{U}^{(3)}$	218
$\mathscr{U}(a)$	219
$\mathscr{T}_{\mathscr{U}}$	220
$(X, \mathscr{U})$	220
$\mathring{B}$	221
$\bar{B}$	222
$\mathscr{U}_A$	224
$\mathrm{V}_{r,d}$	231
$\mathrm{W}_{r,d}$	231
$\mathscr{A}^*$	232
$\mathscr{U}_{\mathrm{d}}$	233
$\mathscr{U}^{(\mathscr{A})}$	233
$\mathscr{T}_d$	233
d_U	236
$\mathscr{U}_{\mathrm{c}}$	238
$\mathscr{C}$	250
$\mathscr{U}^{(c)}$	250
$(\hat{X}, \hat{\mathscr{U}})$	257
$\mathscr{U}^*$	261
$(\widetilde{X}, \widetilde{\mathscr{T}})$	262

INDEX

Absolutely convergent series, 159
Adherence, 16
  in metric space, 130
  of a filter basis, 211
Adherent, 16
Antisymmetric, xi

Baire space, 198
Ball, 110
Banach algebra, 160
Banach space, 158
Basis, 4
  of product topology, 44
Bijective, xi
Bolzano-Weierstrass property (B-W property), 139
Boundary, 20
Bounded subset of a uniform space, 264
Bounds, xii

Cauchy filter, 239
Cauchy inequality, 119
Cauchy net, 243
Cauchy sequence, 133
Closed ball, 110
Closed interval, 49
Closed mapping, 36
Closed set, 5
Commutative algebra, 64
Compact, relatively, 77
Compact spaces, 66, 261
Compactification, 263
  one-point, 83
Complete metric space, 134
Completely regular topological space, 261, 262
Completion, of metric space, 154
  of uniform space, 258
Component (connected), 97
Connected, arcwise, 99
  locally, 98
Connected topological space, 89
Continuous function, 210
  at point, 27
  in metric space, 130
  on set, 31
Convergence, of a net, 213
Convergent sequence, 126
Convex set, 99
Covering, 66
Curve, 99
Customs Passage Theorem, 102

Dense, 19
Diameter of a set, 110
Directed, 213

Elementary sets, 44
Equicontinuous, 166
Euclidean norm (on $\mathbf{R}^n$), 120
Extremely disconnected, 263

Filter, 212
Filter base, 203
  associated with net, 214
  associated with sequence, 204
Filtering, 213
Final topology, 103
Finer covering, 86
Finite intersection property, 73
Frontier, 20
Function
  continuous, 210
    at point, 27
    in metric space, 130
    on set, 31
  Lipschitz, 32, 149
  rational, 65
  semicontinuous, 197

Function (*Continued*)
  uniformly continuous, 245
    extension of, 252
    on metric space, 146
Fundamental system, 9

Generated topology, 3

Hausdorff, 12
Homeomorphism, 35

Inductive, xii
Infimum, xii
Initial topology, 40
Injective, xi
Interior, 15
Interlacing condition (condition (I)), 199
Intermediate value theorem, 101
Interval, 49, 95
Isometry, 150
Isomorphism, 263
  of uniform spaces, 255

Limit, 206, 209
  projective, 56
Lipschitz function, 32, 149
Locally closed, 25, 81
Locally compact, 80
Locally connected, 98
Locally finite, 12

Mapping, 36
Maximal element, xii
Metric, 109
Metric space, 114, 134
  completion of, 154
  continuous function in, 130
  uniformly continuous function in, 146
Metrizable, 115
Minimal element, xii

Neighborhood, 8
Net, 213
Norm, 156
Normal topological space, 180
Normed space, 157
  complete, 158
Nowhere dense, 198

Open ball, 110
Open interval, 49
Open mapping, 36
Open set, 5
Order relation, xi
Ordered set, xii

Paracompact topological space, 88
Part (subset), xi
Partition of unity, 189, 192
  continuous, 193
Polynomial, 64
Premeasure, 57
Preorder relation, 56
Product, of uniform spaces, 227, 249
Product topology, 43
  of regular spaces, 49
Projection, xi
Projective limit, 56
Projective system, 57

Quasi-compact topological space, 66
Quotient space, 105
Quotient topology, 105

Rational function, 65
Reflexive, xi
Regular compact space, 69
Regular locally compact space, 80
Regular space, product topology of, 49
Regular topological space, 12, 260
Regular uniform space, 223
Relation, preorder, 56
Relatively compact subset, 77

Scalar product (on $\mathbf{R}^n$), 118
Semicontinuous function, 197
Semi-metric, 230
  directed families of, 237
Separated topological space, 12
Sequence, 126, 241
  eventually increasing, 131
Series, 159
Set, 5
  convex, 99
  function on, 31
Sigma-compact topological space, 85
Space. See specific kinds of spaces.
Spanned topology, 3
Stone-Čech compactification, 263
Stone-Weierstrass Theorem, 173
  complex form, 178
Subalgebra, 161
Subbasis, 4
Subordinated, 192
Subsequence, 128
Subset, relatively compact, 77
Subspace, 22
  of a vector space, 158
Support of a function, 180

Supremum, xii
Surjective, xi
Symmetric, 217

T_1 space, 14
T_4 space, 191
Tietze, 184
Topological group, 11, 65
Topological space, 4
  completely regular, 261, 262
  connected, 89
  normal, 180
  paracompact, 88
  quasi-compact, 66
  separated, 12
  sigma-compact, 85
  uniformizable, 260
  uniformly locally compact, 264
Topology, 1
  associated with uniform structure, 220
  discrete, 2
  final, 103
  generated, 3
  initial, 40
  quotient, 105
  spanned, 3
Torus (one-dimensional), 107
Totally bounded, 258
Totally discontinuous, 97
Transitive, xi
Tychonoff, theorem of, 76

Ultrafilter, 213
  basis, 213
Uniform space, 220
  bounded subset of, 264
  compact, 221, 258
  complete, 240, 242
  completion of, 258
  locally compact, 221
  product of, 227, 249
  separated, 221
Uniform structure, 218
  associated with family of semi-metrics, 233
  equivalent, 220
  initial, 225, 249
  metrizable, 233
  product, 249
Uniformizable, 224
Uniformizable topological space, 260
Uniformly continuous function, 245
  extension of, 252
  on metric space, 146
Uniformly locally compact topological space, 264
Urysohn, theorem of, 184

Zorn's lemma, xii